Subjectivity and Truth

Also in this series:

SOCIETY MUST BE DEFENDED
(North America & Canada)

ABNORMAL
(North America & Canada)

HERMENEUTICS OF THE SUBJECT
(North America & Canada)

PSYCHIATRIC POWER

SECURITY, TERRITORY, POPULATION

THE BIRTH OF BIOPOLITICS

THE GOVERNMENT OF SELF AND OTHERS

THE COURAGE OF TRUTH

LECTURES ON THE WILL TO KNOW

ON THE GOVERNMENT OF THE LIVING

THE PUNITIVE SOCIETY

Forthcoming in this series:

PENAL THEORIES AND INSTITUTIONS

MICHEL FOUCAULT

Subjectivity and Truth

LECTURES AT THE COLLÈGE DE FRANCE

1980-1981

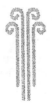

Edited by Frédéric Gros
General Editors: François Ewald and Alessandro Fontana

English Series Editor: Arnold I. Davidson

TRANSLATED BY GRAHAM BURCHELL

palgrave
macmillan

ISBN 978–1–4039–8664–1   ISBN 978–1–349–73900–4 (eBook)
DOI 10.1057/978–1–349–73900–4

Library of Congress Control Number: 2017944763

This Palgrave Macmillan imprint is published by Springer Nature
The registered company is Macmillan Publishers Ltd. London

The registered company address is: The Campus, 4 Crinan Street, London, N1 9XW, United Kingdom

CONTENTS

Foreword: François Ewald and Alessandro Fontana   ix

Abbreviations   xv

one   7 JANUARY 1981   1
        The fable of the elephant in Saint Francis of Sales. ∿ *Versions*
        of the fable in the Middle Ages and sixteenth century. ∿ *The*
        Physiologus. ∿ *Versions of the fable in Greek and Latin*
        antiquity. ∿ *The endpoint with Aristotle.* ∿ *The "subjectivity*
        and truth" relationship: philosophical, positivist, historico-
        philosophical formulations of the problem. ∿ *Subjectivity as*
        historical relationship to the truth, and truth as historical system
        of obligations. ∿ *Principles of monogamous sexual ethics.*
        ∿ *The privileged historical question.*

two   14 JANUARY 1981   25
        Return to the fable of the elephant. ∿ *The arts of living:*
        typology and evolution. ∿ *Mathēsis, meletē, askēsis:*
        relationship to others, the truth, and oneself. ∿ *Notes on the*
        concepts of "paganism," "Judeo-Christianity," "capitalism," as
        categories of self-analysis of Western societies. ∿ *Problem of the*
        pre-existence of "Christian sexual morality" in Stoicism.

three   21 JANUARY 1981   47
        The question of the relations between subjectivity and truth and
        the problem of the dream. ∿ *The oneirocriticism of Artemidorus.*

⌣ *The ethical system of sexual acts through the analysis of dreams.*
⌣ *Distinction between dreams-*rêves *and dreams-*songes. ⌣ *The
economic and social signification of dreams.* ⌣ *The social-sexual
continuum.* ⌣ *Sexual relations in accordance with nature and the
law.* ⌣ *Sexual relations contrary to the law.* ⌣ *Sexual relations
contrary to nature.* ⌣ *Principle of the naturalness of penetration.*

four   28 JANUARY 1981   75
The ethical perception of aphrodisia. ⌣ *Principle of social-
sexual isomorphism and principle of activity.* ⌣ *Valorization
of marriage and definition of adultery.* ⌣ *Modern experience
of sexuality: localization of sexuality and division of the sexes.*
⌣ *Penetration as natural and non-relational activity.* ⌣ *The
discrediting of passive pleasure.* ⌣ *Paradox of the effeminate
womanizer.* ⌣ *Problematization of the relationship with boys.*
⌣ *The desexualized pedagogical erotics.*

five   4 FEBRUARY 1981   97
Process of valorization and illusion of the code. ⌣ *Experience
of the flesh and codification.* ⌣ *The philosophers' new sexual
ethics: hyper-valorization of marriage and devalorization of
pleasure.* ⌣ *Comparative advantages and disadvantages of
marriage.* ⌣ *Should a philosopher marry?* ⌣ *The negative
answer of the Cynics and Epicureans.* ⌣ *The duty of marriage
in the Stoics.* ⌣ *The exception of marriage for the philosopher
in the present catastasis, according to Epictetus.*

six   11 FEBRUARY 1981   123
The kata phusin *character of marriage.* ⌣ *Xenophon's*
Oeconomicus: *study of the speech of Ischomachus to his young
wife.* ⌣ *The classical ends of marriage.* ⌣ *The naturalness
of marriage according to Musonius Rufus.* ⌣ *The desire for
community.* ⌣ *The couple or the herd: the two modes of social
being according to Hierocles.* ⌣ *The relationship to the spouse
or the friend in Aristotle: differential intensities.* ⌣ *The form of
the conjugal bond: organic unity.*

seven 25 FEBRUARY 1981 147
The new economy of aphrodisia. ᔿ *Traditional mistrust of
sexual activity: religious restrictions.* ᔿ *Double relationship of
sexuality: symmetry with death, incompatibility with the truth.* ᔿ
Sexual activity and philosophical life. ᔿ *The medical description
of the sexual act.* ᔿ *Comparison of the sexual act and epileptic
crisis.* ᔿ *Christian transformation of the death-truth-sex triangle.*
ᔿ *Consequences of the conjugalization of sexual pleasure in
the first two centuries* CE *in philosophical texts; the man-woman
symmetry; objectivation of matrimonial sexuality.*

eight 4 MARCH 1981 175
*The three great transformations of sexual ethics in the first
centuries* CE. ᔿ *A reference text: Plutarch's* Erōtikos. ᔿ
Specificity of Christian experience. ᔿ *Plan of* The Dialogue
on Love. ᔿ *The comic situation.* ᔿ *The young boy's place:
central and position of passivity.* ᔿ *The portrait of Ismenodora
as pederast woman.* ᔿ *The break with the classical principles of
the ethics of* aphrodisia. ᔿ *The transfer of the benefits of the
pederastic relationship to within marriage.* ᔿ *The prohibition of
love of boys: unnatural and without pleasure.* ᔿ *The condition
of acceptability of pederasty: the doctrine of the two loves.* ᔿ
Plutarch's establishment of a single chain of love. ᔿ *The final
discredit of love of boys.* ᔿ *The wife's agreeable consent to her
husband.*

nine 11 MARCH 1981 203
The new ethics of marriage. ᔿ *Evolution of matrimonial
practices: the historians' point of view.* ᔿ *Institutional
publicization, social extension, transformation of the relationship
between spouses.* ᔿ *The evidence of writers: the poems of Statius
and Pliny's letters.* ᔿ *Games of truth and reality of practices.*

ten 18 MARCH 1981 227
*The problem of redundant discourse (*discours en trop*).* ᔿ
The Christian re-appropriation of the Hellenistic and Roman

matrimonial code. ⌒ Problematization of the relation between discourse and reality. ⌒ First explanation: representative reduplication. ⌒ Four characteristics of the game of veridiction in relation to reality: supplementary, pointless, polymorphous, efficient. ⌒ Second explanation: ideological disavowal. ⌒ Third explanation: universalizing rationalization

eleven   25 MARCH 1981   249
The spread of the matrimonial model in the Hellenistic and Roman period. ⌒ The nature of the discourse on marriage: tekhnai peri bion. ⌒ Definition of tekhnē and bios. ⌒ The three lives. ⌒ Christian (or modern) subjectivity and Greek bios. ⌒ From paganism to Christianity: breaks and continuities. ⌒ Incompatibilities between the old system of valorization and the new code of conduct. ⌒ Adjustment through subjectivation: caesura of sex and self-control.

twelve   1 APRIL 1981   269
Situation of the arts of living at the point of articulation of a system of valorization and a model of behavior. ⌒ The target-public of techniques of self: competitive aristocracies. ⌒ Historical transformation of the procedures of the distribution of power: court and bureaucracy. ⌒ Re-elaboration of the principle of activity and socio-sexual isomorphism in marriage. ⌒ Splitting of sex and doubling of self on self. ⌒ Cultural consequence: fantasy of the prince's debauchery. ⌒ Problem of the government of self of the prince. ⌒ Subjectivation and objectivation of aphrodisia. ⌒ The birth of desire.

Course summary   293

Course context   301

Index of Concepts and Notions   317

Index of Name   329

FOREWORD

MICHEL FOUCAULT TAUGHT AT the Collège de France from January 1971 until his death in June 1984 (with the exception of 1977 when he took a sabbatical year). The title of his chair was "The History of Systems of Thought".

On the proposal of Jules Vuillemin, the chair was created on 30 November 1969 by the general assembly of the professors of the Collège de France and replaced that of "The History of Philosophical Thought" held by Jean Hyppolite until his death. The same assembly elected Michel Foucault to the new chair on 12 April 1970.[1] He was 43 years old.

Michel Foucault's inaugural lecture was delivered on 2 December 1970.[2] Teaching at the Collège de France is governed by particular rules. Professors must provide 26 hours of teaching a year (with the possibility of a maximum of half this total being given in the form of seminars).[3] Each year they must present their original research and this obliges them to change the content of their teaching for each course. Courses and seminars are completely open; no enrolment or qualification is required and the professors do not award any qualifications.[4] In the terminology of the Collège de France, the professors do not have student but only auditors.

Michel Foucault's courses were held every Wednesday from January to March. The huge audience made up of students, teachers, researchers, and the curious, including many who came from outside France, required two amphitheaters of the Collège de France. Foucault often complained about the distance between himself and his "public" and of how few exchanges the course made possible.[5] He would have liked a seminar in which real collective work could take place and made a

number of attempts to bring this about. In the final years, he devoted
a long period to answering his auditors' questions at the end of each
course.

This is how Gérard Petitjean, a journalist from *Le Nouvel
Observateur*, described the atmosphere at Foucault's lectures in 1975:

> When Foucault enters the amphitheater, brisk and dynamic like
> someone who plunges into the water, he steps over bodies to
> reach his chair, pushes away the cassette recorders so he can put
> down his papers, removes his jacket, lights a lamp and sets off at
> full speed. His voice is strong and effective, amplified by the loud-
> speakers that are the only concession to modernism in a hall that
> is barely lit by light spread from stucco bowls. The hall has three
> hundred places and there are five hundred people packed together,
> filling the smallest free space ... There is no oratorical effect. It is
> clear and terribly effective. There is absolutely no concession to
> improvisation. Foucault has twelve hours each year to explain
> in a public course the direction taken by his research in the year
> just ended. So everything is concentrated and he fills the margins
> like correspondents who have too much to say for the space avail-
> able to them. At 19.15 Foucault stops. The students rush towards
> his desk; not to speak to him, but to stop their cassette record-
> ers. There are no questions. In the pushing and shoving Foucault
> is alone. Foucault remarks: "It should be possible to discuss what
> I have put forward. Sometimes, when it has not been a good lec-
> ture, it would need very little, just one question, to put everything
> straight. However, this question never comes. The group effect in
> France makes any genuine discussion impossible. And as there is
> no feedback, the course is theatricalized. My relationship with the
> people there is like that of an actor or an acrobat. And when I
> have finished speaking, a sensation of total solitude ..."[6]

Foucault approached his teaching as a researcher: explorations for a
future book, as well as the opening up of fields of problematization were
formulated as an invitation to possible future researchers. This is why the
courses at the Collège de France do not duplicate the published books.
They are not sketches for the books, even though both books and courses

share certain themes. They have their own status. They arise from a specific discursive regime within the set of Foucault's "philosophical activities." In particular they set out the program for a genealogy of knowledge/power relations, which are the terms in which he thinks of his work from the beginning of the 1970s, as opposed to the program of an archeology of discursive formations that previously orientated his work.[7]

The course also performed a role in contemporary reality. Those who followed his courses were not only held in thrall by the narrative that unfolded week by week and seduced by the rigorous exposition, they also found a perspective on contemporary reality. Michel Foucault's art consisted in using history to cut diagonally through contemporary reality. He could speak of Nietzsche or Aristotle, of expert psychiatric opinion or the Christian pastorate, but those who attended his lectures always took from what he said a perspective on the present and contemporary events. Foucault's specific strength in his courses was the subtle interplay between learned erudition, personal commitment, and work on the event.

With their development and refinement in the 1970s, Foucault's desk was quickly invaded by cassette recorders. The courses—and some seminars—have thus been preserved.

This edition is based on the words delivered in public by Foucault. It gives a transcription of these words that is as literal as possible.[8] We would have liked to present it as such. However, the transition from an oral to a written presentation calls for editorial intervention: at the very least it requires the introduction of punctuation and division into paragraphs. Our principle has been always to remain as close as possible to the course actually delivered.

Summaries and repetitions have been removed whenever it seemed to be absolutely necessary. Interrupted sentences have been restored and faulty constructions corrected. Suspension points indicate that the recording is inaudible. When a sentence is obscure there is a conjectural integration or an addition between square brackets. An asterisk directing the reader to the bottom of the page indicates a significant divergence between the notes used by Foucault and the words actually uttered. Quotations have been checked and references to the texts used are indicated. The critical apparatus is limited to the elucidation of obscure points, the explanation of some allusions and the clarification of critical points. To make the lectures easier to read, each lecture is preceded by a brief summary that indicates its principle articulations.

The text of the course is followed by the summary published by the *Annuaire du Collège de France*. Foucault usually wrote these in June, some time after the end of the course. It was an opportunity for him to pick out retrospectively the intention and objectives of the course. It constitutes the best introduction to the course.

Each volume ends with a "context" for which the course editors are responsible. It seeks to provide the reader with elements of the bio-graphical, ideological, and political context, situating the course within the published work and providing indications concerning its place within the corpus used in order to facilitate understanding and to avoid misinterpretations that might arise from a neglect of the circum-stances in which each course was developed and delivered.

Subjectivity and Truth, the course delivered in 1981, is edited by Frédéric Gros.

A new aspect of Michel Foucault's "œuvre" is published with this edition of the Collège de France courses.

Strictly speaking it is not a matter of unpublished work, since this edition reproduces words uttered publicly by Foucault. The written material Foucault used to support his lectures could be highly devel-oped, as this volume attests.

This edition of the Collège de France courses was authorized by Michel Foucault's heirs, who wanted to be able to satisfy the strong demand for their publication, in France as elsewhere, and to do this under indisputably responsible conditions. The editors have tried to be equal to the degree of confidence placed in them.

FRANÇOIS EWALD AND ALESSANDRO FONTANA

Alessandro Fontana died on 17 February 2013 before being able to complete the edition of Michel Foucault's lectures at the Collège de France, of which he was one of the initiators. Because it will maintain the style and rigor that he gave to it, the edition will continue to be published under his authority until its completion. - *F.E.*

1. Michel Foucault concluded a short document drawn up in support of his candidacy with these words: "We should undertake the history of systems of thought." "Titres et travaux," in *Dits et Écrits, 1954-1988*, four volumes, ed., Daniel Defert and François Ewald (Paris: Gallimard, 1994) vol. 1, p. 846; English translation by Robert Hurley, "Candidacy Presentation: Collège de France" in *The Essential Works of Michel Foucault, 1954-1984, vol. 1: Ethics: Subjectivity and Truth*, ed. Paul Rabinow (New York: The New Press, 1997) p. 9.
2. It was published by Gallimard in May 1971 with the title *L'Ordre du discours*, Paris, 1971. English translation by Ian McLeod, "The Order of Discourse," in *Untying the Text*, ed., Robert Young, (London: Routledge and Kegan Paul, 1981).
3. This was Foucault's practice until the start of the 1980s.
4. Within the framework of the Collège de France.
5. In 1976, in the vain hope of reducing the size of the audience, Michel Foucault changed the time of his course from 17.45 to 9.00. See the beginning of the first lecture (7 January 1976) of *"Il faut défendre la société". Cours au Collège de France, 1976* (Paris: Gallimard/Seuil, 1997); English translation by David Macey, *"Society Must be Defended." Lectures at the Collège de France 1975-1976* (New York: Picador, 2003).
6. Gérard Petitjean, "Les Grands Prêtres de l'université française," *Le Nouvel Observateur*, 7 April 1975.
7. See especially, "Nietzsche, la généalogie, l'histoire," in *Dits et Écrits*, vol. 2, p. 137; English translation by Donald F. Brouchard and Sherry Simon, "Nietzsche, Genealogy, History" in *The Essential Works of Michel Foucault 1954-1984, vol. 2: Aesthetics, Method, and Epistemology*, ed., James Faubion (New York: The New Press, 1998) pp. 369-392.
8. We have made use of the recordings made by Gilbert Burlet and Jacques Lagrange in particular. These are deposited in the Collège de France and the Institut Mémoires de l'Édition Contemporaine.

This page intentionally left blank

ABBREVIATIONS

The following abbreviations are used in the endnotes:

DE, I-IV,   *Dits et écrits, 195-1988,* ed., D. Defert and F. Ewald avec la collaboration de Jacques Lagrange (Paris: Gallimard, 1994) 4 volumes.

"Quarto," I   *Dits et écrits, 1954-1975,* ed., D. Defert and F. Ewald avec la collaboration de Jacques Lagrange (Paris: Gallimard, "Quarto," 2001).

"Quarto," II   *Dits et écrits, 1976-1988,* ed., D. Defert and F. Ewald avec la collaboration de Jacques Lagrange (Paris: Gallimard, "Quarto," 2001).

EW, 1   *The Essential Works of Foucault, 1954-1984. Volume 1: Ethics, Subjectivity and Truth,* ed. Paul Rabinow (New York: New Press, 1997).

EW, 2   *The Essential Works of Foucault, 1954-1984. Volume 2: Aesthetics, Method, and Epistemology,* ed., James D. Faubion (New York: New Press, 1998).

EW, 3   *The Essential Works of Foucault, 1954-1984. Volume 3: Power,* ed., James D. Faubion, (New York: New Press, 2000).

This page intentionally left blank

one

7 JANUARY 1981

The fable of the elephant in Saint Francis of Sales. ⌒ *Versions of the fable in the Middle Ages and sixteenth century.* ⌒ *The* Physiologus. ⌒ *Versions of the fable in Greek and Latin antiquity.* ⌒ *The endpoint with Aristotle.* ⌒ *The "subjectivity and truth" relationship: philosophical, positivist, historico-philosophical formulations of the problem.* ⌒ *Subjectivity as historical relationship to the truth, and truth as historical system of obligations.* ⌒ *Principles of monogamous sexual ethics.* ⌒ *The privileged historical question.*

[WITH THE] RATHER SOLEMN title *Subjectivity and Truth*, you will forgive me for beginning with some remarks on the life of elephants, and to quote a text that so delights me that I really think I have quoted it to you before, or at any rate referred to it, in one of the last courses[1]: "The elephant is obviously only a huge beast, but it is the most dignified that lives on earth and one with the most intelligence. I want to speak of a feature of its decency. The elephant never changes its mate and loves tenderly the one it has chosen, with which, however, it mates only every third year, and then for only five days and so secretly that it is never seen in this act. However, I have seen it on the sixth day, when, before anything else, it goes straight to a river in which it washes its whole body, not wanting to return to the herd before it is purified. Are these not beautiful and decent feelings on the part of such an animal?"[2]

You will have recognized Chapter 39 of the Third Part of the *Introduction to the Devout Life*. Saint Francis of Sales continues by noting that, of all the examples, all the lessons that nature can give to human kind, that of the elephant is clearly one of the most commendable, and it would be good for all married Christians to be inspired by it. When the sensual and voluptuous pleasures, part of the vocation of married people, are over, says Francis of Sales, the latter, like the elephant, should straightaway purify themselves of those sensual and voluptuous pleasures. They should purify themselves, they should "wash the heart and affection of them" and, Francis of Sales concludes,[3] this counsel, given to men by nature, is remarkably in line with the excellent doctrine that Saint Paul gives to the Corinthians.[4]

Well, I would like to wander for a bit in this fable of the elephant as model and emblem of good conjugal conduct.[5] This idea of an elephant's good conduct and the use of the elephant as emblem, as illumination of good conjugal sexuality, is not just a part of the spiritual rhetoric of the start of the seventeenth century. At any rate, in the same period, someone who is in no way a spiritual author, but a naturalist—Aldrovandi,[6] who, as you know, had quite considerable influence on all the natural sciences and zoology (in the seventeenth century at any rate)—also placed the animal among the highest figures of morality. He sang the praises of the elephant's munificence, temperance, fairness, faithfulness, and indulgence. He noted that the elephant is repelled by anything that is not reasonable, that furthermore it does not like idle discourse and pointless words, and that generally this animal is a veritable document (*documentum*[7]) for morals and virtues. And among all the good qualities that make the elephant a document for human use, the most famous, for Aldrovandi and others, is again and always its chastity, at least in what concerns sex. They practice the greatest chastity, Aldrovandi says.[8]

What is remarkable is that one and a half centuries later, someone like Buffon will still admire the way in which the elephant combines the greatest social virtues, which makes the elephant herd a model for human society: it is prudent, courageous, calm, obedient, and faithful to its friends.[9] To all these grand virtues, which assure the cohesion of the social body in the elephant herd, he adds great constancy, intensity, and restraint in its amorous relationships as well. The elephant may well be bound to the group to which it belongs by the virtues I have just mentioned, but when the females are on heat the great attachment it has for the society of which it is a part gives way, Buffon says, to a more intense feeling. The herd then splits into couples formed beforehand by desire. "The elephants unite by choice; they hide away, and love seems to precede and modesty to follow them in their steps; for mystery accompanies their pleasure. They are never seen to mate; they fear above all being seen by their own kind and perhaps know better than us that pure sensual pleasure of climax in silence, and of being concerned only with the beloved object. They seek out the thickest woods; they reach the deepest solitude to give themselves up without witnesses, disturbance, and reserve to all the impulses of nature: these are all the more intense and durable as they are rare and long-awaited. The female carries for two years: when she is pregnant, the male abstains from her; and it is only in the third year that the season of mating returns."[10]

You see that Buffon's text shifts a certain number of accents in comparison with the texts of Saint Francis of Sales or Aldrovandi I have just quoted. Where in Aldrovandi and Saint Francis of Sales it was above all a matter of the kind of repugnance that the sexual act arouses in elephants, to the extent that they feel the need to purify themselves as soon as they have finished, Buffon puts more emphasis on the heat of desire, the awakening of this desire even before marriage, in a sort of prior betrothal. Buffon also stresses the intensity of a pleasure increased by secrecy. He stresses the values peculiar to the intimacy of the sexual act, an intimacy that the forest protects, but that modesty makes even more intense. And all this plays a much more important role than the somewhat disgusted concern for purification after the act. But anyway, and this is what I want to emphasize, with slightly more specific accents in the eighteenth century, the lesson provided by the good and virtuous elephant is still and always that of good monogamy, of the right and decent sexual behavior of the couple.

It might be thought that this lesson of the elephant appears only with what is thought to be the conjugality peculiar to the modern family and its morality. Now what is remarkable is that this theme of the elephant as emblem of good conjugal sexuality absolutely does not appear with the ethical rigor of the Reformation or Counter

Reformation, nor with the formation of what is generally thought to be the morality of modern conjugality. In fact, if you try to produce what is obviously a purely anecdotal history of the elephant as natural emblem of matrimonial sexuality, you see that the fable is very old, and that this lesson appears to have been passed down through centuries, and even millennia. As pure and simple anecdotal introduction, I would like to go back a bit into the history of this ancient fable and the model of matrimonial sexuality that the elephant is supposed to bring from nature to human society. For example, in the middle of the sixteenth century, a naturalist like Gessner[11] recognized that elephants may possess a great amorous force. But, according to him, these animals were able to maintain a restraint, a quite remarkable continence in their relationships. Notwithstanding the violence of their amorous desire, they did not give way to an immoderate concupiscence. Elephants, Gessner says, do not know adultery, they never think of fighting for a female, and they have relationships only in order to have descendants, ceasing to touch their spouse once she has been impregnated.[12] Further back in history you find again the same indications. Albertus Magnus,[13] Vincent de Beauvais, especially, in his *Speculum naturale*,[14] also saw in the morals of the elephant a sort of animal illumination of conjugal virtue and the virtuous marriage.

In fact, the text of Vincent de Beauvais reproduces almost word for word a work that is very clearly from earlier, the famous *Physiologus*.[15] Throughout the Middle Ages, the *Physiologus* was one of the most widely distributed books, one of the most frequently copied manuscripts, and one of the constant vehicles of animal fables from Late Antiquity onwards. The author and exact date of the composition of this text are not known. There are quite explicit references to the *Physiologus* in several Christian authors: Saint Ambrose refers to it,[16] the *Hexameron* of the Pseudo Eustathius[17] refers to it explicitly, as well as the famous *Homily on Genesis* translated by Rufin and wrongly attributed to Origen.[18] So the *Physiologus* existed in the fourth century. It was probably written in an earlier period that is generally situated between the end of the second and the beginning of the fourth century.[19] In any case, the *Physiologus* is a Christian text and it proposes a Christian and allegorical reading of a number of features, properties, characteristics that ancient pagan naturalists attributed to animals and

to nature in general. It is a sort of allegorical interpretation of items of
Late Antique natural science. The life and character of a series of different animals, around forty in all,[20] are considered and reinterpreted
in terms of the scriptural information available to Christians. And this
was how the *Physiologus* linked scripture and nature, the Bible and the
natural science of antiquity, in an interesting and revealing way.

What, according to the *Physiologus*, would the scriptural blazon of
the elephant and its female be? Quite simply, of all animals, the elephant and its mate tell us about the Fall in the best and clearest way:
the relationship between man and woman before the Fall, the Fall
[itself], and what happened and has to happen after it. In fact, the
Physiologus says, when the elephant wants to procreate, it turns to the
East, that is to say towards Paradise. And it reconstructs, it re-stages
as it were, the original scene of humanity, namely, eating the forbidden
fruit and the Fall itself. In fact, when, in what Buffon called the mating
season, the elephants head towards the East or Paradise, it is in search
of a tree, or rather a plant, the mandrake, whose ambiguous values
you are aware of, which here is like the figure of the forbidden tree.[21]
The elephants reach the mandrake and, of course, the female takes and
eats it first. As soon as she has eaten it she persuades the male to eat
it in turn. And when both have eaten it, they mate. I don't need to
tell you what this fable told by the *Physiologus* represents. Up to the
consummation of the sexual act the elephants reproduce the history
of original humanity. And what happens after refers to humanity after
the Fall and to the battle humanity has to wage for its own salvation.[22]
So, after this union, which follows eating the mandrake, the female is
pregnant. When ready to give birth she heads for a stretch of water.
She plunges into the water and gives birth to her calf only when she is
completely submerged. Why is she submerged? Because, of course, she
has to defend herself against the serpent who is on the look-out, there
on the banks of the river, the shore of the lake, the serpent who cannot
get at the calf because it is born in the middle of the water. Baptismal
water, of course. Diabolical serpent, obviously. And while this scene
takes place (birth of the baby submerged in the salvational water),
the male, who has followed his mate, protects her against the perpetual enemy and struggles with him. You see that, without need of much
further comment, we find in the mating of this admirable species, the

elephant, the image, the reproduction of the history of humanity.[23] The *Physiologus* concludes that, since the elephant represents to us what we were and what we must be, protecting us from the Devil and giving birth to our children only in the salvational baptismal water, it is good to burn in your home a piece of elephant skin and some bones. The smell is enough to drive the serpent from it, as the word of God drives evil from our heart.[24] Here we are present at the origin of the great Christian fable of the elephant, the full and complete expression of which is still given in the seventeenth century by Saint Francis of Sales.

What is interesting, at any rate for what I would like to talk about this year, is that the Christian authors were not the first to seek instruction on conjugal conduct from the old bestiary of morality. In fact, in this chapter as elsewhere, the *Physiologus* obviously only takes up the lessons already explicitly expressed by the authors of antiquity, modifying some of their features and above all adding to them all the scriptural correspondences you will have noted. If one wanted to study the complete and exhaustive history of the bestiary of conjugal chastity, one would have to go back beyond Christianity and cross the borders of what is called Christianity [to paganism]. If we take the plunge and, going back a bit in time, address the authors of Latin and Greek antiquity, we already find this same model of good conjugal behavior in the elephant.

Let's quickly pass over Solinus,[25] a third century compiler of little interest. For him, elephants never fight each other for mates, never commit adultery, copulate only every two years, and, having done so, do not return to their herd before washing themselves. This is what Solinus says in the third century.[26] In the second century, still in pagan literature, in the non-Christian naturalists, Aelian, in *De Natura Animalium*, also extols the elephant's moderation. He applies to the elephant the term that the philosophers use to designate the highest form of wisdom: he calls it the bearer of a *sōphrusunē* (a wisdom) that enables it to keep at bay the two major adversaries of this *sōphrusunē*, that is to say *hubris* (violence) and *hagneia* (debauchery, impurity).[27] Elephants are therefore the image of *sōphrusunē*: they have intercourse with their mate only once in their whole life, when the female consents to it—the lesson is even more strict and rigorous than in later authors. Moreover, again according to Aelian, they perform this sexual act only in order to have a descendant. As for the act itself, they would be ashamed to engage in it anywhere other than deep in a forest where they withdraw so as not to be seen.[28] Before Aelian, in Pliny,[29] there is a description of the elephant's good conduct that we can assume is, with a few variations, the description Saint Francis of Sales reproduces in the *Devout Life*—anyway, of all the ancient texts it is the one closest to Saint Francis' account. Pliny writes: "It is out of modesty ... that they mate only in secret; the male begets at five years, the female at ten; the female lets herself be covered only every second year and, it is said, during five days of each year, no more: on the sixth, the couple bathe in a river and rejoin their herd only after bathing. Adultery is unknown to them and they do not engage in those mortal combats for females that one observes in other animals."[30] But, in Pliny's view, this moderation does not exclude the elephant experiencing the force of love (*amoris vis*).[31] This is, in fact, manifested in the lengthy attachment of the partners, but it is also manifested in stubborn passions. And it is this that gives the elephant its moral value: it has a very strong propensity for amorous passion, but it knows how to control it, or to direct it only to the single one to which it is bound. This is how Pliny recounts an elephant's great, albeit somewhat adulterine love for a flower seller, who was herself the mistress of a grammarian. He also tells the story of an elephant so smitten by a young soldier of Ptolemy's army that it would rather not eat than be fed by someone else.[32] So much for Pliny's accounts of the love-life of elephants.

Can we go back further in this history of the fable of the elephant's love life? Clearly, the elements recounted by Aelian or Pliny no doubt come from earlier authors. Aelian seems mostly to have copied a text of Juba.[33] But what seems remarkable and in need of comment is that we see the fable of the elephant disappear when we reach Aristotle. Aristotle does speak of the elephant and indicates some of the elements that will be found again later, but he makes no attempt to make the elephant a model for human morality. The features of elephant life recounted in, for example, *History of Animals*, are infinitely less close to the eulogy of those cited later. "Elephants," Aristotle says, "go wild when in rut, and this is why those who rear them in the Indies do not let them cover the females: it seems in fact that, going wild at this time, they knock down their master's houses."[34] Here, then, is the amorous force-violence of the elephant's feelings, but without that principle of moderation for which Pliny, Aelian, and later authors praise them. The periodic character of the elephant's mating also appears, but as a purely physiological detail: "The female elephant lets herself be mounted at ten years old at the earliest, fifteen at the latest. But the male mates at five or six years old. Spring is the mating period. After mating, the male mates again after two years ... Gestation lasts two years and only one young is born."[35]

As for the act itself, Aristotle confines himself to describing it thus: "elephants copulate in secluded places, preferably near water"— this is the element that will be found again later, charged with moral values—"and where they usually live."[36] There is only one feature of the elephant's behavior that Aristotle relates explicitly to a sort of moral, almost intellectual quality: this is the respect the elephant is supposed to have for the female once she has been impregnated. In fact, in a text that I think is also in the *History of Animals*, but I am not very sure, he says: "All the senses of the elephant are very developed and its faculty of understanding exceeds that of the [other] animals." And as proof and testimony of this faculty of understanding attributed to elephants, he says: "when the elephant has covered a female and impregnated her, he no longer touches her."[37] This respect for the progeny, or at any rate for the female during gestation, is a sign of reason for Aristotle, but it is the only element he finds that is exemplary for human conduct.

Forgive this somewhat ambling introduction to the history of the elephant and the fable it provides to humankind. I will halt the survey there for a number of reasons. Two reasons, to tell the truth— both of them obvious. The first is that before Aristotle the Greeks quite simply did not know about elephants, because it was, of course, with the expeditions of Alexander that they became acquainted with the animal—there are only one or two references to the elephant in Herodotus, and that is about all.[38] The other reason is more serious and more important. This is that the theme of the moral exemplarity of nature is not very old in the Greeks. Let us be more precise. Of course, there was an old tradition that you find in the Pythagoreans, and in others, too, of identifying something like a moral lesson in this or that story, this or that animal fable. But the idea of making a systematic reading of nature that constantly and in each of its elements provides a moral lesson for human conduct is something that Greek or

Hellenistic culture knew only relatively late on, let us say anyway after Aristotle. At least two fundamental ideas had to be brought together for nature to provide a permanent lesson of conduct for humankind in this way. The first is that nature had to be considered as being governed by an overall and coherent rationality. The idea of a general, omnipresent, and permanent government of nature, that nature was well and truly governed, that is to say traversed by a rationality whose expression and evidence could be found in its different elements, processes, figures, and so on, had first to be accepted. So the idea of nature as general and rational government of the world was necessary to make the different elements of zoology function as permanent examples for human conduct. One also had to have the idea that to be virtuous, to be rational, man did not merely have to obey the particular and specific laws of his city, but that there were much more fundamental and doubtless much more important general laws in nature and in the order of the world than the particular rules that might define this city or State. That is to say, to have the idea of a permanent moral exemplarity of nature, there had to be not only the idea of a rationally governed nature, but also the idea that the law that should govern human behavior had to be derived not only from the City State, but also and even more fundamentally from an order of the world that individuals, human beings, fell under, even before they came under the laws of their own city. You can see what all this refers to: to a sort of moral theme or ideology on the one hand, and to a whole cosmology on the other, that is usually attributed to Stoic thought in general, but which in a more overall sense belongs to Hellenistic thought and the historical, social, and political modifications of the Hellenistic world. That is why there would obviously be no point in looking beyond Aristotle for the roots of this fable of the elephant as emblem of conjugal virtue, even if some anecdotal ones can be found.

Now, with all the excuses I can offer for beginning the course in this way, I owe you some explanations. Why begin a course on "subjectivity and truth" with a fable of animality? There are several reasons. The first concerns the very nature of the question I would like pose; the second concerns the historical domain in relation to which I would like to pose this question; and the third concerns the method I would like to employ.

First: the type of questions that I would like to pose through this theme of "subjectivity and truth." There is, let's say—not to seek a more [specific]* word—a philosophical way of posing the question of "subjectivity and truth." This is the way used from Plato up to Kant at least, and that might be summarized in the following way. In the philosophical tradition, the "subjectivity and truth" problem consists in wondering how and on what conditions I can know the truth, or: how is knowledge as experience peculiar to a knowing subject possible? Or: how can someone who has this experience recognize that it is indeed a matter of true knowledge? Let's say that the philosophical problem "subjectivity and truth" may be characterized in this way: to resolve the tension between two propositions. Obviously there cannot be any truth without a subject for whom this truth is true, but on the other hand: how, if the subject is a subject, can he actually have access to the truth?

To this philosophical formulation of the "subjectivity and truth" question one could oppose what I shall call here—again in a very hasty fashion and for convenience's sake—a positivist formulation, which would be the question the other way round: is it possible to have a true knowledge of the subject, and on what conditions? With regard to the subject, with regard to the form and content of subjective experiences, is it technically possible, is it theoretically legitimate to apply the procedures and criteria peculiar to the knowledge of any object? If you like, the positivist question would be this: how can there be truth of the subject, even though there can be truth only for a subject?

From these two, well-known types of question, I would like to try to distinguish a third. One might in fact formulate, develop the "subjectivity and truth" question in the following direction: what experience may the subject have of himself when faced with the possibility or obligation of acknowledging something that passes for true regarding himself? What relationship does the subject have to himself when this relationship can or must pass through the promised or imposed

---

* M. F. says: specialist.

discovery of the truth about himself? Thus formulated, the question is, I will come back to it shortly, a fundamentally historical question. I would say that it is a factual question. Probably in every culture, every civilization, every society, at any rate in our culture, our civilization, and our society there are a certain number of true discourses concerning the subject that, independently of their universal truth value,[*] function, circulate, have the weight of truth, and are accepted as such. In our culture, in our civilization, in a society like ours there are a number of discourses on the subject that are recognized institutionally or by consensus as true. And the historical problem to be posed is this: given these discourses in their content and form, given the bonds of obligation that bind us to these discourses of truth, what is our experience of ourselves in light of the existence of these discourses? And in what ways is our experience of ourselves formed or transformed by the fact that somewhere in our society there are discourses considered to be true, which circulate and are imposed as true, based on ourselves as subjects? What mark, which is to say as well, what wound or what opening, what constraint or what liberation is produced on the subject by acknowledgment of the fact that there is a truth to be told about him, a truth to be sought, or a truth told, a truth imposed? When in a culture there is a true discourse about the subject, what is the subject's experience of himself and what is the subject's relationship to himself in view of the fact of this existence of a true discourse about him?

The first, philosophical way of formulating "subjectivity and truth" relations is summed up in a word: it is the question of the possibility of a truth for a subject in general. The second way, which I have called positivist, thinks about the possibility of telling the truth about subjectivity. And the third way, which may call, if you like, historico-philosophical, wonders what the effects on subjectivity are of the existence of a discourse that claims to tell the truth about subjectivity.

It is around this third and final way of posing the "subjectivity and truth" question that, for a certain number of years, from a greater or lesser distance, with more or less clarity, I have tried to work. The question can be approached from different [angles]. With regard to madness, illness, and crime, for example, we may think about how types of

---

[*] M. F. adds: and I will not pose this question.

practices are formed involving the existence and development of true discourses about alienated reason, the sick body, or the criminal character, and how the relationship we have to ourselves—and by "relationship to ourselves" I mean not just our relationship to our own individuality, but our relationship to others inasmuch as they are also ourselves—is affected, modified, transformed, and structured by this true discourse and the effects it induces, the obligations it imposes, and the promises it suggests or formulates. How do we stand, what should we do, how should we conduct ourselves, if it is true that there is and must be a certain truth about us, and what's more a truth told to us through what we push furthest away from us, namely madness, death, and crime? It is partly this question that ran through the work I have done until now.

You see anyway that in posing the "subjectivity and truth" question in this way we can keep two or three things in mind—this is all that I will hold onto for the moment, but for us it must serve as the shared basis for the work that remains to be done.

First, in posing the question of "subjectivity and truth" through these historical problems of madness, illness, and crime, subjectivity is not conceived of on the basis of a prior and universal theory of the subject, it is not related to an original and founding experience, it is not related to an anthropology that has universal value. Subjectivity is conceived as that which is constituted and transformed in its relationship to its own truth. No theory of the subject independent of the relationship to the truth.

Second, in posing the problem in this way, truth—I referred to this just now—is not defined by a certain content of knowledge that is thought to be universally valid, it is not even defined by a certain formal and universal criterion. Truth is essentially conceived as a system of obligations,[39] independently of the fact that it may or may not be considered true from this or that point of view. Truth is above all a system of obligations. Consequently it is completely immaterial that something may be considered true at one moment and not at [another]. And we may even accept this paradox[, namely:] that psychiatry is true. I mean that seen as systems of obligations, even psychiatry, even criminology are analyzed as true, inasmuch as the systems of obligations peculiar to true discourse, to the enunciation of the truth, to veridiction, are in actual fact present at the root of these types of discourse. What is important in this question of truth is that a certain number of things actually pass for true, and that the subject must produce them himself, or accept them, or submit to them. So, what has been and will be at issue is the truth as bond, as obligation, and also as politics, and not the truth as content of knowledge or as formal structure of knowledge.

Third, I want to stress this, equally: if all these analyses are necessarily carried out through historical material, the objective of this historical material is not to show the extent to which truth is changeable and definition of the subject is relative, but the way in which subjectivities as experience of self and others are constituted through obligations of truth, through the bonds of what could be called veridiction. The constitution of experiences of self and others through the political history of veridictions is what I have tried to do until now.

I would like now to apply [these same choices of] method or point of view to another domain, to the domain of what has been called—and for a relatively short period of time, moreover, less than two centuries—sexuality.[40] This domain of sexuality presents a certain number of important differences from the previous themes of madness, illness, death, and crime. The first of the differences from madness, illness, death, and crime is our ambiguity [vis-à-vis] the domain of sexuality. Whereas we have a fundamentally negative relationship, a relationship of rejection, to madness, death, illness, and crime, and whereas the question of the truth of madness, illness, death, and crime is posed only on the basis of this rejection and refusal, the problem of sexuality is quite different, since, whatever systems of regulation, disqualification, repression, or even rejection it is exposed to, sexuality is not the object of a systematic, fundamental, and more or less constant rejection. It is the object of an always complex interplay of refusal and acceptance, valorization and devalorization—this is the first difference, consequently, between sexuality on the one hand, and crime, illness, death, and madness on the other.

The second difference is that in the case of madness, illness, death, and crime, true discourse about the subject was essentially pronounced from outside, by another person. It is to the extent that I am not mad, it is to the extent that the doctor as doctor does not have to be ill, it is inasmuch as I am not a criminal and that the person who speaks about the crime is not himself a criminal, that a true discourse could be delivered on illness, madness, crime, death, and so on. You know,

however, that a true discourse about sexuality has been institutionalized in a completely different way. And I refer, as I will, moreover, several times, to last year's lectures on the problem of truth-telling about
oneself in Hellenistic societies and at the beginning of Christianity.[41]
In the case of sexuality, true discourse has been institutionalized, to a
large extent at least, as the subject's obligatory discourse about himself. That is to say true discourse about sexuality is not organized on
the basis of something given as observation and examination according
to accepted rules of objectivity, but is organized around the practice
of confession (*aveu*). Let us say that with regard to sex and desire—
and this is where the history of this problem is completely different
from that concerning madness, crime, death, and so on—true discourse
is organized around a discourse of confession (*aveu*) about a part of
ourselves that we may well detest, or of which we may purify ourselves,
but which is nevertheless inseparable from what we are. Discourse of
confession about an inseparable part of ourselves: it is around this that
the problem of the relationships of "subjectivity and truth" regarding
sex should be understood. In the previous cases cited (madness, illness, death, crime, and so on), the problem was: what experience can
we have of ourselves and others when there is someone who has the
right to say: "That person is mad, you are ill, such a person is criminal?" With sexuality, the problem I would like to pose is this: what
experience can we have of ourselves, what type of subjectivity is linked
to the fact that we always have the possibility and right to say: "Yes,
it's true, I desire?" In sum, in the societies that concern us and to
which we are bound by bonds of kinship whose solidity will indeed
need to be examined, it is a matter of studying the way in which the
subject has been called upon to manifest and recognize himself, in
his own discourse, as being in truth subject of desire. More simply,
I would like to focus this year's lectures [around] the history of the
notion of concupiscence[42] inasmuch as it organized the subjective experience of sexuality, or rather of sex and sexual relations.[43] So that is the
first reason for taking this rather silly little fable about the elephant's
good sexual conduct as our starting point.

The second reason for my choice of this story of the elephant is the
period I would like to study concerning the relations between truth
and subjectivity. Going back into the history of the moral elephant, the

elephant-emblem of good sexuality, you will have noted that a point pulled me up, and I stressed this a bit: this is the moment when, following the ascending line of time, we see this fable pass from pagan to Christian literature. It is this interplay, this communication between pagan and Christian literature that caught and will keep my attention. In the *Physiologus*, a text of Christian allegory on themes of natural history, the fable of the elephant was already completely Christianized. On the other hand, in Aelian, an author from the second century [whose texts are therefore not] much earlier than the *Physiologus* but who echoes a whole pagan literature, we find the same fable, and in Pliny, and so on. Now this passage from the literature of pagan naturalists to Christian literature raises an interesting historical problem. In fact, when one reads the fable of the elephant in Saint Francis of Sales, it is quite evident, as he says himself, that it is a question of presenting a model of sexual behavior for human beings. Saint Francis of Sales says so, and so too does Aldrovandi, a naturalist: the elephant is a document (*documentum*) for human conduct. He employs an interesting expression, saying that the way in which the elephant conducts himself with regard to his female should give men an *idea morum*[44] (a sort of model of conduct). Now we are quite familiar with this model of good sexual conduct. In this mode it is relatively easy to recognize a type of morality whose origin, sometimes the merit, and more often the fault, is attributed to Christianity. It seems that when Saint Francis of Sales and Aldrovandi recount the fable of the elephant all they are doing is reminding us, through this image, of the fundamental principles of what I would call a gloomy sexual ethics, since it is an ethics that conforms to a certain number of principles that are usually attributed to Christianity and that can be summarized as follows.

[First:] the principle of monogamous sexuality, of indissoluble conjugality. One must have one and only one wife, one must keep her throughout one's life, and this bond with the wife must not be broken by adultery. This union presupposes a very strong attachment between the two partners. Second: the principle of strict economy in sexual relations, a strict economy that is established in the first place by, of course, the condemnation of any extra-marital sexual intercourse. But, even within marriage, you see that according to Aldrovandi, according to Saint Francis of Sales, the fable of the elephant directs us to the need for a voluntary and considered rarefaction. Even with one's wife, one must not make love in any way and at any time. A principle of rarity is needed, a principle of economy due to the very temperance of the character, a rarity that has to limit sexual relations to the sole end of procreation (sexual intercourse is permitted simply in order to be able to create, to procreate when one wishes), and a rarity that prohibits touching the female for the whole period of gestation. The third principle found in this fable of the elephant recounted by Aldrovandi and Saint Francis of Sales: the rule of the isolation of sexual activity. Sexual relations must be separated from everyday existence. Elephants withdraw from their familiar place, from their place of existence, for reasons of modesty, linked to the fact that sexual intercourse is permitted [only between husband and wife]* and in the secrecy that prevents any gaze from outside. Sexual intercourse is something one hides, even when it is a legitimate relation between married individuals. And finally, the fourth principle: the principle of purification. After sexual intercourse one cannot return to the light of day and the society of others without an ablution, which points to an intrinsic impurity of the act and some stain that one must get rid of.

So, in this model of the elephant recounted by Aldrovandi and Saint Francis of Sales, we have a sort of description of good sexuality (as matrimonial, monogamous, necessarily infrequent, and nonetheless impure) that we usually attribute to Christian morality and its intrinsic bleakness. Now, through this thin history of the elephant that I have told you, you can see that this model was already conveyed, term for term, by the naturalists of the first to second century CE, outside of any Christian influence. The model, the sexual *pattern* of the elephant presented by the Christian authors, was formed wholly and explicitly well before Christianity. Consequently, we are led to ask ourselves this question: is the code of sexual morality attributed to Christianity actually specifically Christian? Or could it not have been created elsewhere, earlier, and by what is generally called paganism? You will say that maybe the question of whether some of the fundamental principles of our sexual code were formed by or before Christianity is ultimately rather pointless. Why would it be pointless? In the first place

---

* M. F. says: only of couple to couple.

because, after all, is it that important to know if and when this or that element of our morality was formed, if our morality was formed in the philosophy schools of the first centuries before [and] after Christ, or simply in Christian milieus? Maybe the problem is what we should do with our morality rather than where it comes from. And because of this, the question of its origin might indeed be pointless. It might [also] be pointless because the question has something of the naive look of responsibilization. Voltaire's fault, Rousseau's fault? Is paganism or Christianity to blame for our morality? It could be [finally] that the question of the origin of our fundamental moral codes regarding sexuality is rather pointless because the idea of these two (Christianity and paganism) facing each other as perfectly formed and well individualized wholes is obviously very questionable and naive.

Obviously, I can only be in agreement with this last point, namely that Christianity and paganism are not well-formed wholes, perfectly clear cut individualities. And while conducting the analysis that I will develop later, I will attempt to take account, first of all, of the fact that paganism is in no way an unchanging, unitary, and stable system, and also of the fact that whatever institutional unifications or dogmatic systematizations there may have been (due to the organization of Christianity as a Church), it would be imprudent to talk of a single Christianity, as if one Christianity existed. In any case, in the first Christian centuries, it is necessary to take a plurality of experiences into account, all of them Christian but very different from each other. Between the Encratite[45] currents on the one hand, and then the canonical decisions on the problem of divorce or remarriage, there is a difference of at once context, objective, and thought. Again, there is a whole world between the Alexandrians of the end of the second century, like Clement of Alexandria, who repatriates a whole morality of marriage into Christianity that is very much in line with what might have existed in previous centuries, and then the Cappadocian Fathers of the fourth century, who are completely impregnated with a new mysticism of virginity, and so on. Paganism cannot be treated as a single unit, and Christianity perhaps even less so. So the question I will try to pose is not that of whether our morality is Christian or pagan, but: when, how, within what processes, within what milieus, within what practices or institutions, pagan or otherwise, were the principal elements of our sexual morality formed. It remains true nonetheless that this question of when our morality was formed, and what took place exactly in that period situated at the end of what is called the ancient world and [the beginning] of Christianity is important. One cannot dispense [with this study] if one wants to pose the theoretical question, the practical question, and the political question of morality, and of sexual morality [in particular] of sexual morality.

I would say that most major theoretical developments, most of the major theoretical questions posed, always have a privileged associated historical question. For example, currently I do not think one can conduct any kind of political reflection without recognizing the need to associate it with a certain privileged historical question. [This is:] what is the French Revolution? What took place between 1789 and 1800? There cannot be a political reflection, however theoretical and general, nor can there be any political question, however actual, however contemporary, that [does not include], as its projected shadow, the historical question of what took place at the moment of the Revolution. I would say the same with regard to the theory of science. It seems to me that one cannot undertake a theory of what science is in general without linking it with an historical question that is its projected shadow, as it were. And the historical question that is the projected shadow of general reflection on science is: what took place when mathematical physics was formed? What took place with Kepler, Galileo, Newton, and so on? Just as there cannot be any political reflection without the question of what the French Revolution was, so there cannot be any theoretical question about science that, in one way or another, does not take up the question of what took place historically with, and around Kepler, Galileo, and Newton.

Well, I would say the same with regard to morality: it seems to me that no moral reflection, however theoretical and general, no moral question, however contemporary, can avoid an associated historical question that is its projected shadow, and that would be: what took place in the first century CE, at the turning point of what is called pagan ethics and Christian morality? In the history of our morality, this historical problem is associated with any general question or any political question concerning our morality, as the question of the foundation of mathematical physics is associated with any reflection on science, as the question of the French Revolution is associated with any political reflection.

So it is around this historical question, associated with any problem of morality, that I would like in part to focus my reflections on "truth and subjectivity." Let's stop there, and, next week I will sketch out some remarks on problems of method.

1. The "model of the elephant" is not referred to in any of Foucault's previous courses.
2. Saint Francis of Sales, *Introduction à la vie dévote* [1609], III, 39, text established and presented by Charles Florisoone (Paris: Les Belles Lettres, "Collection des universités de France," 1961) vol. II, p. 117; English translation by Fr. Antony Mookenthottam, Fr. Armind Nazareth, and Fr. Antony Kolencherry, *Introduction to the Devout Life* (Bangalore: S.F.S. Publications, 1995, 2nd revised edition) Part 3, ch. 39, 154-155: "The elephant is only a huge beast but the most dignified and most intelligent animal which lives on earth. I wish to tell you an instance of its excellence. It never changes its mate and loves tenderly the one it has chosen. However, it does not mate with it except every third year, and that for five days only, and so secretly that it is not seen doing this act. Nevertheless, it is seen on the sixth day on which before anything else, it goes straight to some river. There it washes completely its whole body without any wish to return to the flock before it is purified. Are not these beautiful and chaste characteristics of such an animal an invitation to the married?"
3. Ibid., Fr., "It [the elephant] invites husband and wife not to remain fondly attached to the sensual and voluptuous pleasures they will have practiced according to their vocation, but to purify themselves as soon as possible, in order afterwards to practice the purer and more elevated actions with freedom of spirit. This advice consists in the perfect practice of the excellent doctrine that Saint Paul gives to the Corinthians"; Eng., "They [the married couple] are not to remain entangled in attachment to sexual pleasures which they have experienced according to their vocation. When they are over, they are to wash their hearts and affection for them, and purify themselves as soon as possible. Thus, soon after, they can practice with freedom of spirit other purer and more noble actions. The perfect practice of the excellent teaching which St. Paul gives to the Corinthians consists in this counsel."
4. *Corinthians*, I, 7:29.
5. Foucault will take up this example of the elephant-model in his Introduction to the *Histoire de la sexualité*, vol. II: *L'Usage des plaisirs* (Paris: Gallimard, "Bibliothèque des histoires," 1984) "Un schéma de comportement," pp. 23-24; English translation Robert Hurley, *The Use of Pleasure. Volume 2 of The History of Sexuality* (New York: Pantheon Books, 1985) "An ideal of conduct," p. 17.
6. Ulisse Aldrovandi (1522-1605) left a work of encyclopedic ambition in the domains of botany, zoology, mineralogy, and medicine. He was the first professor of natural history at Bologna. In *Les mots et les choses* (Paris: Gallimard, "Bibliothèque des sciences humaines," 1964) Foucault uses his works to describe the Renaissance episteme (references to *Monstruorum historia* in ch. II: "La prose du monde," § "Les quatre similitudes," pp. 32-40; English translation A. Sheridan, *The Order of Things. An Archaeology of the Human Sciences* (London: Tavistock Publications, 1970), ch. 2, "The prose of the world, I. The four similitudes," pp. 27-25).
7. I. Aldrovandi, *De quadrupedibus solidipedibus. Volumen integrum*, ch. IX: "De elephanto" (pp. 418-479), Bononiae, apud Nicolau Thebaldinum, 1649 [1639], p. 418: "In eo [omni vitae tenore] enim excellit elephas, quod a se natura suggerente & morum & virtutum documentum praebeat, quod homines nisi doceantur vix assequi posse videantur."
8. Ibid., in particular the section, "Coitus. Partus," pp. 454-456.
9. G.-L. L. Buffon, *Histoire naturelle* [1749], in *Œuvres complètes*, t. XV (Paris: Eymery, Fruger et Cᵉ, Libraires, 1829) pp. 60-166.
10. Ibid., pp. 74-75.
11. C. Gessner, *Historiae animalium*, book I: *De quadrupedis viviparis* [1551] (Francofurti: in bibliopolio H. Laurentii, 1620) pp. 376-403.
12. "Nec adulteria nouere, nec ulla propter foeminas inter se praelia, caeteris animalibus pernicialia: non quia desit illis amoris vis (nam & homines ab eis amatos constat ...). Ab omni immoderata libidine castissimi sunt. Numquam enim, neque et construpratores, neque ut item valde lasciui societatem Veneris cum foemina faciunt, sed tamquam generis suecessione carentis, liberis procreandis dant operam sic hi, sua stirps ut ne deficiat, complexu Venereo iuguntur" (ibid., p. 382).
13. Albert le Grand, *De animalibus* [1258; Rome ed., 1478; Venice, 1495] (Münster: Aschendorff, 1916 and 1920); English translation Kenneth Kitchell, Jr. Albertus Magnus, *On Animals: A Medieval "Summa Zoologica"* (Baltimore: John Hopkins University Press, 1999).

14. Vincentius Bellovacensis, *Speculum Naturale* [1624], bk. XIC, ch. xliv (Graz: Akademische Druck-und Verlagsanstalt, 1964) pp. 1406-1407). The first part of *Speculum majus* was written between 1240 and 1260.
15. The *Physiologus* is a Christian bestiary. It offers an allegorical interpretation of animal behavior. Written in Greek but quickly spread across the Latin world, it is one of the reference texts for the Middle Ages. Foucault probably used the Droz edition with the Latin text, *Physiologus latinus*, ed. and trans., Francis James Carmody (Paris: Droz, 1939).
16. In particular in Saint Ambrose *Hexaéréron* (*Divi Ambrosii Hexameron*, c. 389), in *Patrologiae cursus completus omnium SS. Patrum, doctorum scriptorumque ecclesiasticorum sive Latinorum, sive Graecorum, Patrologia Latina*, ed., Jean-Paul Migne, t. 14, 1885; see, for example, 6, 13, 3; English translation John J. Savage, *Hexameron, Paradise, and Cain and Abel. Fathers of the Church* (Washington, DC: The Catholic University of America, 1961) vol. 42.
17. Eustathius of Antioch, *Ancienne Version latine des neuf homélies sur l'*Hexaéméron *de Basile de Césarée* [*Eustathii in Hexaemeron Basilii Caesareae Cappadociae episcopi latina translatio*], ed., Emmanuel Amand de Mendieta and Stig Yngve Rudberg (Berlin: Akademie Verlag, 1958).
18. Origen, *Homélies sur la Genèse* [*In Genesim Homiliae*], trans., Louis Doutreleau (Paris: Éd. du Cerf, 1944); English translation Ronald E. Heine, *Homilies on Genesis and Exodus. Fathers of the Church* (Washington, DC: The Catholic University of America, 2001) vol. 71.
19. The oldest mentions of the text are from the fourth century (the *Hexameron* of Saint Ambrose, that of the Pseudo Eustathius, the treatises of Rufinus of Aquileia). But it could have been known already in the second and third centuries by authors like Origen and Justin. As it uses a method of interpretation formulated by the school of Clement of Alexandria, following Philo, the text was probably written in the second century, in what was then the center of Christian philosophy and culture: Alexandria.
20. 49 chapters in the first anthology (ed., Offermans, 1966, from the lion to the cuckoo) to which can be added nine chapters in the second (from the water horse to the hare), and six in the third (from the wolf to the pheasant—ed., Sbordone, 1936).
21. *Physiologos. Le bestiaire des bestiaires*, translated from the Greek, with introduction and commentary by Arnaud Zucker (Genoble: Éd. Jérome Million, 2004) p. 234: "The mandrake, the Hebrew name of which (*dudha'im*) has the same root as love, is actually nothing other than the tree of temptation. Its aphrodisiacal properties, no doubt necessary to awaken the ardor of the cold and prudish animal, are attested to by Rachel (*Genesis*, 30, 14), who cured her sterility with the plant of red or white berries, also called 'love apples'; they are also referred to by the wife in the *Song of Songs* (7, 14). It is the face of the other glimpsed in paradise, whose root resembles a human face and that Pythagoras calls 'anthropomorphous' (Theophrastus, *History of Plants*, 9, 8, 8); this miniature silhouette that it offers innocently makes it a panacea (Dioscorides, *De Materia Medica*, 4, 75), suitable for treating all affections: 'for whatever member causing pain, eat the corresponding member of the plant and one will bear it better' recommends the holy Abbess Hildegarde (*Plantes*, 56)." See also, J. Bersez, *Les Extraordinaires Pouvoirs et vertus de la mandragore: plante magique de tous les temps ...* (Vaduz: Éd. du Lion d'or, 1990).
22. "The male elephant and his mate correspond to the figures of Adam and Eve: when they were in the delights of paradise, before their transgression, they did not yet know the carnal relationship and did not even have the idea of coupling; but when the woman had eaten from the tree, that is to say spiritual mandrakes, and had given it to her spouse, *then Adam knew his wife, and she gave birth to Cain* on the waters of reprobation": *Physiologos. Le bestiaire des bestiaires*, p. 235; emphasis in the text.
23. "This animal [the elephant] is wholly ignorant of sexual desire. When he wishes to have an offspring he withdraws east towards paradise. The female accompanies him. There is a tree there called 'mandrake.' The female tastes the tree first and then offers it to her mate, and she pesters him until he tastes it too. When he has tasted it, the male comes to the female and mates with her; and she conceives straightaway. When the time arrives to give birth, she heads for a marsh of water and enters it up to her udders. It is then that she gives birth to her young in the water. The latter rises on his legs and sucks its mother's udders. The male elephant watches over her during the birth, because of the serpent, for the serpent is the elephant's enemy. And if he encounters a serpent, he tramples it underfoot and kills it" (ibid., p. 234).

24. "When one burns some elephant hair or bone somewhere, no evil spirit draws near to that spot, and no evil is found there" (ibid.).

25. Iulius Solinus [Gaius Iulius Solinus], *Collectanea rerum memorabilium* (25, 1-15), ed., Theodor Mommsen (Berlin: In aedibus Friderici Nicolai, 1895 [1864]).

26. Ibid., 25, 5-6, p. 111: "Venerem ante annos decem feminae, ante quinque mares nesciunt. Biennio coeunt quinis nec amplius in anno diebus, non prius ad gregarium numerum reversuri quam vivis acquis abluantur. Propter feminas numquam dimicant: nulla enim noverunt adulteria."

27. Aelian, *La Personnalité des animaux*, VIII, 17, ed. and trans., Arnaud Zucker (Paris: Les Belles Lettres, CUF, 2001) t. I, p. 215: "It is perfectly appropriate to say how their sexual continence is explained (*sōphrosunēs des opōs meteilēkhasin*). During mating, they [the elephants] do not approach the female with lust (*os hubrizontes*) and as if they were going to rape her (*ōs lagnoi*), but as beings who desire an heir of their blood and aim at procreation, in order not to be deprived of a common offspring and to leave a descendant"; English translation A. F. Scholfield, *On the Characteristics of Animals*, vol. II, Books 6-11, (Cambridge, MA: Harvard University Press, Loeb Classical Library, 448, 1959) Book VII, 17, pp. 200-201: "it is most fitting to state that they have been gifted with temperance. For they seek intercourse with the female not as though minded to commit an outrage or from lust, but like men desiring a succession to their family and to beget children, in order that their common offspring may not fail but that they may leave their seed after them."

28. Ibid., Fr., "Also, they pursue the pleasures of Aphrodite only once in their life, when the female agrees to it. Every elephant, when he has made her pregnant, no longer has relations with his companion. What is more, they do not copulate without shame in view of others, but in a withdrawn spot. They shelter behind the thick trees, in a dense coppice, or in a deep hollow that enables them to pass completely unseen"; Eng., pp. 200-203: "At any rate once only in a life-time do their thoughts turn to love, when the female herself submits. Then when each one has impregnated its mate, thereafter it knows her no more. And they do not couple without reserve or in the sight of others but withdraw and screen themselves in thick trees or in some close-growing forest or in some deep hollow, which affords them ample means of hiding."

29. Pliny the Elder, *Histoire Naturelle*, VIII, 1-11, trans., Alfred Ernout (Paris: Les Belles Lettres, CUF, 1952) pp. 23-24; English translation H. Rackham, *Natural History* (Cambridge, MA: Harvard University Press, Loeb Classical Library 353, 1940) vol. III, Book 8, 1-11, pp. 11-13.

30. Ibid., VIII, 5, 13, Fr., p. 27; Eng., p. 12.

31. Ibid., VIII, 5, 31, Fr., p. 28: "Not that they are unaware of the power of love (*amoris vis*)"; Eng., p. 13: "they are [not] devoid of strong affection."

32. Ibid., VIII, 5, 13-14, Fr., pp. 27-28: "An elephant is referred to that was enamored of a flower seller; and don't think this was a random choice: she was the favorite of Aristophanes, the famous grammarian. Another was smitten by Menander, a young Syracusan who served in the army of Ptolemy, and when he did not see him, he showed his regret by not eating"; Eng., p. 13: "it is reported that one elephant in Egypt fell in love with a girl who was selling flowers, and (that nobody may think that it was a vulgar choice) who was a remarkable favorite of the very celebrated scholar Aristophanes; and another elephant is said to have fallen in love with a young soldier in Ptolemy's army, a Syracusan named Menander, and whenever it did not see him to have shown its longing for him by refusing food."

33. The King Juba II of Mauretania (c. 52 BCE—23 CE) was a great scholar. His works on the history of Rome, sadly lost, were a reference for historians like Titus Livy and Pliny the Elder. The latter cites him at several places in his *Natural History*. See J. Lahlou, *Moi, Juba, roi de Maurétainie* (Paris: Éd. Paris-Méditerranée, 1999).

34. Aristotle, *Histoire des animaux*, VI, 18, 571b-572a, ed. and trans., Pierre Louis (Paris: Les Belles Lettres, CUF, 1969), t. II, p. 103; English translation d'A, W. Thompson, *History of Animals*, Book VI, 18, 571b-572a, in *The Complete Works of Aristotle*, ed., Jonathan Barnes (Princeton: Princeton University Press, Bollingen Series LXXI.2, 1984) vol. 1, pp. 897-898: "Male elephants get savage about pairing time, and for this reason it is stated that men who have charge of elephants in India never allow the males to have intercourse with the females; on the ground that the males go wild at this time and turn topsy-turvy the dwellings of their keepers."

35. Ibid., V, 3, 540a, Fr., p. 5; Eng., pp. 862-863: "The female elephant becomes sexually recep-tive when ten years old at the youngest, and when fifteen at the oldest; and the male is sexu-ally capable when five years old, or six. The season for intercourse is spring. The male allows an interval of three years to elapse after commerce with the female ... The period of gestation with the female is two years; and only one young animal is produced at a time."

36. Ibid., Fr., V, 3, 540a, p. 6; Eng., V, 2, 540a, p. 853: "Elephants ... copulate in lonely places, and especially by river-sides in their usual haunts."

37. Ibid., Fr., IX, 45, 630b, t. III, p. 136: "Of all the wild animals, the elephant is easiest to tame and domesticate. For it learns many things and understands them since one even trains it to prostrate itself before the king. Its senses are highly developed and on the other hand its fac-ulty of understanding is superior to that of other animals. When it has covered a female and made her pregnant, it no longer touches her"; Eng., IX, 46, 630b, p. 980: "Of all the wild animals the most easily tamed and the gentlest is the elephant. It can be taught a number of things, and it understands them; as, for instance, it can be taught to kneel in presence of the king. It is very sensitive, and possessed of an intelligence superior to that of other animals. When the male has had sexual union with the female, and the female has conceived, the male has no further intercourse with her."

38. Herodotus, *The Persian Wars*, vol. II, trans., A. D. Godley (Harvard, MA: Harvard University Press, Loeb Classical Library, 1921) Book IV, 191 (on the elephants in Eastern Libya).

39. On the truth as obligation, see M. Foucault, *Du gouvernement des vivants. Cours au Collège de France, 1979-1980*, ed., M. Senellart (Paris: EHESS-Gallimard-Le Seuil, "Hautes Études," 2012), p. 92; English translation Graham Burchell, *On the Government of the Living. Lectures at the Collège de France 1979-1980*, English series editor Arnold I. Davidson (London: Palgrave Macmillan, 2014) pp. 94-95.

40. See M. Foucault, *Histoire de la sexualité*, t. I: *La Volonté de savoir* (Paris: Gallimard, "Bibliothèque des histoires, 1976) p. 91; English translation Robert Hurley, *The History of Sexuality. Volume I: An Introduction* (London: Allen Lane, 1979) p. 68: "nearly one hundred and fifty years have gone into the making of a complex machinery (*dispositif*) for producing true discourses on sex: a deployment (*dispositif*) that spans a wide segment of history in that it connects the ancient injunction of confession to clinical listening methods. It is this deployment (*dispositif*) that ena-bles something called 'sexuality' to embody the truth of sex and its pleasures."

41. M. Foucault, *Du gouvernement des vivants*; *On the Government of the Living*.

42. Foucault envisaged devoting the second volume (*La Chair et le Corps; The Flesh and the Body*; fragments of this partially destroyed manuscript are cited in P. Chevallier, *Michel Foucault et le Christianisme* (Lyon: ENS Éd., 2011) pp. 149-150) of the first *History of Sexuality* announced in *La Volonté de savoir*; *The History of Sexuality. Volume One: An Introduction* in 1976, to "a genealogy of concupiscence through the practice of confession (*confession*) in Western Christianity and of spiritual direction, as it developed from the Council of Trent" (D. Defert, "Chronologie," in *DE*, I, p. 53; "*Quarto*" vol. I, p. 73). In the subsequent research Foucault will take this genealogy of concupiscence back to the first centuries CE: "in order to under-stand how the modern individual could experience himself as a subject of a 'sexuality,' it was essential first to determine how, for centuries, Western man had been brought to recognize himself as a subject of desire ... the slow formation, in antiquity, of a hermeneutics of the self" ("Introduction" to *L'usage des plaisirs*, p. 11; *The Use of Pleasure*, pp. 5-6). The fourth volume of *The History of Sexuality*, as realized in the eighties (*Les Aveux de la chair; Confessions of the Flesh*), remaining unpublished, regarding the construction and the use of the notion of concu-piscence in Foucault, one can read: *Les Anormaux. Cours au Collège de France, 1974-1975*, ed., V. Marchetti and A. Salomoni (Paris: Gallimard-Seuil, "Hautes Études," 1999) lectures of 19 and 26 February 1975, pp. 155-215; English translation Graham Burchell, *Abnormal. Lectures at the Collège de France 1974-1975*, English series editor A.I. Davidson (New York: Picador, 2003) pp. 167-230; "Le combat de la chasteté" (given by Foucault as to be read as a chapter from *Les Aveux de la chair; Confessions of the Flesh*), *DÉ*, IV, no. 312, pp. 295-308; "*Quarto*," vol. II, pp. 1114-1127; English translation Anthony Forster (amended by editors), "The Battle for Chastity," in Michel Foucault, *EW*, I, pp. 185-198; and "Sexuality and Solitude" (analysis of the notion of concupiscence in Saint Augustine), *EW*, 1, pp. 175-184.

43. In fact, as often happens, the movement of research will lead rather to Foucault studying, through the examination of the arts of living of the Hellenistic and Roman period, a new

problematization of sexuality, of desire, and of the couple, that Christianity will broadly inherit.

44. U. Aldrovandi, *De quadrupedibus solidipedibus*, ch. IX, p. 418: "Multa sunt, quae ab eo [ab elephante] tamquam a morum idea homines desumere & imitare coacti fuerint."

45. The Enkratites (from the Greek *egkrateia*: continence, self-control) formed a particularly ascetic and rigorist current of Christianity (formal prohibition of sexual relations, eating meat, and drinking alcohol—wine prohibited even for the Eucharist, celebrated with water) that arose in Syria in the second century and that took on a certain importance in Rome in the fourth and fifth centuries. It was finally proscribed by Theodosius I. According to Irenaeus of Lyon, it was founded in the second century by Tatian the Assyrian, a disciple of Justin.

two

14 JANUARY 1981

[
Return to the fable of the elephant. ∽ *The arts of living: typology*
and evolution. ∽ Mathēsis, meletē, askēsis: *relationship*
to others, the truth, and oneself. ∽ *Notes on the concepts of*
"paganism," "Judeo-Christianity," "capitalism," as categories of
self-analysis of Western societies. ∽ *Problem of the pre-existence*
of "Christian sexual morality" in Stoicism.
]

SO I HAVE BEGUN this course, entitled a bit pretentiously
Subjectivity and Truth, with the account of the fable of the elephant as
it is found equally in the naturalists, the moralists of antiquity, and
Christian authors: a fable in which this sympathetic animal is presented as the model of good sexual conduct (model of temperance,
chastity, modesty, and so on). This elephant, moral emblem of nature,
[gave] then five great lessons to attentive humanity: the principle, first
of all, of monogamy and conjugal fidelity as the general framework of
sexual relations; second, within this framework of monogamy and conjugal fidelity, infrequency of the sexual act; [third,] reproduction and
solely reproduction as the aim of this sexual act; [fourth,] the need
for secrecy, discretion, modesty (the sexual act is not part of ordinary,
public, visible, everyday life); and finally, the obligation to purify oneself after this act. This is what the elephant recounted to men.

There are several reasons for choosing this little fable to introduce a
subject with quite other pretensions. I indicated two of them last week.

On the one hand, I said that this fable seemed to me to be significant for the following reason. This year, I would like to raise the question of subjectivity and truth as I have raised it in earlier studies, that is to say [by taking] the other side of the classical philosophical problem, namely: how, to what extent, up to what point, on what grounds, and according to what rights can subjectivity ground knowledge of the truth? This is the classical philosophical question. And the problem that I would like and have already tried to pose is in a way the opposite. It is a question of the experience we may have of ourselves, the field of subjectivity that may be opened to the subject for himself, when there exists in fact, historically, in front of him, in relation to him, a certain truth, a certain discourse of truth, and a certain obligation to be bound to this discourse of truth—either so as to accept it as true, or so as to produce it himself as true. What, therefore, are the effects of the existence of this truth and of this discourse of truth for our experience of ourselves? What is our relationship to reason and madness when there exists a science of or knowledge of madness that claims to be true? How may we experience ourselves, and how may we experience our own sexuality, or how may sexuality appear as a field of subjective experience, when with regard to sexual practice, to sexual activity, there exists a certain knowledge that claims to be true? This is the type of question I would like to pose. This was the first reason for choosing to begin with this fable of the elephant.

The second reason was of an historical order. It seems to me that if one wants to recover the formation of this experience of sexuality in the least bad way, understood as the mode of relationship that may exist between our consciousness of ourselves and the discourse of truth, there is an important, particularly fertile period, [namely:] between what one calls Christianity and paganism, that is to say in the last centuries of Hellenistic and Roman history and the first centuries of what one calls the Christian era. It is in that framework, within these historical reference points, that I would like to situate myself. And the fable of the elephant, [which] passes from the discourse of the naturalists (like Pliny or Aelian) into Christian discourse, seems to me to be a point to consider in detail in order to understand this passage, to understand this moment.

The third reason I chose to begin these lectures with the fable of the elephant is one of method. Obviously, this fable of the elephant is at once ordinary, banal, flat, and of little interest. At any rate, it does not offer any great theological, anthropological, or ethical insights that could ground sexual morality in Christianity or earlier philosophical thought. So it provides no insight into the great fundamental principles of morality. Nor any insight, you will have noted, into what, after all, constituted the great system of prohibitions in sexual practice and life in our societies. There is nothing, for example, about the prohibition of incest, nothing on homosexuality, as if the elephant's image could not even be soiled by such references. In this fable of the elephant one has the flattest, most mediocre form of a series of little pieces of advice (advice for life, rules of conduct). This fable of the elephant belongs to a minor genre, that of the arts of conducting oneself, of the arts of living, of advice for existence.

However insignificant it may appear, I think this literature harbors a quite considerable interest; it seems to me that it should enable us to clarify some of the problems I would like to pose. This literature concerning the way to conduct oneself, modes of life, ways of being, was very extensive throughout antiquity, and in particular in the Hellenistic and Roman period, in the first centuries of Christianity that precisely I would like to talk to you about. This literature on the arts of living, on the art of conducting oneself, lasted for a very long time and has now disappeared. No one today would dare write a book on the art of being happy, the art of not getting angry, the way to lead a tranquil life or achieve happiness, and so on. Which does not mean that in a society like ours models of conduct are not formed, proposed, circulated, more or less imposed or absorbed. There is an art of conduct in our kind of society, but it has absolutely lost its autonomy. One no longer finds these models of conduct except, of course, invested, packaged within the big, thick, massive practice of pedagogy. It is through pedagogy that a major part of these instructions of existence is conveyed. There is also all of what may be called social stereotypes, which offer models of good behavior through the intermediary of literature, of word or image. It needs to be said again that what are called the human sciences, at whatever level of utilization one takes them, from the lowest to the highest, also more or less explicitly convey schemas that are thought to be good schemas of existence, good models of conduct. But it is a fact that since the seventeenth-eighteenth centuries, although this needs to be studied more closely, there is no longer an autonomous and specific literature with the intention of saying, basically, how to live. Now, it is clear that in this fable of the elephant I have been talking about, we have, reduced to its simple expression, as a simple little anecdotal schema, a small fragment of the art of living, of the art of behaving.

I would like to linger on this notion of art of living as it existed, as it was recognized in antiquity, and in any case in the first centuries CE. In reality the arts of living made up a whole domain covering completely different objects. For example, the art of living might focus on what were looked upon as significant moments of existence. The art of dying: how to prepare for it, how to think about it in advance, how to make arrangements for the moment of one's death. Also, when one feels death is coming, how should one comport oneself, behave, what should one say, how should one be at that precise moment of the arrival of death? It might also be a moment of misfortune, exile, or ruin. The art of living also consists in knowing how to behave, to comport oneself, to do and think what is right when one experiences a bereavement, when someone passes away; so an art of living applied to those [particular] moments of life. [It is] interesting to study them to see precisely what, in a given society or in the kind of civilization I am referring to, might be the significant moments, the essential moments of existence, those through which an existence might acquire its meaning. Death, exile, ruin, bereavement, and so on. It is very interesting to see this dissemination, this distribution of the essential moments of life.

Some formulae of the art of living focus not on moments of life, but on activities, particular activities. For example, there is the art of rhetoric that teaches the activity of speaking in public—you will see shortly moreover that in reality it involves something completely different and much more than this. There are arts that are even more circumscribed: the famous art of memory, which to some extent is part of the art of rhetoric, the art of memory* that Yates has studied in a book that should obviously be read most urgently, although it was published in France at least fifteen years ago and few people refer to it.[1]

---

* The manuscript adds: "this is a very precise technique, but it is also a whole knowledge (*savoir*), a whole attitude."

A whole part of these arts of living is not addressed to moments of life, nor to this or that particular activity, but to what could be called the general regime of existence. The regime of the body, of course, and here these arts of the body are bound up with what is called medicine. Or rather: a large part of Greek and Roman medicine was essentially devoted not so much to saying how to treat illnesses, as to defining an art of living that is, as it were, a physical, bodily, physiological, and psychological art of living.[2] The regime of the soul also: how to control, to master one's passions, and so on. Innumerable treatises on anger,[3] for example, come under this art of living regarded as general regime of existence. But even more broadly I think that some regimes of life, and to some extent all, even the most specific, refer to a certain conception that enabled the description and definition of different modes of life to which individuals might desire access and the means of gaining access to that life: public or private life, active life or life of rest, contemplative life, and so on. I will try to come back to this later when we talk about Christianity and the transfer of the rules of the contemplative life to the monastic institution. The arts of living therefore cover a quite considerable domain. But I think it is necessary to be a bit more specific.

These arts of living, which extend from the art of confronting something specific in existence to the art of gaining access to some (general, complete, definitive) mode of life are distinctive in that it is less a matter of them teaching people how to do something as teaching them above all how to be, how to succeed in being. Simplifying a great deal, and keeping it as hypothesis—all that I am telling you is not the result of work, but a possible program for possible work—, one could say that with Christianity — and this becomes quite clear in the Renaissance—the emphasis in these arts of living, which continue to exist then up to the seventeenth-eighteenth centuries, increasingly focused on the "activity" side, on the side of "how to do something." If you read books on the arts of living in the sixteenth century, from Erasmus to the moralists of the seventeenth century (for the seventeenth century moralists are still people who write arts of living), it seems to me that the problem is increasingly, and in an increasingly marked way over time, the question of doing: how to comport oneself, how to behave, what to do in order to acquire this or that aptitude, what type of relationship one should establish with others, how to appear in public, how to behave decently, and so on. For example, the particular study of the art of dying is quite revealing regarding this evolution. With Christianity, in the Renaissance, you see the arts of dying, which were so important from antiquity up to the thirteenth century, becoming increasingly collections of advice, manuals, little practical books concerning gestures, attitudes, and postures, the clothes one should wear, the words one should utter. It is not so much knowing how to be in relation to death, how not to fear it, how not to miss those who are dead. Above all, and increasingly clearly, these arts of living are concerned with defining good behavior towards others when death becomes the main problem for oneself or others. Consequently they are arts of doing, arts of behavior. Now it seems to me, and this is what I would like to stress, that the arts of living of antiquity, those found in the Hellenistic and Roman period, and also in early Christianity, focus less on the question of doing, although they also deal with this, than on the question of being, on the way of being. One learns from them not so much, or not only behavior, the performance of certain actions, conformity to a certain social model, but one learns to change one's being, to modify or model one's being to give oneself an absolutely specific type of experience.

Take, for example, the problem of the art of rhetoric. It is true that in a sense the art of rhetoric is a very technical art by which one teaches people to speak. It is a professional apprenticeship: one teaches them to speak in public, what gestures they must make, how to employ good arguments, how to make what is false appear true, and what is true appear false, and so on. All of this comprises technical formulae. But in antiquity the arts of rhetoric, the rhetor's apprenticeship, learning the rhetor's life, is nevertheless something else as well. One has to teach the person who has to become a rhetor a whole style of public life, a whole use of language in political and judicial institutions, a whole relation to others in collective life, a whole morality to be promoted in public relationships; one teaches him to be a public man, a man linked to others in a certain relationship of public life, of publicity, of politics, of social life, and so on.

Let us say—here again, as hypothesis, schema, something to be gone into much more—that the arts of living we see developing in Greek and Roman antiquity, at the beginning of Christianity, focus essentially on the being one is. They do concern the things one can do, but essentially and above all, insofar as one can transform what one is through the things one can and must do, through the actions one has to accomplish. Whereas, it seems to me that in their evolution from the end of the Middle Ages until the seventeenth-eighteenth centuries, the arts of living tend increasingly to define what needs to be done and to revolve around what could be called, roughly, schematically, a professional apprenticeship. The transition from art of living to professional training is obviously one of the major developments one can observe and one of the factors that led to the disappearance of the art of living as an autonomous genre of reflection and analysis.

In what do these arts of living (mainly focused around the question, not of how to do, but of how to be) consist? Here too, still in rapid overview, I would say this: viewed from the perspective of its objectives, the art of living should permit the acquisition of certain qualities that are neither abilities nor exactly virtues in the moral sense that these will be understood later, [but] rather qualities of being, qualities of existence—what I would call modalities of experience, qualities that affect and modify the being himself. Tranquility, for example (achieving the *vita tranquilla*[4]) is not just a certain style of behavior, it is not just a way of reacting minimally, or with the greatest possible control in the face of this or that accident. It is this, it is true, but tranquility is really something else. Tranquility is a certain quality of being, a certain modality of experience that means that the events that occur around one, in existence (favorable, unfavorable accident, and so on) produce the least possible effect on the individual and enable him to maintain his autonomy and independence in relation to them. Beatitude also is a quality that means that the entire being is happy whatever happens to him.[*] In short, these arts of living involve enabling the individual to acquire a certain ontological status that opens up to him a modality of experience describable in terms of tranquility, happiness, beatitude, and so on. Modification of being, transition from one ontological status to the other, opening up of modalities of experience: this is what is involved in the arts of living. You see that we are a

---

[*] The manuscript mentions a third objective of these arts of living: reputation, "that brilliance that radiates from the existence and that, even after death, remains in men's memory."

long way from the kind of learning an ability we find later. The methods by which these arts of living can thus bring about a change in the ontological status of the individual and give him qualities of existence might be summarized by saying that it is a matter of defining the complex work by which one can attain this ontological status of experience through: first, a certain relationship to others, second, a certain relationship to the truth, and third, a certain relationship to self.

Relationship to others means that the arts of living are learned. They are learned by a teaching, a listening (learning, teaching). That is to say, the presence of the other, his speech, and his authority are clearly indispensable in these arts of living. The arts of living are passed on, are transmitted, are taught and learned in and through a certain relationship of master to disciple. The activity of the master's direction of the disciple, the way in which he teaches and forms a conduct, are absolutely indispensable and constitutive of the art of living. One cannot learn the art of living on one's own, one cannot accede to the art of living by one's own means, without this relationship to the other, this relationship to direction, to the other's authority, without this at least provisional power relationship that means that one submits to the other and to his teaching until one has attained the ontological status that enables one to deploy, for oneself and in full autonomy, the mode of experience to which one aspires. After the relationship to the other, the second element in these arts of living is the relationship to the truth. That is to say that every art of living entails not only that one learns it, but, as we would say in our vocabulary, that one interiorizes it. In any case it is necessary that one thinks it oneself, that one reflects on it, that one meditates on it. Knowledge has to be brought into play, a knowledge that one receives, of course, but that will then become a sort of permanent reference in one's existence. There is no art of living without the necessity for a periodic resumption of the teaching received from the master, a periodic resumption that one carries out in and for oneself, reflecting, recalling precepts given, examining what is taking place either in or around oneself, reading books, and gathering one's thoughts, so that the teaching and the truth it brings really become our own thought or our permanent and constant relationship to the truth. Relationship to others through learning, relationship to the truth through permanent reflection, and finally relationship to self (this is the third element of the work peculiar to the art of living) that entails a whole ascesis, a whole series of exercises. Attempts to do this or that, checking what one has done, self-examination, examination of the faults one may have committed during the day, examination of what one has to do to arrive at a particular outcome, progressive, increasingly difficult trials, until finally, through all this series of tests, one ends by recognizing that one has really attained the ontological status that one sought, that one really is affected by the quality of being one was aiming for. All this constitutes the element of the ascesis of oneself, of the exercise, the work of self on self.

Relationship to others, teaching: this is what was called *mathēsis* in the Greek terminology of the arts of living. Relationship to the truth, that is to say: permanent reflection and ceaseless resumption of what has been taught and what must be considered to be true, is what the Greeks called *meletē* (meditation, reflection on). And finally this work of testing, of successive trials to see where one is and if one has really progressed, is what the Greeks called *askēsis* (ascesis). *Mathēsis, meletē,* and *askēsis* are the three elements found [in the arts of living]. Think of the third discourse of Epictetus,[5] for example, in which you have a definition of the path to wisdom, that is to say of what someone must be who learns this art of living as a sage proposed by philosophy. You will see that Epictetus regularly gives you these three elements, according to a formula that is his but that is generally that of the Stoics, and we can say [that these elements are to be found] in all the great arts of living, even if [their theorization is especially Stoic]. *Mathēsis, meletē, askēsis*: teaching (*mathēsis*), relationship to the other; meditation (*meletē*), relationship to the truth; exercise (*askēsis*), work of self on self, relationship of self to self.

Despite the dull and commonplace character of the advice they offer, these arts of living nevertheless constitute a highly structured practice that has an important sense, signification, and a relatively dense, coherent, rich, and so on internal architecture. I want to say one more word regarding them. I have employed several words to designate what these arts of living are about. I have spoken of art of existence, of transformation of being, of action of self on self. In fact the Greeks, the Greeks not the Latins, have a word that designates quite precisely what these arts of conducting oneself must be brought to bear on.

This is the word *bios*.[6] You know that for a Greek there are two verbs that we translate with one and the same word: "to live." You have the verb *zēn*, which means: to have the property of living, the quality of being alive. Animals live, in this sense of *zēn*. Then you have the word *bioūn*, which means to pass one's life and is related to the way of living this life, the way of leading it, conducting it, the way in which it may be described as happy or unhappy. *Bios* may be good or bad, whereas the life one leads because one is a living being is simply given to you by nature. *Bios* is life that can be qualified, life with its accidents, its necessities, but also the life one may make oneself, decide oneself. *Bios* is what happens to us, of course, but from the angle of what we do with what happens to us. It is the course of existence, but in the light of the fact that this course is inseparably linked to the possibility of managing it, transforming it, directing it in this or that direction, and so on. *Bios* is the correlative of the possibility of modifying one's life, of modifying it in a rational fashion and according to the principles of the art of living. All these arts, all these *tekhnai* that the Greeks, and the Latins after them, developed so much, these arts of living bear on *bios*, on that part of life that falls under a possible technique, a considered and rational transformation. They have, moreover, a word to express all this, an expression that should be taken quite literally. They called this art of living (you find the expression in Epictetus, but you also find it in many others): *tekhnē peri bion*—the *tekhnē* that is applied to life, the technique that concerns existence understood as life to be led, the technique that enables this life to be fashioned.[7]

Of course, it could be said that these arts of living (*tekhnai peri bion*) are quite precisely bio-techniques. But clearly the word could hardly be used, for the meaning one gives to it now shifts us towards something else entirely.* So rather than use the term bio-technique to designate what is at issue in these arts of living, I would prefer to employ the expression technique or technology of the self, since in all these

---

* The manuscript offers the following development:
    "Biopoetics would be justified because it is indeed a sort of personal fabrication of one's own life (note that in these arts the question often arises of whether or not an act is beautiful). One could thus follow the problem of sexual conduct: biopoetics where it is a matter of the aesthetic-moral conduct of individual existence; biopolitics where it is a matter of the normalization of sexual conducts according to what is considered politically as the requirement of a population."

practices it is a matter of thought out, elaborated, systematized procedures taught to individuals in such a way that, through the management of their own life, through the control and transformation of self by self, they can attain a certain mode of being. This is the meaning that needs to be given to these arts of living. And the little fable of the elephant that I have just spoken to you about is basically just a miniscule fragment of this major genre of reflection, analyses, prescriptions, control, elaboration, and change of conduct that the Latins and Greeks practiced under the heading of *tekhnai peri bion* (techniques of life). I would like to pose the questions of these lectures on the basis of these techniques of life, these arts of life. They will serve as the guiding thread, or rather as the basic material for trying to pose them.

Actually, the "subjectivity-truth" problem is absolutely central in these arts of life. As I have tried to show you, by art of life we understand a technique by which the individual, not without a relationship to others but even so by himself, by exerting himself and acting on himself, tries to acquire a certain quality of being, a certain ontological status, a certain modality of experience. This acquisition of a modality of experience cannot be produced without an action of self on self, a relationship to the other, and a relationship to the truth. And in this sense, it seems fairly clear to me that by studying these arts of living more closely we could doubtless identify the way in which, in the Hellenistic and Roman period, a certain mode of connection between their relationship of self to self and to the truth was proposed to individuals. How could the truth, true discourse, the obligation to acknowledge a truth, the need to seek out a truth modify the [individual's] experience of himself? The subjectivity-truth connection is particularly legible in these arts of living. I am not saying that it is legible only there, [but it] is legible in them in big letters, as if through a lens. The arts of living are essentially methods and procedures [by which] individuals, by action on themselves, may modify and transform their experience of themselves by [referring] to a true teaching, truthful speech, the discovery or search for a certain truth. For this reason, the arts of living seem to me to constitute a particularly interesting and rich resource for the general problem I would like to pose.

[Furthermore,] it is quite clear that problems of marriage and of sexual activity are very important elements in these arts of living.

[In fact, first,] marriage is one of the most discriminating elements for dividing, separating, distinguishing different modes of life. For a very long time—[this] will be set out in a different form in Christianity—one of the features that distinguish the philosophical life, or the life of study (what will later be called the contemplative life) from the life of everyone else, is precisely that the sage, the person who is called to the contemplation of the truth or to a privileged relationship with God, does not need this human type of marriage relationship. The person capable of leading the life of pure contemplation of the truth is alone. In the set of prescriptions concerning the set of modes of life, the question of whether or not one should marry is clearly, if not central, at least one of the important questions and an unavoidable one. Second, the question of sexual relations and sexual activities is obviously important in this literature of the arts of living insofar as the problem of the economy of pleasures, of self-control, of mastery of the passions, of the limitation of involuntary impulses by which one is carried to this or that, is of course quite fundamental since the ontological status, the essential modality of experience towards which these arts of living are oriented, is obviously the total and perfect mastery of oneself. The problem of the mastery of oneself necessarily raises the question of sexual activity, of the economy of pleasures, and of the different modes of life* [...]. For all these reasons, the analysis of these arts of living is a precious, interesting, significant resource, and it is to this that I will address myself† [...].

And now let us return to the historical problem I would like to pose, the problem suggested by the fable of the elephant that I have recounted to you.‡ One is intuitively aware of the massive differences between what are called pagan ethics and Christian morality.

---

* A barely audible passage. One can hear only: contemplation, studies [...] public life.
† Gap in the recording. One hears only: general prohibitions.
‡ The manuscript offers the following development:
"And since in this problematic of kinds of life, the problem is not: how to be moral, obey the law, avoid committing errors, but: how to conduct oneself, how to establish a relationship of self to self that passes through the obligation to know the truth, it is understandable that it is here that one can most find the matrix of the experiences of sexuality. The working hypothesis is this: it is true that sexuality as experience is obviously not independent of the codes and system of prohibitions, but it needs to be recalled straightaway that these codes are astonishingly stable, continuous, slow to change. It needs to be recalled also that the way in which they are observed or transgressed also seems to be stable and repetitive. On the other hand, the points of historical mobility, what no doubt change most often, what are most fragile, are modalities of experience."

A whole kind of more or less confused historical memory, whose status should really be questioned, seems to indicate that, despite everything, between pagan morality and Christian ethics as it developed historically, there is a total difference, a gulf that is very difficult to cross and yet one that our societies crossed, and, it seems, quite quickly.* There is a kind of scansion that appears quite evident to us in the terms of our familiar knowledge, our quasi-intuitions. And then, when we look [further forward] (and it is precisely because of this that the fable of the elephant interested me and held my attention), a whole series of historical indications show that in fact the division, transition, or discontinuity that seems so massively obvious to us, undoubtedly did not arise like that. The history of the elephant [is significant in this respect]. Saint Francis of Sales presented [it] at the beginning of the seventeenth century as the classic example of good Christian morality regarding sexuality. [It is enough to trace back the history of this lesson to see] that it was already formulated in absolutely the same terms in people like Pliny, where it seems that Saint Francis of Sales had more or less directly looked for it. It is about this that we should say something now: how to establish this division, how to make the cartography of this "watershed," as Peter Brown expressed it,[8] between what one calls Christianity and what one calls paganism?† [...]

An interesting study could be made of this historiographical theme of paganism and Christianity. To start with, on the way in which the theme of paganism was formed, in which the kind of historical obviousness [was imposed] that there is a series of things that one can logically, that one can legitimately, as much from the theoretical as historical point of view, call paganism. In fact, when we look a little at

---

* M.F. adds this sentence: We may question whether they crossed it quickly, but it doesn't matter.
† Passage difficult to hear. One hears only:
> Before entering into a bit more erudite detail on all this, I would like all the same—again, forgive the very programmatic character of all this [...] but [...] problem of [...] Basically, what I am doing here [...].

M.F. adds:
> You know that this is the only "public" establishment in the strict sense in France, since everyone can come, without any formal inscription, without criteria of level or whatever. One speaks to everyone. Speaking to everyone is not easy, it raises many problems, it entails a whole work of uncertain adjustment to a public with fluid borders, such that it means that I am summarizing the relatively detailed work that I have been able to do too quickly for you. It also means, and this is what I am going to do again this time, that I am giving a kind of possible program for possible work, assuming that, after all, some of you are here for the encouragement of possible work.

what is put into the category of paganism (of pagan thought, pagan morality), we find things as incredibly different as: Pythagorean religiosity, the juridical-moral rules common in Roman society, Neoplatonist mysticism, the philosophical and abstract monotheism of the Stoics. All this is called paganism. So what is this idea of paganism? Where does it come from? It is quite clear that paganism was designated as a unified entity very early on. In actual fact—we have evidence of this in thought and institutions—from the second century, at any rate from the third century CE, Christianity saw itself as in a conflictual relationship with [both] paganism and heresy. Reading all the great heresiological literature of the second and especially the third-fourth centuries is very interesting for seeing how the division is made, how, confronting heresy, paganism, which is clearly a very different notion, was constituted and makes it possible to define a whole series of things that are supposed to be external to Christianity, not only as errors, but ontologically as it were, inasmuch as it is not a matter of Christianity at all. But let us leave to one side the constitution of the notion of paganism, of the pagan field, in the polemics and discussion of the first Christianity. It is a complex question and I am not the one who could talk to you about it and it is not what I would like to talk about here. On the other hand, it would be very interesting to see how this notion of paganism reappeared and how it was utilized, it seems, from the classical age (the seventeenth and, especially, the eighteenth century); how this notion functions in the analyses of the philosophers, historians, historians of religion, and so on, from the eighteenth century; how it is distinguished from the notions of barbarian, infidel, and primitive; how it functioned massively throughout the nineteenth century, the themes it covered. Paganism does include the Stoics and their monotheism, but the way in which this notion was made to function proves that paganism was connected above all to polytheism [and] to what was judged to have been, in a certain type of civilization, the extreme proximity, the kinship, almost the cohabitation of gods and men. A privilege granted to matter, to the body, and to everything making up the sensory world, is also attributed to paganism. An ethical tolerance concerning pleasures, and in particular concerning sexuality is attributed to paganism. Now all these themes that, explicitly or implicitly, permeate, animate, make vibrate and

shimmer this theme of paganism in the course of the nineteenth century, have a very specific role in relation to the analysis made of ourselves and others. On the one hand, paganism was in fact perceived in
the nineteenth century as something absolutely external to our world,
to our non-pagan world, to the civilization in which we are and from
which we would try to free ourselves, inasmuch as non-pagan civilization, Christian culture would have covered over a paganism that was
plural, polytheistic, warm, close to the divine, open to the sensory, tol-
erant of pleasures, and so on. This world would have been our world,
[but] would have been covered over and forgotten by Christianity. And
it is what we should strive for. [On the other hand,] we have a way
of criticizing our own society, analyzing what we are, what we have
to be, programming our liberation in terms of this theme of paganism
as it developed. Paganism is the other, and then it is a certain ground
of ourselves. And if we want to return to the ground of ourselves, it
is this paganism, this absolutely other, absolutely lost paganism that
we have to find again. Paganism is to historical consciousness, to the
consciousness we have of our irreversible history, a bit as nature has
been since the eighteenth century in relation to our consciousness of
our necessary technology. Beneath technique, nature. Behind our history, paganism. So it would be interesting to study the history of this
theme of paganism.

As [a complement]—[and] I do not think one could separate the
two studies—one would have to study the history of the theme that
is its vis-à-vis, that is to say Christianity, or more precisely Judeo-
Christianity. The notion of Judeo-Christianity is even more paradoxical, more astonishing than that of paganism. For if paganism somewhat
problematically covers a series of extremely different things, the notion
of Judeo-Christianity, which was also so important in the nineteenth
century and linked to the notion of paganism, is even more paradoxical, inasmuch as it was totally unthinkable for centuries and millennia.
Let us take as reference point, as point of departure, the first major
anti-Judaic and anti-Semitic texts that can be found in the Christian
literature of the fourth-fifth centuries; Saint Augustine is a marvelous
example.[9] Let's say that, from that moment up until practically the
end of the seventeenth century—thinking of Judeo-Christianity—a
sort of historical, trans-historical, meta-historical identity of Judaism

and Christianity was something strictly impossible. And then this notion became possible. Not only did it become possible, but it was one of the most frequently used categories in Western society's historical analysis of itself in the nineteenth century. What happened? It is very interesting to see that this notion of Judeo-Christianity, like that of paganism, always served critical intentions—always very confused, very ambiguous critical intentions, moreover, that should be deciphered. It is certain that in the historians or philosophers of the nineteenth century the coupling "Judeo-Christianity" was often a way of discrediting Christianity, of bringing to Christianity all the negative connotations that might affect Judaism, without saying so too directly. And the claim that Christianity was ultimately only the heir and continuation in us of whatever was the Jewish destiny. Jewish misfortune, Jewish abstraction, and Jewish legalism operated very strongly in the [nineteenth]* century. From the texts of Hegel on Jewish consciousness[10] to the *Genealogy of Morality*,[11] you have reference points for the constitution of this strange entity of Judeo-Christianity. [But this] does not mean that this notion has always received, or has continually carried negative values. Rather, it has carried positive values and it has been in the name of a Judeo-Christian tradition that the equally fictional tradition, or rather that completely fictional historical field called paganism, has been opposed.

All this is to tell you that the use of these notions ("paganism," "Judeo-Christianity") clearly becomes impossible as soon as one tries to filter it a bit and cast a critical eye on the history of thought in the nineteenth century. They are categories that were forged at that time, categories that have had interesting, important critical, political functions that are worth deciphering. But it is not possible for us currently to adopt either the notion of paganism or that of Judeo-Christianity. A few more words on the history of these two themes. If one were to study the history of this couple: paganism/Judeo-Christianity, it would be interesting to see how this theme intersects with another major category of the self-analysis of Western societies: capitalism.

---

* M.F. says: "eighteenth," but all the examples in the following sentence refer to the nineteenth century.

It could be said that in the nineteenth century two great categories of the self-analysis of Western societies were formed: the socio-
economic category with capitalism, and the socio-religious category
with Judeo-Christianity. And it is interesting to see how they have
functioned in relation to each other; for example, how French socialism, formed essentially from an analysis, pretty crude in comparison
to others, of the problem of capitalism, very quickly diverged in its
interests, in its sharp polemics, by according more and more importance to the notion of Judeo-Christianity. And when, around the years
1860, 1880, 1890, 1895, 1898, French socialism was permeated, if not
dominated by anti-Semitism,[12] this was the divergence of the category
of self-analysis defined in economic terms by capitalism towards this
other category of self-analysis, that of Judeo-Christianity. The problem
is why, in terms of which reasons did this drift take place in the ideological or social field; in what respect did the German approach differ?
For if you take the Hegelian or post-Hegelian movement up to around
1840-1848, you realize that the Young Hegelians' self-analysis of the
West is above all an analysis in Judeo-Christian terms. What about religious consciousness, what is the status of our religious consciousness,
in what respect can this religious consciousness effectively account
for what we are? Feuerbach represented exactly the summit of this
form of analysis focused on these categories of Judeo-Christianity and
paganism,[13] and then the evolution takes place in exactly the opposite direction to that of the French. That is to say, through Marx and a
certain number of others in German socialism, it is the economic categories, notably those of capitalism, that finally prevailed in this self-
analysis of the West. And that is why it may be thought that, at the
end of the nineteenth century, the greatest work of synthesis carried
out in Western thought to bring about its own self-analysis is found
in Max Weber. Max Weber is the person who precisely tried to combine in the most rational way possible, and according to an historical
knowledge and analysis as positively grounded as possible, the categories of religious [and economic] self-analysis. What is the situation
of Christianity, of Judaism, of the specificity of our religious civilization in relation to economic questions?[14] And conversely: how have
economic processes actually made possible the filtering, the incrusta-
tion, the implantation of this or that type of religious consciousness?[15]

It is Max Weber who tried to compress together as much as possible in one and the same framework of self-analysis, the category of Judeo-Christianity and the category of capitalism as modes of consciousness, analysis, and decipherment of Western societies.

This is all a bit long-winded. So let's drop paganism and Judeo-Christianity, telling ourselves that [these categories] are not part of an historical methodology, but may be at most the object of an historical study. Let us ask how to make this analysis that I would like to propose to you now, namely: what are the transformations regarding what is called sexual morality that can be observed between a period that historiography has called pagan and a period that historiography has called Christian, following these transformations through the arts of living?

So two or three more minutes to have done with all the questions of method and be able to tackle the content of the analysis next week. Let's say—this is a fact that historians have recognized for almost a century now*—that the sexual morality attributed to Christianity had already given perfectly clear and well-defined proofs of existence before the spread of Christianity, and even before its appearance within the ancient world. There is a pre-existence of this so-called "Christian sexual morality" within so-called pagan thought and morality. And this pre-existence is usually attributed in a privileged way to the Stoics and to a whole series of texts, some well-known, others less so, like those, for example, of Seneca, of course, but also of Musonius Rufus, Hierocles, Antipater, and so on. This is absolutely right and all the examples I will try to show you next week confirm this impression. It is absolutely true that the Stoics played a fundamental and decisive

---

* The manuscript clarifies:
  "For almost a century, three major moments. End of nineteenth century: role of the university: secular morality, morality for a State; Kulturkampf: Zahn; Epictetus; Bonhöffer. 1930: sage, saint, hero Stelzenberger: preparation. 1957: Spanneut establishes that a pre-existing Stoic morality is taken up by the Christians; this enables him to found a 'Christian universalism,' a Christian humanism, not on an Aristotelian type of philosophy of knowledge, nature, and science (which is what Gilson does), but on a moral philosophy."

role in this whole affair of the arts of living and of the economy of sexual relations.

Simply, two cautions for this kind of analysis. First, it should be stressed that with these texts it is always a matter of late Stoicism. If you take Stoicism as a coherent form of philosophy, born, developed on the basis of the great principles of Zeno and Chrysippus, you risk making a mistake, inasmuch as in Zeno, and very often in Chrysippus, you do not find these elements of sexual morality. They represent rather an inflection, if not a deviation in relation to early Stoicism. Paul Veyne says that in reality it was a veritable interpolation of sexual morals within Stoic philosophy itself.[16]

Second, it should be stressed that many authors who are not Stoics, who are even explicit adversaries of Stoicism, like Plutarch for example, give exactly the same type of advice in their arts of living and consequently convey the same sexual morality as the Stoics. Adversaries of the Stoics like Plutarch, neo-Pythagoreans like Sextius,[17] the Cynics too, and on many points also the Epicureans, give the same type of application of their art of living to the question of sexual relationships and sexual activities. Consequently, the arts of living that I will study are to a great extent of Stoic origin, but they are texts from late Stoicism, inflected by a certain number of processes. [Furthermore,] there will be a certain number of, and even not bad texts that are not of Stoic origin. That is to say that ultimately we will see emerging a certain art of leading one's life with regard to sexual relationships that is more or less common to most of the philosophical schools in the centuries preceding or which initiate precisely what we call our era. Next week I will talk about sexual relationships in the arts of living at the end of Antiquity, in the last centuries of what is called pagan Antiquity.

1. Francis Yates, *The Art of Memory* (Chicago: University of Chicago Press, 1966); French translation Daniel Arass, *L'Art de la mémoire* (Paris: Gallimard, "Bibliothèque des histoires," 1975).
2. On this definition of medicine as art of living, see M. Foucault, *Histoire de la sexualité, t. III: Le Souci de soi* (Paris: Gallimard, "Bibliothèque des histoires," 1984) pp. 121-126; English translation Robert Hurley, *The Care of the Self* (New York: Random House, Pantheon Books, 1986) pp. 101-104; as well as *L'Herméneutique du sujet. Cours au Collège de France, 1981-1982,* ed., F. Gros (Paris: Gallimard-Seuil, "Hautes Études," 2001) pp. 93-96; English translation Graham Burchell, *The Hermeneutics of the Subject. Lectures at the Collège de France 1981-1982,* English series editor Arnold I. Davidson (New York: Palgrave Macmillan, 2005) pp. 97-100.
3. See, for example, Seneca, *De la colère,* trans., Abel Bourgery, in *Dialogues, t. I* (Paris: Les Belles Lettres, CUF, 1922); English translation J. W. Basore, "On Anger," in *Moral Essays* (Cambridge, MA: Harvard University Press, Loeb Classical Library, 1928) vol. I; Plutarch, *Sur les moyens de réprimer la colère,* trans., Vicotor Bétolaud (Houilles: Éd. Manucius, 2008); English translation W. C. Helmbold, "On the Control of Anger," in *Plutarch's Moralia* (Cambridge, MA: Harvard University Press, Loeb Classical Library, 1939) vol. VI; Philodemus of Gadara, *De ira liber [Philodemi De ira liber],* ed., Carolus Wilke (Lipsiae: B. G. Teubner, 1914). For a history of treatises on anger in the Hellenistic-Roman period, Foucault cites in *The Hermeneutics of the Subject,* p. 392, n.4; Paul Rabbow's work, *Antike Schriften über Seelenheilung und Seelenleitung auf ihre Quellen untersucht, Bd. I: Die Therapie des Zorns* (Leipzig: B. G. Teubner, 1914).
4. See Seneca, *On Tranquility of Mind,* trans., John W. Basore, in *Moral Essays* (Cambridge, MA: Harvard University Press, Loeb Classical Library, 1979) vol. II.
5. Foucault is doubtless referring here rather to the fourth discourse of Book I (*Peri prokopês:* Of progress), which actually describes the advance towards wisdom, emphasizing the requirements of teaching (by a master or books) as much as the importance of exercises and meditations.
6. In the lecture of 25 March (below p. 253) Foucault will endeavor to give a definition of the Greek *bios* in contrast with the Christian and modern "subject."
7. "For as wood is the material of the carpenter, bronze that of the statuary, just so each man's own life is the subject-matter of the art of living (*houtos tês peri bion tekhnês hulê ho bios autou hekastou*)": Epictetus, *The Discourses as Reported by Arrian,* trans., W. A. Oldfather (Cambridge, MA: Harvard University Press, Loeb Classical Library, 2000) Book I, 15, 2, pp. 104-105.
8. Peter Brown, "A Debate on the Holy," in *The Making of Late Antiquity* (Cambridge, MA and London: Harvard University Press, 1978) p. 2: "... what is at stake ... is not whether, at some point between Marcus Aurelius and Constantine—two emperors who have been conventionally considered to sum up in their own persons and in the quality of their reigns the opposite poles of a pagan, classical world and the Christian Late Roman Empire—a watershed was passed. The real problem is what it is like for a great traditional society to pass over a watershed." Peter Brown takes the expression "watershed" from a passage in W. H. C. Frend, *Martyrdom and Persecution in the Early Church* (Oxford: Blackwell, 1965) p. 389: "this watershed between the Ancient World and the European Middle Ages."
9. See Aurelius Augustinus, *Tractatus adversus Judaeos* (Migne 8, 29-43, PL 42, 5, 1, 64), which is however not strictly speaking an anti-Semitic text (on this point see the studies of B. Blumenkranz, *Die Judenpredigt Augustins. Ein Beitrag zur Geschichte der judisch-christlichen Beziehungen in den ersten Jahrhunderten* (Basle: Helbing & Lichtehahn, 1946; Paris: 1973) and, id., "Augustin et les Juifs; Augustin et le judaïsme," *Recherches augustiniennes, t. I,* (*Supplément à la Revue des Études augustiniennes*), 1958, pp. 225-241).
10. These are those from the Frankfurt period (1797-1800), grouped together in G. W. F. Hegel, *L'Esprit du christianisme et son destin [Der Geist des Christentums und sein Schicksal, 1798-1800],* preceded by "L'Esprit du judaïsme," ed., Olivier Depré (Paris: Vrin, "Bibliothèque des textes philosophiques," 2003). For a general presentation see B. Bourgeois, *Hegel à Francfort. Judaïsme, christianisme, hégélianisme* (Paris, Vrin, "Bibliothèque d'histoire de la philosophie," 1970). See also J. Cohen, *Le Spectre juif de Hegel* (Paris: Galilée, 2005). [The editions cited in the bibliographic references mentioned in this course are those that we have consulted; F. G.]

11. F. Nietzsche, *On the Genealogy of Morality*, trans., Carol Diethe and ed., Keith Ansell-Pearson (Cambridge: Cambridge University Press, Cambridge Texts in the History of Political Thought, 1994).

12. On "social racism," see the end of the lecture of 17 March 1976 and more specifically on anti-Semitism, the lecture of 4 February 1976 in M. Foucault, *"Il faut défendre la société." Cours au Collège de France, 1975-1976*, ed., M. Bertani and A. Fontana (Paris: Gallimard-Seuil, "Hautes Études," 1997) pp. 76-77 and pp. 232-234; English translation David Macey, *"Society Must Be Defended." Lectures at the Collège de France 1975-1976*, English series editor Arnold I. Davidson (New York: Picador, 2003) pp. 88-89 and pp. 207-209.

13. L. Feuerbach, *The Essence of Christianity. A Philosophy and Critique of Religion*, trans., George Eliot (New York: Cosimo, Inc., 2008).

14. M. Weber, *Sociology of Religion*, trans., Talcott Parsons and ed., Ephraim Fishoff (Boston: Beacon Press, 1993).

15. M. Weber, *The Protestant Ethic and the Spirit of Capitalism*, trans., Sephen Kalberg (New York: Oxford University Press, 2011).

16. P. Veyne, "La Famille et l'amour sous le Haut-Empire romain," *Annales ESC*, no. 1, 1978; republished in P. Veyne, *La Société romaine* (Paris: Seuil, 1991).

17. Quintus Sextius the Elder, often cited by Seneca in his *Lettres à Lucilius*, trans., Henri Noblot (Paris: Les Belles Lettres, 1957); English translation Richard M. Gunmere, *The Epistles of Seneca* (Cambridge, MA: Harvard University Press, Loeb Classical Library, 3 volumes, 1989), was a Roman philosopher of the first century BCE, mixing Stoic and Neo-Pythagorean influences. See Seneca, *Epistles 66-92*, Letter LXXIII, p. 113: "Let us therefore believe Sextius when he shows us the path of perfect beauty, and cries: 'This is "the way to the stars"; this is the way, by observing thrift, self-restraint, and courage!'"

This page intentionally left blank

three

21 JANUARY 1981

The question of the relations between subjectivity and truth and the problem of the dream. ∽ The oneirocriticism of Artemidorus. ∽ The ethical system of sexual acts through the analysis of dreams. ∽ Distinction between dreams-rêves and dreams-songes. ∽ The economic and social signification of dreams. ∽ The social-sexual continuum. ∽ Sexual relations in accordance with nature and the law. ∽ Sexual relations contrary to the law. ∽ Sexual relations contrary to nature. ∽ Principle of the naturalness of penetration.

I SHALL CHOOSE THE study of a text that bears on the dream (*rêve*). After all, we should recall that the dream is obviously a strategic point, a privileged test for the question of relations between truth and subjectivity. You know that the dream (its fleeting, illusory images, and so on)—this is a principle absent from few cultures—is a surface of emergence for the truth. More precisely, in most cultures the illusion by which the subject is enchanted, captured by the dream, [and] from which he frees himself in the spontaneous movement of waking, is supposed to tell a subject's truth, anyway to tell him a truth that, most often, concerns him. Indeed, sometimes this illusion, by which the subject is enchanted and from which he frees himself by waking, is charged with telling the subject the truth of what he is, the truth of his nature, of his state, and also of his destiny, [since] the dream tells him what he is already and what he will be in a time that he does not yet see. So the dream as strategic point, as privileged test in the truth-subjectivity relationship, is no doubt not a trans-cultural constant but an idea, a recurrent theme in a considerable number of cultures.

For the problem that concerns me, obviously I situate myself only in the West, and from the angle of the constitution of a knowledge with scientific status, claims, presumptions. And I will say that Western knowledge has encountered the problem of the dream in two or three of its main moments, and at the precise moment when what was at issue was the redistribution, the reevaluation of the apparatus (*dispositif*) of the relations between truth and subjectivity, truth and subject. Thus, when an answer was needed to the question: how is it possible to conceive that the truth (the true truth of the world, the truth of the object) comes to a subject? When the question had to be posed of how the subject can be certain of having access to the truth, how he can possess the truth of the truth, this question, contemporary with the foundation of classical science, could be answered only by way of the problem, the obstacle, and the threat of the dream. The subject had to be freed from the possibility of the dream, he had to be guaranteed that he is not in actual fact dreaming when he has access to the truth, or that, at any rate, the access he has to the truth cannot be threatened or compromised by the possibility of the dream. To become the fundamental element in the development of knowledge, to become the subject of *mathēsis* (a *mathēsis* valid everywhere, for everyone, a universal *mathēsis*), the subject, insofar as he is capable of thinking the truth, still has to be freed from the dream. This was not just Descartes' problem, but Descartes[1] gives it what is clearly its most radical expression in the seventeenth century. The problem of the dream is found again, but in a different form, in the critical age, from the end of the eighteenth and throughout the nineteenth century. For, after all, the question posed then—it is found obviously very implicitly in Kant,[2] much more in Schopenhauer,[3] and very clearly in Nietzsche[4]—was: but is the truth of the truth true? Might it not be thought that the truth of the truth is not true, that at the root of the truth there is something other than the truth itself? And what if the truth were true only on the ground [of this] rootedness in something like illusion and dream? And if the truth were ultimately only a moment of something that is only the dream? And then you find again [this theme of the relationship between] subjectivity-truth and dream with Freud,[5] when the question raised was: how can one know the truth of the subject himself, what is the situation of the truth of the subject, and might it not be that the subject's most secret truth is expressed through what is most manifestly illusory in the subject? So that the problem, the theme of the dream reappears when it is a matter of founding the subject's access to the truth, of wondering about the truth of the truth, or again of searching for what is the truth of the subject. Explicitly or mutedly, the question of the dream has run through the history of the relations between subjectivity and truth, with particularly marked moments when the overall apparatus (*dispositif*) of the relations between subjectivity and truth were reorganized and modified.

It is because of what I believe to be the dream's strategic position in relation to this problem of subject-truth relations that I wanted to begin with the study of a text that concerns the dream, a text that of course relates to the period I would like to study, that is to say the period in which what is called paganism and what is called Christianity—with, of course, all the question marks and scare quotes I have emphasized—encounter, meet, confront, define, and above all become entangled with each other, so a text that dates from the second century CE. This text also has the advantage of paradoxically being the only complete surviving text of a genre, oneirocriticism (the interpretation of dreams), that was, however, familiar to Antiquity. This is Artemidorus' famous text, which was translated by Festugière a few years ago,[6] and which is a sort of little manual, or to tell the truth, encyclopedia of the interpretation of dreams.

The second reason for opting to begin with this text is that* the type of document to which I would like to address myself in this study of the relations between subjectivity and truth is, in sum, the arts of living. It is the arts of living (arts of governing one's conduct, of taking oneself in hand, technologies of oneself as it were) that will have to form the domain of investigation. Oneirocriticism is not, of course, an art of living, it is not exactly an art of conduct. But, as you will see when we enter into a bit more detail, oneirocriticism is not just a way

* M. F. adds: I told you this the first or second week [see above, lecture of 14 January, pp. 26-45].

of deciphering the small or big part of truth that may be hidden in the illusions of the dream, it is also a certain way of defining what to do with one's dream, what to do, when awake, what to make of that obscure part of ourselves that is illuminated in the night. When we are awake we cannot not be the same person we were when we were sleeping, so how can I insert the dreamer subject that I was, how can I integrate it, give it meaning and value in my waking life? Ancient oneirocriticism is this: a way of living, a way of living inasmuch as, for at least a part of one's nights, one is a dreamer subject.

The other further reason for choosing Artemidorus' text is that the author, about whom little is known, is even so someone about whom we have two types of information.[7] The first is that he was more or less inspired by Stoicism. Certain passages—some of which will have to be studied a bit more closely—indicate his membership of the Stoic school, or at any rate his closeness to philosophical work, to philosophical reflection on life and morality that belongs precisely to the domain that I particularly want to study. And then, while being a philosopher, while being immersed in Stoic philosophy, while presenting himself effectively as, basically, a theoretician of the dream, of the fundamentals of oneirocriticism, Artemidorus also provides many elements of what could form the traditional and as it were popular interpretation of dreams. A passage in the preface gives a very interesting account of the method he employed to construct this oneirocriticism in which he shows himself conducting, across the whole of the Mediterranean basin, a veritable quasi-ethnological inquiry into oneirocriticism, into the way in which dreams were interpreted in his time. In explanation he says: For myself, there is no book of oneirocriticism that I have not have obtained, and at the cost of great and lengthy searches. But more, although serious people may generally disparage public soothsayers and treat them as charlatans and impostors, despite their bad reputation, I have spent many years in their company and agreed to listen to old dreams (*songes*) and the way in which they were fulfilled (that is to say the way in which they actually demonstrated their prognostic value), and this in Greece, Asia, Italy, and many towns. There was no other way, Artemidorus concludes, to be skilled in this discipline.[8] So, in this text we have both the collection of an oneirocritical tradition that was no doubt very old and anyway very widespread in the second century, and at the same time a reflection, a philosophical type of elaboration around this practice, on the meaning it had and the way in which it could be grounded and justified. To that extent, this text looks in two directions with regard to time: on the one hand, it certainly, or very probably, bears witness to something very old and doubtless going back some centuries. Up to a certain point it may therefore be valid as evidence of a thought and morality that was much older than those of the time in which its author lived. On the other hand, it also represents something like a reflection that is relatively modern in relation to that tradition; it attempts a philosophical, Stoic, theoretical reflection on morality, on the different problems or aspects of life that are evoked through these dreams.

Yet another reason for studying this text is quite simply that four whole chapters are devoted to dreams with a sexual content.[9] And insofar as it is with regard to this "sexuality"[*] that I would like to study the relations between subjectivity and truth, you will quite understand me wanting to talk about these chapters. In the Greek and Roman literature we presently possess, these chapters are in fact the only document offering us a more or less complete picture of sexual acts, sexual relations that are real, possible, imaginable, and so on. It is a systematic presentation of the whole of sexual life ("sexual" in the contemporary sense of the term) in its most elementary form, a picture of acts, gestures, and relations in a relatively objective tone inasmuch as there is obviously no question of Artemidorus practicing literary elision to avoid saying things too crudely. He tells things as they are; it is not a literary work. Nor is it a work of moral philosophy, at least not directly—we will come back to this. No indignation, no outcry, except on certain points of repulsion. He says things such as they appear in the dream. The fact that he is describing not so much real conducts as their dreamed representation allows him to develop fairly dispassionately a relatively complete picture, inasmuch as one can be complete in this domain. Nonetheless there remains the question that even if we have here a picture of a number of possible dreams concerning a relatively significant number of sexual acts, is it a sensible method to look in this text for evidence of the sexual ethics of the second century CE, and possibly of an earlier

---

[*] M. F. makes clear: in quotes.

period? If indeed this text really does reflect what was said and thought, not only in the second century, but no doubt also in an earlier tradition handed down from generation to generation.

In fact I have focused on this text because through its analysis of dreams with a sexual content, through what is in principle an objective interpretation (this is what one may dream/this is what these dreams mean), it reveals an evaluative system. Not of course that Artemidorus makes explicit judgments on each of the acts he mentions because one may encounter them in a dream. But even so there is an interpretive system that appears in two ways. First of all, quite simply from the fact that for Artemidorus a dreamed sexual act will have a favorable or unfavorable (diagnostic or prognostic) value according to whether the act represented has a positive or negative moral value. In clear terms, take any sexual act, and if this sexual act is morally good it will have a favorable prognostic value (the event for which it is the sign will be favorable for the subject). On the other hand, if it has a negative moral value, the event announced will be unfavorable. One has here a general principle of Artemidorus' interpretation, which he did not have to invent, moreover, but takes from an old tradition. At any event, he says it very clearly: it is a general principle that, [in] dream visions, all the acts that are in accordance with nature, the law, and custom, all the acts [also] that are, as he says, "in keeping with the time" (by which he means: fitted to the moment at which they should be performed, in accordance with the principle of the *kairos*, the occasion)— all those in keeping with the names (by which he means: acts which have a favorable name), well, all these acts are auspicious. While the opposite visions—any representation consequently of something that is not in accordance with nature, law, custom, time, and name—are on the contrary of ill-fated value and without advantage.[10] He adds of course, and we will see this moreover in the analyses of dreams with sexual content, that this is not quite a universal principle and that sometimes the representation of something absolutely morally bad will nevertheless, at the cost of a certain number of supplementary circumstances, of additional facts, signify, announce something positive.[11] But all told, this is relatively rare. And, for Artemidorus, when what is morally good announces something favorable, when one reads his interpretation of dreams with sexual content, it suffices to see what the sexual acts are that announce something favorable to see that for Artemidorus, for the moral universe to which he belongs, for the tradition he represents, such a sexual act is morally good or anyway morally acceptable. So, through the very meaning that Artemidorus gives to each dream one has the possibility of finding the moral distributions and hierarchies of the sexual acts he is talking about.

The second reason why this analysis [gives access to] a sort of ethical system of sexual acts, is the fact that, apart from their localized, individual value, the way in which Artemidorus groups or classifies the different sexual acts makes it possible to find a sort of hierarchy of values or anyway an overall distribution of the major values in sexual activity. For example, the division[*] he makes between what is or is not in accordance with law or nature enables us to reconstruct partly, at least in a general way, the system of values assigned to sexual behavior. Let us say in a word that obviously Artemidorus' text does not directly represent a legal or moral code. Moreover, in the Greek or Roman literature there is no legal or moral code dividing up the set of what we call sexual conducts. This idea, the very possibility of a general moral code dividing up sexual conducts is something that appears ([and late] in the history of Christianity) only with the treatises on confession—not even the first, but those that we see developing and multiplying from the twelfth century. There is no general code of sexual acts before this. So Artemidorus' text is not a legal or moral code. It is an indirect document that, while its project is not to lay down the law [but] merely to tell the meaning, to tell the truth, maybe enables us to see an evaluative system as it existed at the time of Artemidorus, and also probably well before him, since he reflects a tradition.

So how is the interpretation of dreams, and more specifically the interpretation of sexual dreams carried out in Artemidorus' text? Artemidorus distinguishes two general forms—I am taking the words employed in Festugière's translation—of "nocturnal visions" (what we call dreams [*rêves*]), which enable us to make the necessary distinctions. So, two categories of nocturnal visions. [First,] the category of *enupnia*, of things that take place in sleep, which Festugière translates as actual "dreams-*rêves*." Actual dreams-*rêves* quite simply translate a

---

[*] M. F. adds: to which we will have to return.

certain present state of the subject, the affects, passions, states of the soul and body of the dreaming subject at the very moment he dreams. For example, if a subject is hungry he will [have a] dream that will immediately translate the sensation of hunger he is experiencing. Similarly if he is afraid, or even if he is in love. Let's say that these *enupnia*, these actual dreams-*rêves*, have simply a diagnostic value for deciphering the present, and possibly transient state of the subject who is dreaming.[12] The second category is the *oneiroi*, [a term] that Festugière translates as "dreams-*songes*." *Oneiroi* (dreams-*songes*) also show what is (*to on*, being). But they do not tell of what is in the sense of a transient state of the subject's soul and body, but of what is inserted in the inevitable thread of time, of what is in the future, of what is inscribed in the general unfolding of the order of the world. So the *to on* consequently contrasts with the affect, the state of the subject, however real this state or affect may be. Inasmuch as the dreams-*songes* (the *oneiroi*) express being (*to on*), they act on the soul, they push it towards the future they announce, insofar in fact as the individual and the soul itself are part of this general sequence of the world. There is like a sort of ontological call of the dream,* a call, coming from the world itself and from the elements in the world, to the soul. The dream is nothing other than this sort of call of future being to the soul to which it makes itself known. And Artemidorus justifies, or supports, if you like, his explanation of the mechanism of the dream, of the *oneiros*, [with] a series of plays on words which explain quite precisely what he understands by *oneiros*. The word *oneiros*, he says, comes first from: *to on eirein*. That is to say the *oneiros* (the dream) expresses being, says what is. Second, the word *oneiros* comes from *oreignein*, which means: to attract, call to, pull, or push. And then there is the noun *Iros*, which in Homer, in the *Iliad*, designates a messenger. The dream is a messenger that draws the soul by telling it the truth, by telling it what is.[13]

Hence, two consequences for oneirocriticism. The first form of nocturnal vision I talked about (*enupnion*: dream-*rêve*) is no more than the translation of the state of the soul and the body. This *enupnion* has

---

* From this paragraph Foucault frequently uses *rêve* to refer to both types of dream, *enupnion* and *oneiros*, making the distinction *rêve/songe* only when necessary or when comparing the two types; G.B.

above all a medical value, for oneself and for the physician. It enables
one to know the illness one is suffering from. We have examples from
elsewhere than Artemidorus that show that this kind of oneirocritical
analysis existed in Antiquity. In Galen, for example, there are several
examples of medical diagnostics given by a physician [thanks to] the
analysis of a dream in the sense of *enupnion*, inasmuch as the dream is
an immediate translation of the state of the individual's soul or body.[14]
This is the first, medical use. But clearly the analysis will not be con-
ducted in the same way in the case of an *oneiros* rather than an *enupnion*, when it is a matter then of that something that announces being,
the truth to the soul, and draws the soul towards it. This is no longer
a diagnostic analysis, it is a prognostic analysis. And this dream-
messenger, which addresses the soul in order to tell it what is, must
be deciphered. There are two possibilities for this. Either the dream
says everything clearly, and shows, by pointing out as it were, by direct
image, what will happen. This is the so-called "theoreomatic" dream-
songe, which basically needs no interpretation. It is enough to record
it and remember it, one knows what will happen. But there are many
other cases in which this announcement to the soul is made through a
dream that is not the direct representation of what will happen.[*] There
is simply an analogical relationship, that is to say the dreamed image
resembles what will happen.[15] Artemidorus says so expressly: in this
case the interpretation of dreams-*songes* consists in nothing other than
placing similar things alongside each other (the representation of the
dream and the event to which it [refers][†] [...]).

This domain of analogy is quite clear, quite coherent. What does a
sexual dream generally refer to, what type of event does it announce,
to what type of being does it refer? To the individual's destiny (health,
illness, death), of course, but this is still the rarest case. The domain of
truth, reality, and being to which the sexual dream refers, the reality
it announces, in fact, is essentially what we would call social, political life, economic life in the old sense of the word. The sexual dream
speaks of the management of [our] own existence in the city, in the
household, in the family or the political body. Prosperity or financial

* The manuscript clarifies: "allegorical dream."
† Gap in the recording. All that can be heard is: Artemidorus reports the sexual dreams.

reverse, success or failure in private or public affairs, good fortune or misfortune in a political career, increase or loss of goods, family affluence or impoverishment, an honorable marriage or instead one of little advantage or of which one would be ashamed, change in social status: the sexual dream speaks of all this. Quite clearly, the sexual dream-*songe* has an economic and political signification. It speaks of the way in which life unfolds in the space of the city as well as in that of the household, the home, the family. So a projection of what we call the sexual* onto the social.

Obviously a number of things need to be taken into account to explain this. On the one hand, [the fact] that this is Artemidorus' general style of analysis in this oneirocriticism; sexual dreams are not the only dreams to have this value—they have this value in a slightly more accentuated way than others for [a] reason I will come back to. [On the other hand,] one must [also] take into account the fact that Artemidorus' oneirocriticism inevitably takes the point of view of the man, the adult male, father of a family, of the male with a social, political, and economic activity. In all of Artemidorus' oneirocriticism, the dreamer is always the adult male, father of a family, in charge. This, of course, relates to the status of most discourses on sex which [adopt] the man's and only the man's point of view[16]—a general cultural feature. But it also and especially relates no doubt to the role of those sorts of works [of which] Artemidorus' [is representative]. As I said to you at the start, oneirocriticism should not be thought of as mere curiosity or speculation. It is not a marginal art that some more or less credulous individuals practice or by which they let themselves be seduced. Oneirocriticism is a practice that should and actually does help men to conduct themselves. To understand one's dreams, to know why one has dreamed and what consequences one may legitimately draw from it in daily life, to manage the affairs of reality while taking account, each for oneself, of the part of night that belongs to and is inscribed in the heart of our existence, is the raison d'être of oneirocriticism. Telling one's dreams, understanding one's dreams, is indispensable for a good family father. The surrealists said: "Parents, tell your dreams to your children!"[17] Well, Artemidorus would say: Fathers,

---

* M. F. adds: we will have to come back to this.

never forget your dreams, do not forget to tell them and consult those who know how to interpret them, for you will not be able to manage your affairs well unless you really know what you have dreamed, why you dreamed it, what it announces, and what you should make of it in your daily life. It is a father's, a good family man's book, and of course due to this it is normal that all dreams, sexual or otherwise, be related to that reality of social, political, and economic life.

But I think sexual dreams are related to this domain of reality more precisely and emphatically than others. It is in fact quite significant that regarding these sexual dreams, and not others of course, Artemidorus makes use of a fairly clear and obvious linguistic fact, that is to say the polysemy, the economic-sexual semantic ambiguity of many Greek words. In his analysis of dreams, Artemidorus exploits, for example, the fact that *sōma* means the body, the body with which one has sex, but also wealth, the goods one possesses.[18] *Ousia* is substance. Substance may also mean the very substance of wealth, the fortune one possesses, as well as the fundamental substance of the human being, the organism, the seed, (produce a seed).[19] *Blabē* means an injury, a reverse of fortune, a loss of money, as well as a setback in a political career. But it is also the fact that one has been sexually assaulted and even more generally the fact that one has been more or less voluntarily passive in a sexual relationship.[20] The *ergasterion* is the workshop, but it is also the house of prostitution, the brothel.[21] The word designating constraint, as when one is obliged to pay a debt to someone to whom one is a debtor, is the same as the word used to designate the fact of experiencing a constraining need to make love and to free oneself in any case of excess sperm that the body has retained and kept in itself for too long. One clears oneself of debt as one gets rid of one's sperm, and so on.[22] Artemidorus rests a large part of his analysis of dreams on the economic-sexual polysemy peculiar to the Greek language.

So we can say that we see something emerging through Artemidorus' analysis that we can hold on to as a hypothesis for the moment, and this is what could be called the co-naturalness for the Greeks of the sexual and the social. Having a sexual relationship in a particular form or with a particular partner, and having an activity in society within one's family or in the city, acquiring a certain wealth, making a certain profit, suffering a reverse in private or public affairs, are for the Greeks, or anyway

for a whole tradition that Artemidorus in the second century still echoes, almost the same thing. In any case, it fell in the same dimension, it
belonged to the same reality. It was as though there was an immediate
communication, a natural analogy, since in this type of interpretation it
is indeed a matter of natural analogy between sexual activity and social
activity in general. I think it is very important to grasp this, inasmuch as
it is very different from a perspective like ours, in which, of course, in our
oneirocriticism—which developed obviously throughout the nineteenth
century, but which had [also] begun to develop before—the social always
tends to be a metaphor of the sexual. And of any dream with a social
content, speaking of a reverse of fortune, of political success, we ask what
sexual truth it conceals. It is the exact opposite in the oneirocriticism
of Artemidorus. One asks for the political, economic, social truth that a
sexual dream expresses. In our oneirocriticism, the possibility of decod-
ing social contents of the dream in sexual terms rests on the postulate
that there is a natural discontinuity between the sexual, or anyway desire,
and the social.* It needed a barrier between [dreams] and their content,
it needed heterogeneity of principles and all the mechanisms of repression and conversion for it to be possible to make one speak the truth of
the other. On the contrary, in Greek oneirocriticism one is dealing with a
sort of continuum of the social and the sexual, a continuum in which the
polysemy of the vocabulary serves as permanent exchanger.† This is the
general framework in which we can question Artemidorus' text.

---

* The manuscript notes: "The articulation of the social on the sexual is thought either in the form
of the law (prohibition), or in that of the institution (marriage)."
† The manuscript notes:
  "In any case, and whatever this 'continuity' involves, this establishment of a relation between
  the sexual and the social, this interpretation of the sexual relationship as social success or
  failure makes it possible to pick out the value that Artemidorus and Oneirocriticism accords
  it. The interpretation is at the same time an indication of value. This permits the analysis
  to be used as an indirect but quite effective moral indicator. Let's say: we have acquired the
  habit of questioning morals in terms of the truth of desire that they convey and conceal. With
  this text by Artemidorus we can analyze in moral terms an interpretation that claims to be
  decipherment of truth. What is very distinctive regarding this text is a way of posing the same
  question. Not: what is the truth of morality, but: what is the morality of the truth, what is
  the will in which the truth is rooted? Or more precisely: what is the morality of the truth of
  morality, in what will is this search to link together the discourse charged with speaking the
  law and the discourse that is supposed to tell the truth rooted? If the genealogy of morality
  consists in telling the truth of all morality and if the function of joyful knowledge is to liber-
  ate the will that is striving for truth, let's say that it is a matter of making the genealogy of
  knowledge a joyful knowledge."

So I would like now to move on quickly to the content of this interpretation whose general outlines I have just given. Regarding dreams with a sexual content, Artemidorus distinguishes three types of dreams-*songes*. Those that represent acts in accordance with the law; those representing acts contrary to the law; and finally those representing acts contrary to nature.[23] In accordance with the law, contrary to the law, contrary to nature: you see straightaway how obscure this structure is. To start with, it is obscure because neither of the two terms (*nomos* and *phusis*) is explicitly, or even implicitly defined by Artemidorus, neither in themselves nor in relation to each other. You can see that this way of simultaneously using two criteria to arrive at a tripartite division cannot stand up. There may be that which is in accordance with the law and that which is contrary to the law, that which is in accordance with nature and that which is contrary to nature. But to draw three categories from these two criteria is clearly logically impossible. To which it should be added that certain things that are in accordance with nature are found in those that are contrary to the law, but there are also things contrary to nature in those contrary to the law, and so on. The text is very confused as to this *nomos/phusis* division, leaving us with the impression—to which moreover we will have to return—that actually this division is both fairly important for constituting the general framework of the distribution of these sexual acts, and of sufficiently little importance that one does not need to define them, or need to classify strictly this or that act in this or that region. So let us take this material as it is, that is to say somewhat confused, and follow the three chapters.

Sex in accordance with the law. What will the general mechanics of analysis be? Towards the middle of the paragraph or chapter devoted to acts in accordance with the law, Artemidorus gives the general principle of his interpretation. He says: "One must consider unknown women" (that one sees in a dream with sexual content) "as images of activities that should concern the dreamers."[24] This puts to work what I was just telling you. In a sexual act with a partner, the partner in question is to be taken as image of the dreamer's future activities (familial, political, social, and so on). "So that whatever the woman may be, whatever her condition, this is also the condition in which his activity will place the dreamer."[25] What appears here is that the relevant element for extracting the meaning of the dream is the partner's condition. What matters is not the nature and form of the sexual act, it is not so much what one will do with anyone, it is who she is, her condition, her social status, what social but also sexual partner she constitutes. It is the condition of the other person that determines the prognostic meaning of the dream. According to the general principle of equivalence, of correspondence at any rate between prognostic and moral values, one will be able to define the moral value of a sexual act not so much by the form of the act, what it consists in, but by the condition of the person with whom one performs it. Moreover, the whole development of this first chapter, which may at first sight appear completely confused and muddled, can be reconstructed quite logically on the basis of this principle. It is not an analysis of different possible sexual acts; it is an analysis of the different persons with whom one may have sexual relations, not in terms of what one does with them, nor even of their sex, but in terms of their [social] condition. And in this way one gets the following development, which can be reconstructed in a way that is not too arbitrary. The first character to appear is, of course, one's wife. And dreaming that one has sexual intercourse with one's own wife is obviously very favorable. It is very favorable because the wife signifies the dreamer's occupation. The wife is that through which and on which one exercises one's activities, privileges, and rights, and it is that from which one draws profit, benefit, and consequently pleasure.[26] This little analysis by Artemidorus is very interesting because [on the one hand] the sexual refers of course to the social, on the other hand the wife is not so much the metaphor for the social but an element, a part of it. It is inasmuch as the wife is part of the domain of activity, of the domain of sovereignty, inasmuch as she belongs [to], falls under the exercise of the individual's rights, that dreaming of having sex with her is a favorable sign. And you see [also] that the pleasure is deduced from the profit. It is because it is profitable to have a wife, because making love with one's wife refers to this profit, to the social advantages or gains, to the value of the individual's status, and so on, that one can draw pleasure from all of this. So the social refers to the sexual and the advantage refers to the pleasure. Not that the sexual or the pleasure would be the secret or truth of the social and the gain, but because the sexual and pleasure are part of the whole constituted by the social network, the individual's rights, his advantages, profits, gains. It is as [partial]* that the wife and the pleasure one can get from her constitute a favorable prognostic element in the dream. An important thing to underline: in at least one case, dreaming of making love with one's wife is not very favorable, and this is when one dreams that the wife does not consent, thereby showing that, in reality, the wealth, goods, rights, and everything over which one asserts one's authority risk being taken from one.

So the wife is the first and initial character. And then the different conditions, the different places in which one may have sexual intercourse are listed. [With] the concubine first of all. According to a well-known situation in Antiquity, there is only an ultimately fairly slim juridical difference between the concubine and the married wife, and Artemidorus says: There is no difference between a dream concerning the concubine and a dream concerning the wife.[27] [Then] comes the *ergasterion*, that is to say the house of prostitution. Dreaming that one enters a house of prostitution is not too unfavorable, but it is not very good even so. It is not very good because there is after all something a bit shameful in going to an *egasterion*; one makes expenditure, of money, of course, which is useless, and of sperm also, which serves no purpose; in short, and this is one of the rare elements that diverge slightly from Artemidorus' general analysis, it is not really very favorable to dream of these *egasteria* because if one considers how an *egasterion* is constructed, that is to say, small rooms, small booths separated from and alongside each other, it bears a terrible resemblance to a cemetery. Consequently, dreaming that one goes to the *ergasterion* may well be a sign of death.[28] This is one of the rare elements that does not have the same organization as the whole of Artemidorus' analysis.

After the *ergasterion*, quite naturally, comes another place where sexual intercourse can be practiced easily, which is the house itself, the household, that is to say with slaves. And dreaming that one has sex with slaves is obviously a good [sign], because it indicates that the dreamer will be able to get pleasure, profit, good, and so on, from his possessions, from his goods. This indicates that his possessions will become greater and more magnificent. Obviously, it goes without saying, there is no difference between boy and girl slaves. The

---

* M. F. says: It is as part (*C'est a titre de partie*).

only problem that arises, in the case of the father of the family making love with a male slave, is the position occupied by the former. If he is active in relation to his male slave it is of course a positive sign: one takes pleasure from one's possessions. But if he is passive in relation to his male slave it indicates that he will suffer a *blabē* (injury, sexual aggression, state of passivity, of constraint, and so on).[29] After the household, there is the general circle of relations and friends. If one dreams of an unmarried woman, the dream is favorable on condition that she is rich. It is much less favorable if one dreams that the woman is not rich. It is a bad sign if she is married. It is a bad sign because the married woman is "in the power of the husband": the law prohibits taking her, and consequently the dream of adultery announces that the individual will suffer the penalties [foreseen by] the law on adultery.[30] You see again here the direct communication of the sexual and the social. Finally, Artemidorus says, still regarding relations in accordance with the law, there is the dream in which one dreams that one lies with a man and is passive with him. Here there are two possibilities. Either the dreamer is a woman (this is the only time in the whole text when the woman appears as a possible subject of dreaming), then, if the woman dreams she is penetrated by someone, this is obviously an advantage for the woman for it is natural for her to be penetrated. If the dreamer [is a man, and] he dreams that he lies with a man [while being] passive, the value depends on the social status of each of them: to be possessed by someone older and richer than oneself is obviously good; but to be possessed by someone younger and poorer is bad.[31]

So much for sex in accordance with the law, and consequently also in accordance with nature. Now let us move on to sex not in accordance with the law. In fact, the two chapters devoted to these relations *para nomon* (foreign to the law more than contrary to the law) are oddly devoted to just one subject, as if there were only one possibility of sexual relation in this category of contrary to the law. And this sexual relation not in accordance with the law, foreign to the law, is incest, and incest in an extremely restrictive form since it is a matter of incest between parents and children: father, mother-son, daughter, with a small addition we will talk about later regarding brother-sister or brother-brother relations.[32] This seems to indicate that the *nomos* to which these two chapters refer evidently does not designate particular

laws of the city—for example in the chapter on what is in accordance
with the law, you have seen that there was the question of adultery,
which at the same time was said to be condemned by the laws of the
city. The *nomos* involved here does not designate the city's laws but a
certain general principle that separates, not even exactly what we
would call incest, but [rather] the parent-child relation as having a
particular status and constituting a particular dangerous zone, or at
any rate one to be identified and isolated. Schematically, we can say the
following about this incest. The first chapter being devoted mainly to
the problem of the father, it is understood, apart from one or two very
marginal exceptions, that the relation between the father and one of
his children (son or daughter) is always negative. What is very interesting, on the other hand, is that in the following chapter, mother-son
incest, which is developed at length (mother-daughter incest is not
even mentioned, we will come back to this shortly), almost always
has a favorable signification. First of all, it generally signifies that the
father will die. But, well, let's leave this, because I specifically do not
want to Oedipalise this history, on the contrary. On the other hand, all
the significations given spontaneously by the language itself, by language, to the mother character, mean that dreaming incest with the
mother has to be considered as a positive prognosis. And, according to
the general principle I spoke to you about at the start, this therefore
means that the moral value is at any rate not so unfavorable. For the
mother signifies one's occupation: to dream that one lies with one's
mother is therefore a sign that one will make profits in one's profession. Dreaming that one will sleep with one's mother is to dream that
one has positive relations with one's homeland and that one will be
successful if one has a political career. The mother is nature, and as a
result dreams of incest with the mother will assure the individual good
health and long life. The mother is also the earth, and as a result, if
one is a peasant this is a sign that one will have a good harvest. If one
is involved in legal proceedings with someone, it is a sign that one will
win one's case and obtain possession of the land one is after. There
is only one case in which dreaming that one is sleeping with one's
mother is not very good, and this is if one is ill: this is a sign that
one will be thrown into the mother earth, that is to say that one will
die. But, apart from this case, practically all the examples of dreams

with the mother cited by Artemidorus are favorable, as if ultimately, in all this history of sexuality, there was something that did not worry the Greeks overmuch, and that is the story of Oedipus. The story of Oedipus does not appear so terrifying when one reads it through these documents. Now if you see the relatively numerous allusions to dreams with the mother in Greek literature,[33] you well know that this positive value of the dream with the mother is something that recurs insistently, and which, consequently, would perhaps call for a certain revaluation of the fundamentally dramatic interpretation given to mother-son incest on the basis of the tragedy by Sophocles.

So much then for what concerns that which is contrary to the law. So [we have] a [chapter devoted to what is] in accordance with nature, two chapters to what is not in accordance with the law and a final chapter on what is contrary to nature—*para phusin*, that is to say again: contrary, but especially aside from nature, outside nature. In fact, Artemidorus returns twice to this notion of outside nature. In the chapter devoted to incest, and precisely in the paragraph devoted to incest with the mother, he speaks in a rather curious way of what is natural and not natural in sexual relations, as if it were regarding this question of the mother that the differentiation between natural and not natural had to be made. Artemidorus says: "That the chapter on the mother appears under various aspects, with many parts, and that it is susceptible to numerous divisions, has escaped many oneirocritics. Sexual union is not enough by itself to show the things signified, but, as the couplings and positions of the bodies are diverse, this is also what makes the fulfilments diverse."[34] In this rather faithful and literal translation, [the meaning] is quite clear: if one wishes to extract the signification of a sexual dream, one must take into account not just with whom one makes love, but also the way one does so. Whereas, in the chapter devoted to what was in accordance with the law it was a question only of the partners, with regard to the mother the differential morphology of the sexual act will appear.

This diversity of sexual acts is presented in the following way. First, for every species of animal, says Artemidorus, nature has of course determined one and only one mode of union. Thus for the horse, the goat, the cow, and so on, the females are covered from behind. Vipers, doves, and weasels make love with the mouth. And in the case of fish, the female is content to gather the sperm spread in the water by the male. As for humans, like every species they have a certain natural position. This positon is, of course, face to face, the superimposition of the male on the female, and so on. This form, Artemidorus says, has been fixed by nature, and it is what gives man in fact most pleasure and least pain. As for all the other positions, whatever they may be, there's no point listing them—although Artemidorus does so, moreover—they are the inventions of immoderation, intemperance, and excess induced by drunkenness.[35] Consequently you see that this normalization of a certain type of sexual relation does not date from Christianity. It is already absolutely established in the Greek, Hellenistic, and Roman tradition. And in this same paragraph, which is in the [passage concerning] incest with the mother, he adds, however, that among all the things that are the invention of immoderation, intemperance, and excess (all that is not the absolutely normal position), there is even so a particularly bad domain with an always ill-fated meaning indicating a future useless expenditure, a loss: this is oral sex. Oral sexual relations, the sole point of Artemidorus' manifest and clear indignation in the whole of this list, are of fearsome prognostic and detestable moral value, for, Artemidorus says, after such sex it is no longer possible to exchange a kiss or share a meal with the person with whom on has had this kind of relation.[36] So that is what Artemidorus says regarding [what is] against nature in this chapter devoted to incest with the mother.

And then, at the end of the journey, there is a chapter devoted entirely to what is contrary to nature. There are five categories of acts considered by Artemidorus to be contrary to nature: sexual intercourse with gods, animals, cadavers, those one has with oneself, and those two women may have with each other.[37] Let's say that the first three categories are relatively self-evident, that there is no point commenting on the unnatural character of these three forms of intercourse. The two others are more interesting. Why is union with oneself considered to be contrary to nature? What first of all must be understood by this union with oneself? It is not a matter of masturbation—in fact there are a few lines devoted to this problem in the very first chapter (on what is in accordance with the law), where it was said, which is rather strange and interesting, that it happens in fact that a man dreams he is masturbating.[38] What does this dream mean? Quite simply that he will have sex with a slave. For his own body is the signifier of what he possesses in his house, namely his slave.[39] This is the only time in Artemidorus that a sexual dream is interpreted as having a sexual content. All sexual dreams have a social signification, but this sexual dream (dreaming that one is masturbating) has a sexual signification, as if the fact of masturbating was not really in itself exactly sexual, [but is situated] on the border of the sexual and was really sexual only by what it signifies. If one adds to this that in classical Greek ethics masturbation was considered to be an activity without importance—to which obviously slaves resort to, but no free man would have thought of doing—you see that the relation masturbation—sub-sexuality—slave sexuality, and so on is fairly clear in this short passage. So, it is not a question of masturbation when Artemidorus speaks of union with oneself. He is designating self-penetration (the Greek terms show this [with great evidence]). That is to say, obviously, an impossible act, but in which the essential element is penetration. This act, [of which one may dream,] is contrary to nature because it is obviously impossible. As for the union between women, it is equally contrary to nature; which may seem bizarre if one thinks of what Artemidorus said about sex between men: in Artemidorus' description, even sex between father and son (even a son less than five years old),[40] is not considered to be contrary to nature. On the other hand, two women, yes, that is contrary to nature. Here again we need to refer to the Greek word. The word employed is *perainein*, that is to say what is involved is not just an undifferentiated, sexual, erotic relationship, but well and truly a relation of penetration.[41] So it is a matter of a relation that is unnatural or para-natural due to the fact that one of the women imitates the man's role and usurps the penetration that is the man's privilege.

❧

What, in broad terms, can we retain from the whole of Artemidorus' text and all its interpretations? [On the one hand,] it clearly does not make a clear division between nature and contrary to nature. On the other hand, there is fairly clearly the presence and effectiveness of a domain of what one could call naturalness. The main element of this is clearly penetration. The male sexual organ, with its ability to penetrate, defines the universal and constant naturalness of the sexual act. Where there is no penetration, as in masturbation, well, it doesn't count. Where, on the other hand, there is penetration by artificial, non natural means, as between women, one is clearly in the realm of the unnatural. If penetration by the masculine organ is the general form of what is natural, [when] there is penetration—whether of one's own wife, of a slave boy, all this is unimportant—one is in the domain of the natural. It remains the case that this general field of naturalness defined by penetration by the male organ nonetheless had one form that was as it were perfectly and completely natural, more intensely natural, or natural with even more density than the others: this is the relation of man and wife, in a lying down, horizontal position and so on. And around, on the basis of this form, which had in effect a natural privilege, one sees the rest deployed in increasing indifference to nature, up to the point of two women making love with each other, and who completely depart from nature because they have escaped the law of penetration, or at any rate have wished to usurp its form.

On the other hand, apart from this field of naturalness, you see that one of the elements with the most influence in modifying the value of a sexual dream is the status of the partner. Here again, in this field of sociality, there is an organization that is both unsystematic and relatively clear. An absolutely privileged type of relation is that with the spouse herself, the consenting spouse. At the other end, there is a strongly prohibited region: the incestuous relation (especially between father and children). And then, between the privileged form with the consenting spouse and the highly negativized form of the father-child relation, you have a whole multiplicity of more or less acceptable relations according to the reciprocal point of view of each and in which the most important elements are age, social status, wealth, and so on. So that, in short, sexual relations appear as integrated within a game of more or less valid, more or less respectable social combinations in which the oppositions old/young, active/passive, rich/poor, above/beneath combine to form something more or less good.

Penetration as the highest form of sexual naturalness; definition of sexual relations in terms of the social position of each: these two criteria combine very easily inasmuch as penetration is effectively the sexual act that manifests most directly the position and social privilege of the man who is dreaming. And if in fact he dreams of a sexual act in which penetration is perfectly acceptable in terms of the social game within which it takes place, one has a favorable dream. Naturalness (defined by penetration) and sociality of sexuality (characterized essentially by the social position of each of the partners, their role in society): these two elements combine perfectly to produce a sexuality that is fundamentally, essentially, a sexuality of the family father.

Consequently, whatever fundamental differences of code there may be between what has been called pagan ethics and what will be called Christian ethics—again, with all the uncertainties this involves—there are two things to keep in mind. First, the idea of a sexuality entirely valorized around the family relationship, the male position, the position of the family father, already exists in the ethics represented by Artemidorus. But—and perhaps the most important difference is here, where no doubt elaboration by Christian thought will play the main role—this sexuality is thought in a sort of continuity with the social relation, such that social relation and sexual relation are realities of the same type, the same category, coming under an absolutely continuous ethics. The social and the sexual are not distinguished. To have a good sexuality is to have a socially recognized sexuality. For example, for a family father it is possible, legitimate, to have sexual intercourse with his spouse to give him children, and then immediately after with a young (male) slave. In Greek eyes, according to the ethics represented by Artemidorus, these are clearly not two different forms of sexuality. The individual who does this plays legitimately on two aspects of a same social role. It is the same social role that means that he is married, rich, has a wife and wants children in order to have heirs, and then that he has slaves. Having sex with his wife and taking his pleasure with his slave are both part of a same social role. The idea that it is a matter of two different sexualities, articulated in two types of desire that are foreign to each other, is [alien to] the tradition, the ethics, the form of thought represented by Artemidorus. It is a pleasure that there is no reason to think was different in one case or the other. The different [factors] that will transform the Greek experience of what were called *aphrodisia* into the Christian experience of the flesh will act much more on this continuity of the social and the sexual than on the privilege of the family father. But this development will not be the work of Christianity alone, and I will try to show you how within ancient philosophy—philosophy anyway in the Hellenistic and Roman period, in the Stoics but also in the Epicurians and most of the philosophers—this new form of experience begins to develop in which the *aphrodisia* (the pleasures of sex) will be thought in a new form and with a different type of relation to the truth.*

---

* In the manuscript there are some indications relating to marriage that will be developed in the lectures of 4 and 11 February:

"1/ Marriage: old discussion that goes back to the classic centuries: should one marry? Now in this traditional way of posing the debate around this question, two characteristic elements: (1) it was a matter of putting two series of arguments for and against marriage in parallel; (2) in these texts a profound distinction was made between ordinary life and the philosophical life: in ordinary life the question of marriage was one of pure appropriateness, in the philosophical life it raised the more essential question: can one practice philosophy when one is married? Now in the texts of Musonius Rufus, Seneca, Epictetus, Antipater, Hierocles, marriage is apparently recommended; but looking at it more closely the transformation is more important on two points: marriage is defined as *proēgoumenos* ("absolute," Bonhöffer); [for the] Cynics, marriage is entered into according to circumstances (love, according to Epictetus), for the Stoics it is rather an "absolute" act that must be performed whatever happens. Why? Because it is useful (*sumpheron*) for oneself, the family, and the city. Because it is completely necessary. Marriage should therefore be conceived of as [*stoikheion*]; the elements of the city are not men but households; but households are not completed (*teleios*) if there is no couple; hence the fact that marriage is the root of everything: city, humanity (Cicero).

2/ But an important consequence is drawn from this: if marriage is part of human nature insofar as it is rational, the philosopher, inasmuch as he gives himself the end of fulfilling human nature in its perfection, must marry; as for the ordinary man, there is no difference. On the contrary, by marrying, he takes interest, but without too much knowing it, in the rational ends of every human being: marriage, which was one of the criteria of the difference between the sage and the ordinary man, becomes rather one of the essential points through which they communicate.

3/ But at the same time one sees that marriage, as rational part of life, enters the field of those exercises by which the individual tries to fulfil his rational essence. Philosophy has three dimensions, three aspects: *mathēsis, meletan, askeinai*, three forms of the philosophical practice having to be applied to all aspects of the philosophical life. One must learn, one must reflect, one must practice in married life. The problem is no longer: should one? but: how?"

1. It may be recalled that, in his controversy with Michel Foucault, regarding the interpretation of a passage from Descartes' *Meditations* ("But so what? Such people are insane") put forward in *Folie et Déraison. Histoire de la folie à l'âge classique* (Paris: Plon, 1961); English translation J. Murphy and J. Khalfa, *History of Madness* (London and New York: Routledge, 2006), Jacques Derrida tried to show that if Descartes abandons the hypothesis of madness, as vector of deepening doubt, in favor of that of the dream, it is because the dream makes possible a more radical ontological suspicion, and not because of the exclusion of madness by classical reason, an exclusion evidenced by the confinement of the mad in the *Hôpitaux généraux* at the same time. On this controversy, see the text by J. Derrida, "Cogito et histoire de la folie," in *L'Écriture et la Différence* (Paris: Seuil, 1967); English translation by Alan Bass, "*Cogito* and The History of Madness," in *Writing and Difference* (Chicago: Chicago University Press, 1980), and Foucault's replies: "Mon corps, ce papier, ce feu," *DE*, II, no. 102, pp. 245-268/ "Quarto," vol. I, pp. 1113-1136, English translation J. Murphy and J. Khalfa, "My Body, This Paper, This Fire," Appendix II in M. Foucault, *History of Madness*; and "Réponse à Derrida," *DE*, II, no. 104, pp. 281-295/ "Quarto," vol. I, pp. 1149-1163; English translation J. Murphy and J. Khalfa, "Reply to Derrida," Appendix III, in *History of Madness*.
2. On this point see Kant's text on Swedenborg's theosophy: I. Kant, "Dreams of a Spirit-Seer Elucidated by Dreams of Metaphysics," in *Theoretical Philosophy 1755-1770*, trans. and ed., David Walford in collaboration with Ralf Meerbote (Cambridge: Cambridge University Press, "The Cambridge Edition of the Works of Immanuel Kant," 1992).
3. Arthur Schopenhauer, *The World as Will and Presentation* [*Di Welt als Wille und Vorstellung*, 1819], trans., R. E. Aquila in collaboration with D. Carus (New York: Longman, 2008).
4. F. Nietzsche, *The Birth of Tragedy* [*Die Geburt der Tragödie*], trans., Ronald Speirs, in *The Birth of Tragedy and Other Writings*, ed., Raymond Geuss and Ronald Speirs (Cambridge: Cambridge University Press, "Cambridge Texts in the History of Philosophy," 1999).
5. Sigmund Freud, *The Interpretation of Dreams* [*Die Traumdeutung*, 1900], trans. and ed., James Strachey (New York: Basic Books, 2010).
6. Artemidorus, *La Clef des songes. Onirocriticon*, trans., André Jean Festugière (Paris: Vrin, 1975); English translation Daniel E. Harris-McCoy, *Artemidorus' Oneirocritica: Text, Translation, and Commentary* (Oxford: Oxford University Press, 2012). The study of this text forms the first chapter of *Souci de soi*: "Rêver de ses plaisirs," pp. 15-50; *The Care of the Self*: "Dreaming of One's Pleasures," pp. 1-36.
7. In his "Introduction," A. J. Festugière attests that "what is known about him is indicated to us only by his work" (*La Clef des songes*, p. 9). Artemidorus of Ephesus was born in Lydia in the second century CE, in the small city of Daldis where Apollo Mystes was honored. It is this god who invited him to compose his book, for the creation of which the author undertook numerous voyages and many searches "in order to be as complete and instructive as possible."
8. We give A. J. Festugière's translation here, ibid., p. 16: "For me, not only is there no book of oneirocriticism that I have not acquired, employing great research to this end, but more, although public soothsayers are greatly disparaged, are called charlatans, impostors, and buffoons by people who adopt a solemn air and frown, I, scorning this description, have had dealings with them for many years, agreeing to listen to old dreams (*songes*) and their fulfilment, in towns and general assemblies in Greece, in Asia and Italy, and in the most important and populous islands: there was no other way in fact of being skilled in this discipline"; *Artemidorus' Oneirocritica*, pp. 44-47: "But there is no book on dream-interpretation which I have not acquired, expending much zeal in this regard, and I have also consorted for many years with the much-maligned diviners of the marketplace, whom the high and mighty and the eyebrow-raisers call beggars and charlatans and altar-lurkers, though I have rejected their slander. And in Greece, in its cities and festivals, and in Asia and in Italy and in the largest and most populous of the islands, I have listened patiently to old dreams and their outcomes. For in no other way was I able to gain practice in these matters."
9. Actually, it is three chapters: 78, 79, and 80 in Book I, Fr., pp. 84-93; Eng., pp. 136-151.
10. Ibid., IV, 2, Fr., p. 222: "Now it is a general principle that all dream visions in accordance with *nature* or *law* or *custom* or *art* or *names* or *time* augur well, that all the opposite visions are harmful and without advantage"; Eng., p. 307: "And so it is a fundamental principle that all

things that are in accord with nature or law or custom or craft or words or time are regarded as good, but things that contradict them are grievous and unprofitable." It will be noted however that in "the chapter on sexual relations," Artemidorus announces a more summary division; Book I, 78, Fr., pp. 84-85: "we will deal first of all with relations in accordance with nature, law, and custom, and then with relations contrary to law, and thirdly with relations contrary to nature"; Eng., p. 137: "one ought to speak first about intercourse that is in accordance with nature and law and custom; then about that which is contrary to law; and, third, about that which is contrary to nature."

11. Ibid., IV, 2, Fr., p. 222: "Remember however that this principle is not absolutely universal but applies only to most cases. For there are many things seen that are turned to a good end even though they follow neither nature nor any of the other fundamental facts as contrary to what takes place in real life"; Eng., p. 307: "But keep in mind that this doctrine does not apply in all cases, but for the most part holds thus. For in fact many things observed to be discordant with nature or another of the elements have turned out to be good contrary to the nature of these items."

12. Ibid., I, 1, Fr., p. 19: "The vision of the dream-*songe* differs from the dream-*rêve* in this, that the former signifies the future and the latter present reality. You will understand this more clearly in this way. Some of our affects are arranged by nature to accompany the soul in its course, to go alongside it and thus give rise to dreams-*rêves*. For example, the person in love inevitably dreams that he is with his beloved, and the hungry dreams that he eats, the thirsty that he drinks, and what is more the person who is full of food dreams that he vomits or that he is suffocating. It is therefore possible to have these dreams-*rêves* because the affects are already the basis for them, these dreams themselves do not comprise an announcement of the future but a memory of present realities. Things being thus, you can have dreams that concern the body alone, or dreams that concern the soul alone, or dreams concerning body and soul together"; Eng., p. 47: "For the *oneiros* differs from the *enhypnion* insofar as it is characteristic of the former to be significant of things in the future, and of the latter to be significant of things in the present. And you might come to understand this more clearly in the following way. Certain types of affections are inclined to rush up and marshal themselves in the mind and to bring about nocturnal emissions. For example, it is inevitable that, in a dream, a lover imagines that he is with his boyfriend, and a frightened man observes the things he fears and, again, that a hungry man eats and a thirsty man drinks, and, moreover, one who is stuffed with food either vomits or chokes ... It is therefore necessary to regard these dreams, where the affections are already accounted for, as containing not a warning of things to come but a recollection of things that are. And, when this is the case, you will see that some dreams are particular to the body and some are particular to the soul and some belong to both ..."

13. Ibid., Fr., p. 19: "The vision of the dream (*songe*) on the other hand not only affects like the 'dream (*rêve*) *during sleep*' in that it brings us to attend to the announcement of what will come, but again, *after sleep*, by bringing about the enactment of undertakings, its natural property is to excite the soul and put it in motion (*oreinein*), the name itself *oneiros* having been applied to it from the start due to this, or because it 'states,' which means 'tells' 'what is' (*to on eirei*), according to the word of the poet, '*I say what is truthful*' (*Od.* 11, 137). And the people of Ithaca call the beggar Irus, '*because in his journeys he conveyed messages when he was commanded to*' (*Od.* 18, 7)"; Eng., p. 49: "the *enhypnion* ... is active as long as sleep lasts but, when sleep stops, it disappears. But the *oneiros*, which is also an *enhypnion*, makes us observe a prophecy of future events and, after sleep, it is by nature inclined to rouse and stir the soul by inciting active investigations. And its name was originally given to it for these reasons or else because it 'tells' ⟨'the truth'⟩, which is to say, as the poet renders it: 'I have spoken infallible things.' And the men of Ithaca named the beggar 'Irus' 'because he, running about, used to deliver messages whenever someone prompted him.'"

14. Galien [Galen], *De dignotione ex Insomniis*, in *Claudii Galeni Opera omnia*, ed., Karl Gottlob Kühn (Leipzig: C. Cnobloch, 1821-1833) 22 volumes: see vol. VI, pp. 832-834.

15. Artemidorus, *La Clef des songes*, I, 2, p. 20: "Among the dreams-*songes*, some are *theoreomatic*, others allegorical. Theoreomatic dreams are those whose fulfilment fully resemble what they have shown. For example a navigator has dreamed he is shipwrecked and this is what has happened ... Allegorical dreams on the other hand are dreams that signify certain things by means of others: in these dreams-*songes* it is the soul that, according to certain natural laws,

obscurely lets an event be understood"; *Artemidorus' Oneirocritica*, p. 49: "Moreover, within the category of *oneiroi*, some are 'directly perceived' and some are 'allegorical'. Directly perceived dreams are those [whose outcomes] are identical to their appearance. For example, someone who was sailing imagined that his ship was wrecked and it occurred just as it was presented in sleep ... But allegorical dreams are those that signify different things through different images, since in these dreams the soul [in fact] naturally speaks in riddles."

16. See, on this point also, *L'Usage des plaisirs*, p. 29; *The Use of Pleasure*, p. 22 regarding ancient ethics: "It was an ethics for men: an ethics thought, written, and taught by men, and addressed to men—to free men obviously."

17. Declaration carried on one of the ten "surrealist butterflies" printed in December 1924 for the *Bureau de recherches surréalistes*.

18. Artemidorus, *La Clef des songes*, I, 78, p. 86: "It goes without saying that the person who offers his body to someone must also provide him with the goods he possesses"; *Artemidorus' Oneirocritica*, p. 139: "For she, making her own body available to someone, would rightly also furnish things in addition to ⟨her⟩ body." See A. J. Festugière's note on this sentence: "The 'things relative to the body' (*ta peri to sōma*) here are the *chrēmata*, wealth, goods" (ibid).

19. Ibid., Fr., I, 78, p. 87: "If the father is poor, he will make his secretions in his son in the sense that he will send him to the schoolmaster and pay the latter's salary; if the father who sees this dream is rich, he will make his loss of substances (*eis auton apousias poiēsetai*) in his son in the sense that he will give him great gifts and transfer his fortune to him by contract." See A. J. Festugière's commentary in his note on this sentence: "*ousia* = wealth ... the dream of 'secretions' for a rich father signifies that, by gift to his son, he will have great losses of wealth." Eng., p. 141: "if, first, the father is poor, by sending the son to a teacher and paying wages, he will make 'expenditures' [and outlays] upon him. And should a wealthy man observe a thing of this sort, by giving many gifts to his son and allotting them to him, he will lavish 'outlays' upon him."

20. Ibid., Fr., I, 78, p. 88: "In the case of a friend, if one penetrates him he will conceive a hatred of you, for he will have been first of all raped by you (*blabenti*)"; Eng., p. 143: "one who penetrates a friend will fall into enmity with him since he has injured him without provocation."

21. Ibid., I, 78, Fr., p. 85: "I know of one who dreamed that having entered a brothel he could not leave it and died a few days later, this dream having had for him its fulfilment in just fashion: this is because one calls the brothel, like the cemetery, a 'common place.'" In the note devoted to this passage, A. J. Festeguière recalls that *ergasterion*, as well as "house of prostitution," "brothel," signifies also "shop"; Eng., pp. 137-139: "And I know of a certain man who imagined that he entered a brothel and was not able to get out, and he died after just a few days. [And] logically did this happen to him. For the place is called 'open to all,' as is that place which receives dead people."

22. The Greek word here is *anagkaion*. Ibid., I, 79, Fr., p. 91: "If the woman is richer than the man, she will have to settle many debts for her husband, ands if she lives in marriage with a slave, contributing for her part with money, she will have to free her husband, and it is thus found that the husband's 'constraint'—this is the name given to the male organ—that is to say the constraint that pressed him, will have been freed of all difficulty"; Eng., p. 147: "Moreover, a woman with greater wealth than her husband will satisfy many loans on her husband's behalf, and a woman who lives with a slave, furnishing money from her own resources, will free the man, and thus it will occur that the man's 'necessity' (for the genitals are called this)—that is, his compulsion to work—will be wiped away."

23. Ibid., I, 78, Fr., pp. 84-85; Eng., p. 137.

24. Ibid., I, 78, Fr., p. 86; Eng., p. 139: "For it is necessary to consider unrecognized women as signs of future activities that will relate to the observer."

25. Ibid., Eng., "And so, whatever sort of woman she is and however she is disposed, in this same way, too, will these activities affect the observer."

26. Ibid., I, 78, Fr., p. 85: "To have sexual intercourse with one's wife, if she agrees to it willingly and without resistance, is good for both because the wife is either the craft of the dreamer or the profession from which he draws his pleasure or over which he presides and commands, as he does to his wife. This dream thus indicates the profit that may be expected from the craft or profession: for men draw pleasure from the sexual act, and they draw pleasure also from these profits. If on the other hand the wife resists and does not give herself, this indicates

the opposite"; Eng., p. 137: "To have sex with one's wife, if she is willing and submissive and does not resist intercourse, is good for all in common. For the wife is, in fact, the craft of the observer or his business, from which he derives pleasure or over which he is positioned and rules, just as he does over his wife. ⟨And⟩ the dream signifies that one will benefit from these things. ⟨For⟩ people delight in sex, and they also delight in benefits. But if one's wife should resist or not make herself available, it signifies the opposite."

27. Ibid., Fr., "Same interpretation in the case of the mistress"; Eng., "And let the same explanation stand for a lover."

28. Ibid., Fr., "Having sexual relations with prostitutes established in brothels indicates a slight shame and a slight expense: for to approach these women entails both shame and expense. This is good for every kind of enterprise: for some name these women 'workers,' and they give themselves without refusing anything. One should judge good also, once in the brothel, being able to leave it, for not being able to leave is bad. I know of one who dreamed that after entering a brother he could not leave it, and he died a few days later, the dream having had for him its fulfilment in just fashion: this is because one names the brothel, just like the cemetery, a 'common place,' and a great loss of human sperm takes place there. So it is for good reason that the brothel is associated with death"; Eng., pp. 137-139: "And to have sex with female courtesans stationed in brothels signifies a small amount of shame and a small expense. For men both feel shame and spend money when they consort with these women. But they are good for every undertaking. For in fact they are called 'working girls' by some and, never refusing a client, make themselves available. And going into a brothel and being able to get out again is good, since not being able to get out is grievous. And I know of a certain man who imagined that he entered a brothel and was not able to get out, and he died after just a few days. ⟨And⟩ logically did this happen to him. For the place is called 'open to all,' as is that place which receives dead people, and the destruction of much male seed also takes place there. And so this place logically resembles death."

29. Ibid., I, 78, Fr., p. 86: "Having sexual intercourse with his slave, woman or man, is good: for slaves are the dreamer's possessions. Also this indicates that the dreamer gets pleasure from his possessions, and that probably they will increase and become more sumptuous. Being penetrated by a servant, on the other hand, is not good: this implies contempt and injury on the servant's part"; Eng., p. 139: "And to have sex with one's slave, female or male, is good. ⟨For⟩ slaves are the property of the observer. They therefore signify that the observer delights in his property, which will fittingly increase in quantity ⟨and⟩ become more valuable. But to be penetrated by a household slave is not good. For it signifies being both despised and harmed by that slave."

30. Ibid., Fr., "Having sexual intercourse with a woman who is known and familiar to you, if one sees this dream in a state of erotic tension and desire for the woman, does not predict anything due to the tension of the desire. If one is not in a state of desire for the woman, it is good for the dreamer when the woman is rich: in any case in fact there will be some advantageous successful enterprise close to and thanks to the woman one has seen. For it goes without saying that she who offers her body to someone must give him what she possesses ... But to penetrate a legally married woman in the power of the husband is not good because of the law: whatever may be in fact the punishments to which the law subjects someone caught in adultery, it is to these same punishments that the dream also leads"; Eng., p. 139: And to have sex with a woman whom one knows and with whom one is well-acquainted, if the person observing this dream is sexually attracted to that woman and desires her, it foretells nothing due to his desire having been aroused. But if he does not desire the woman, it is good for him, that is, when the woman is wealthy. For he will accomplish something that is wholly profitable directly or indirectly through this woman whom he saw in his dream. For she, making her own body available to someone, would rightly furnish things in addition to ⟨her⟩ body ... And to penetrate a lawfully wedded woman is not good due to the law. For whatever the law demands of one who has been arrested for adultery, the dream does the same."

31. Ibid., Fr., "As to being penetrated by one of your acquaintances, for a woman on the one hand it is an advantage, whoever penetrates her. For a man on the other hand, if he is penetrated by someone older and richer, this is good, for from such people one usually receives; but if it is by someone younger and poor, this is bad; for usually one gives to such people

a part of what one has"; Eng., "And to be penetrated by a familiar person is, for a woman, [pleasant and] profitable, whatever the man may be like. And for a man to be penetrated by a wealthier man and an older man is good. For it is customary to receive things from men of this sort. And to be penetrated by a man who is younger and poor is grievous. For it is customary to give to men of this sort."

32. Ibid., I, 78-79, Fr., pp. 87-92; Eng., pp. 141-149.
33. On this point see F. Héritier-Augé, "L'inceste dans les textes de la Grèce classique et post-classique," *Mètis. Anthropologie des mondes grecs anciens*, vol. 9-10, 1994, pp. 99-115.
34. Artemidorus, *La clef des songes*, I, 79, pp. 88-89; *Artemidorus' Oneirocritica*, p. 143: "And the section on [sex with] one's mother, ⟨being⟩ intricate and manifold and allowing for several distinct analyses, has eluded many dream interpreters. And it holds as follows. The sex itself is not enough to reveal what the dream signifies, but in fact the combinations and positions of the bodies, when different, create different outcomes."
35. Ibid., I, 79, Fr., p. 91; Eng., p.147:
36. Ibid., I, 79, Fr., pp. 91-92; Eng., pp. 147-149.
37. Ibid., I, 80, Fr., pp. 92-93; Eng., pp. 149-151.
38. Ibid.,, I, 78, Fr., p. 86; Eng., p. 141.
39. Ibid., Fr., "One dreams that one makes the member erect with one's hand, one penetrates a male or female slave, because the hands one brings to the member have been in the service of this member"; Eng., "And if one should suppose that he masturbates, he will penetrate a male or female slave because his hands, brought to the genitals, are its servants."
40. Ibid., I, 78, Fr., p. 87; Eng., p. 141.
41. Ibid., I, 80, Fr., p. 92; Eng., p. 149. See A. J. Festugière's note: "the Greek keep *perainein*, *perainesthai* ('to penetrate' and the converse) as in the case of the man: it is a matter of the tribade [lesbian who adopts the male role; G.B.]."

28 JANUARY 1981

The ethical perception of the aphrodisia. ∽ *Principle of social-sexual isomorphism and principle of activity.* ∽ *Valorization of marriage and definition of adultery.* ∽ *Modern experience of sexuality: localization of sexuality and division of the sexes.* ∽ *Penetration as natural and non-relational activity.* ∽ *The discrediting of passive pleasure.* - *Paradox of the effeminate womanizer.* ∽ *Problematization of the relationship with boys.* ∽ *The desexualized pedagogical erotics.*

*WHAT CAN WE RETAIN from this text by Artemidorus, which is then an oneirocriticism (a method of the interpretation of dreams), and which devotes three chapters to dreams with a sexual content? It is not possible to find and it would not be good method to seek in a text like this the way in which the Greeks codified—in their own way—a supposedly absolutely common, trans-cultural, trans-historical domain of sexuality. We should not look in Artemidorus' text for the

---

* The lecture begins in this way:
    Today I would like to return to the text of Artemidorus that I spoke to you about last week, because I get the impression of having been a bit confused, insufficiently precise, of not having exactly made the objective clear, at least at the end of last week's lecture. I would like to return to this because I think it is important for reasons of method, and then for the articulation of the set of things that I would like to propose to you this year. So forgive me if there is a bit of repetition, but once again I will try to re-state these things from a slightly more precise perspective, before passing on as quickly as possible to the rest.

distribution into good or evil, permitted or forbidden, recommended or prohibited, of a raw material, sexuality, that would be both indifferent and universal. I would like to try to show you something rather different in this text: what I would call the ethical perception that characterizes a quite specific Greek, Greco-Roman, Hellenistic, and Roman experience for which the Greeks had a precise word: *aphrodisia*. In his text Artemidorus says that dreams with a sexual content concern the *sunousiai aphrodisiōn*, that is to say the coming together, the unions of those things that I will call, simply and for want of anything better for the moment, the *aphrodisia*, using the Greek word itself. I will not [attempt] to define this notion of *aphrodisia* now for the good reason that its meaning is precisely what is at stake in the analysis. The meaning of *aphrodisia* is what we have to try to bring out in order to distinguish it from two other experiences (defined in contrast with it): the Christian experience of the flesh and the modern experience of sexuality: I not think that these are three domains of separate objects, but rather three modes of experience, that is to say three modalities of the relation of self to self in the relation that we may have with a certain domain of objects related to sex.

As for the notion of ethical perception, since it will thus be a matter of bringing out the ethical perception peculiar to the Greek experience of *aphrodisia*, what I mean by this notion is this. In his method of the interpretation of dreams, Artemidorus does not make directly a positive or negative judgment on the different sexual acts. He does not say: this is good, this is not good. His problem is how to determine the prognostic value of a dream with sexual content, that is to say how to determine whether this dream announces something favorable or unfavorable. So it is not a question of extracting the direct moral value of the act, but rather its prognostic value, this being understood as signification of the event to come and, at the same time, of the value of this event (is it favorable or unfavorable?). Consequently, you see that the objective of Artemidorus' analysis, when it focuses on the sexual acts represented in dreams, is to elaborate the content of these dreams while trying to retain, to reveal, to bring out the elements, the features that may reveal precisely those favorable or unfavorable values of future events. Consequently, it involves singling out the elements, the features that resemble evil, hardship, misfortune. In sum, what is

it in the very nature of the sexual act represented, in the basic components of the dreamed sexual act, that links it to the negative values thereby announced? What Artemidorus reveals spontaneously in
the way he deploys his analysis, without this being of course the actual
objective of this analysis, is not therefore a code, a table of prohibitions
and things permitted, but rather the general mechanisms that determine the formation of positive or negative judgments, of more or less
favorable valuations that one might fairly commonly bring to bear on
such and such a sexual act. What can be extracted from Artemidorus'
text is, as it were, the formation of the evaluative system of sexual acts,
and not a catalog of prohibitions and authorizations. What is involved
is the formation of evaluative mechanisms, and this is what I mean by
"fundamental principles of the ethical perception of these *aphrodisia*."
What are the major formative mechanisms of the evaluative system put
to work in order to determine spontaneously whether a sexual act is
more or less good, more or less bad, of more or less favorable value in
itself and in its diagnostic significance? What are these principles?

[Returning] a bit to what I said last week, I think that one can
extract two principles from Artemidorus', which are the most fundamental principles of [this] ethical perception of *aphrodisia*. Once
again, I remind you that this text is from the second century, was
written by someone who was learned to a certain extent, a philosopher clearly inspired by the Stoics. But I remind you also that
Artemidorus' text was—he explains it himself—composed on the basis
of, and as it were in the current of a whole tradition collected in books
(he says he has read many of them) and equally in an oral tradition
(he travelled through the towns, the markets, the crossroads of the
Mediterranean world to listen to the readers of dreams and soothsayers).[1] Artemidorus thus developed his oneirocriticism in the current of
this whole tradition rethought philosophically. Through this text one
can extract the principles of Artemidorus' ethical perception, but no
doubt also, much more broadly, of the ethical perception of a culture to
which he is both heir and witness.

This ethical perception of *aphrodisia* is governed by a first principle that I will call the principle of isomorphism. This is what I mean.
You recall that sexual acts in dreams have a prognostic value for events
that are essentially events of social life. And I have stressed the fact

that in Artemidorus' analysis this projection of the sexual onto the social does not take place because the sexual is, as it were, the symbol of the social. There is not a symbolic code between the sexual and the social. The sexual refers to the social because they are in continuity with each other. The sexual acts of which one may dream are basically of the same nature, the same substance, and, one must go further, the same form as the social relations, social acts, social events to which they refer.

That is the point that I would now like to stress. What really needs to be understood—in Artemidorus' perspective and, it seems, in the ethical perspective to which he bears witness—is that sexual relations are as it were the physical, corporal, intense forms of social relations themselves. They are directly bound up with them and it is as if they are their major point of intensification. First, it should not be forgotten that in this general category of *aphrodisia* that the Greeks talk about, sexual acts are designated by certain words like: *homilia, sunousia, sumplokē*. Now what do these words mean? *Homilia* is above all a meeting, an assembly, an encounter with someone. It is the intercourse one may have with someone. *Sunousia* is to be with, be familiar with, live with—and may also designate a meeting, an assembly. *Sumplokē* is the intertwining of two things, or rather of two individuals, it is combat, struggle, the entanglement of bodies in a physical confrontation. In short, most of the words that designate the sexual act, the sexual union have above all, fundamentally, a social signification. But obviously this is not the main thing.

For Artemidorus it is a matter of extracting from the different components of the sexual act what has favorable or unfavorable prognostic value for events. When he analyzes these sexual acts, the way in which he breaks them down reveals what could be called a social dramatics as though this were intrinsic to the sexual act. When Artemidorus describes a sexual act as prognosticating something favorable, and so belonging to the domain of the "rather good," he makes it appear as an initiative, a trouble taken by someone. He describes it also as an expenditure made by the person who performs the sexual act and which takes the very precise form of that part of oneself, of one's being, one's body, one's substance that one parts with, that one emits, projects outside, divests oneself of in order to profit from it, in either the pleasure or the descendants that result from it. In this sense, the sexual act is an investment. It is also, of course, an act of possession, an act by which one appropriates something, by which one marks one's rights over someone. The sexual act is also defined as a constraint, either that one imposes on the other person (the partner), or an internal constraint that forces us to commit this act. And by constraint must be understood the same type of juridical-economic constraint—the terms used are the same—as that to which one is subject because one is in debt.[2] One has a debt to oneself that one is obliged to pay because one is pushed by the need of the sexual act, or one makes the other person pay a debt by constraining them to have sexual intercourse with oneself. So there is a whole social dramatics that in a way forms the very fabric of this act. This social dramatics unfolds on a stage and with socially marked characters, between partners whose social value is absolutely essential for evaluating the prognostic meaning and value of the act itself. This sexual theater in Artemidorus is essentially made up of young and old, rich and poor, free men and women, slave men and women. The elements of the social setting are not removed in the sexual act in favor of bodies, in favor of anatomies for which the only rule of appropriateness would be their modesty. The sexual partners are and remain throughout social characters, and consequently the judgment brought to bear on these sexual acts is inseparable from the social marking of the individuals involved in them.

So, the problem now is the respect in which these sexual acts are inseparable from social marking. Here we need to analyze things a bit more closely. In this ethical perception to which Artemidorus bears witness, a sexual act will have a positive value inasmuch as its intertwinings, its intrications, its *sumplokai* extend or reproduce the same model of relation as that linking individuals involved in the whole social field. So one can say that if there is in actual fact this continuity and isomorphism between the sexual relation and the social relation, then the act can be considered as rather good. On the other hand, the sexual relation will be considered rather bad, it will be morally discredited if it overturns, reverses, upsets, or more simply if it strays, deviates, or diverges from the social relations of which it is part. The true division, consequently, in this field of *aphrodisia*, is clearly not that of homo- or heterosexuality, it is the problem of socio-sexual iso- or heteromorphism. The fundamental characteristic of this principle of isomorphism produces a whole series of effects in the field of *aphrodisia*, in the ethical perception of *aphrodisia*, that would be interesting to analyze. I would like to focus on just two or three of them.

First, take an adult, powerful, rich, virtuous man, full of moral and intellectual qualities, and so on. He has relations with a young boy in which, on the one hand, he will make that expenditure that constitutes the activity itself of the sexual act. He will also make that other expenditure which is the attention he pays the boy, the care he gives him, possibly the gifts he [offers] him. And at that point his sexual relationship, thanks to this whole set of care, attention, gifts, watchfulness, and so on, will effectively be isomorphic to the social relationship that every adult in the Greek city must have with regard to the young, since he must help them, serve as model for them, be as it were their patron, their teacher of virtue, of citizenship, their political teacher, and so on. We have an isomorphic type of relationship. On the other hand, the relationship will be heteromorphic if a man who is rich, powerful, and so on, is passive with his slave, that is to say has sexual intercourse with his slave in which he does not occupy the social position of the master that he is, but rather the inferior position of a slave that he is not, consequently putting his slave in a position, calling on him to adopt a position of superiority, which is not the slave's position in society. So we have here this very clear division between isomorphism and heteromorphism. The first conduct is good, the second is bad. You see [then] that the major line of division is not that of homosexuality or heterosexuality, but that of socio-sexual isomorphism and heteromorphism.

The second consequence of this fundamental principle of homomorphism is that marriage and the sexual relationship between husband and wife is the very model of the isomorphic relation. In the conjugal relationship in fact, the husband asserts his rights, he exercises his superiority, he takes possession of precisely what he possesses, he expends energy, he profits (pleasure, a descendant), he is on top, and so on. Matrimonial sexuality, the type of sexual relations that a man and a woman may have as a couple, as spouses, is exactly isomorphic to the social relationship defined by marriage. But at the same time, and this is an important point, you see that this very high valorization of marriage, which places it at the head, at the summit of all possible isomorphisms, is absolutely not a principle of the exclusive localization of sex within marriage. One already sees the development of the principle that sex is actually legitimate and acceptable only within marriage in the philosophy contemporary with Artemidorus, and later of course in Christianity. Anyway, it will become fundamental in later societies. But this is not at all the role that marriage plays in the organization of Greek *aphrodisia*. Marriage is valorized, but not as the only possible place in which sexual intercourse can take place. It is simply the most perfect form of that isomorphism that is one of the fundamental principles of the evaluative mechanisms of the sexual act. And it is on the basis of this form that one will be able to see other forms of reduced isomorphism extending to the point of complete heteromorphism.

Hence a very interesting and important definition of adultery, or rather a [very] revealing [aspect] of the way in which extra-marital relations are evaluated. Since marriage is nothing other (which is already not bad) than the highest form of socio-sexual isomorphism, you quite understand that extra-marital relations are not condemned in themselves. However, several cases may arise. For example, when a married man has sexual relations with his servant and takes her in a dark corner, it is clear that this is a case of a perfectly isomorphous sexual relationship, since in doing this with his servant or slave, in his physical relationship with her, the master does no more than bring into play the same type of social relationship that he has in the rest of existence. Consequently, what we would call "deceiving his wife" with his servant cannot be considered adultery, and certainly not a serious act. It is not adultery since the definition of adultery is much more precise—in any case, if the sexual relationship with [his] wife is valorized, it is not as [the only one], but simply as the best possible. It is not an act that is blameworthy in itself since, once again, it reproduces the social relations within the act itself. On the other hand, when the same married individual has sex with his neighbor's wife, the relationship is then heteromorphous and adultery. First, because adultery is not defined by the fact of one of the couple deceiving the other, but by the fact that one man takes the wife of the other, or that a married woman has sexual relations with a man who is not her husband. This juridical definition of adultery makes it possible to specify when there is a legal offense. But at the same time there is a moral fault and the act is judged bad; it is seen as [bad] quite apart from specifically legal considerations, quite simply because making love with one's neighbor's wife contradicts the type of social relations validated in society. One encroaches on the rights of one's neighbor, one takes his goods from him, one attacks his sovereignty, or at any rate the authority he exercises. And because of this, it is not just an act that is legally prohibited, and legally punished moreover, but also a morally bad, typically heteromorphic act, whereas having sex with one's servant cannot be bad because it is an isomorphic act.

Consequently, let us say that in an ethical perception with which we are familiar, since it is the one we have or anyway the one passed on to us by our societies, by our culture, the fundamental elements are [the following]. First, the problem of what could be called localization: where are legitimate sexual relations localized? For a very long time— since Christianity no doubt, but even earlier, in the moral philosophy I will talk about shortly—the answer to this question of localization was: legitimate sexual relations must be localized in marriage, and exclusively in marriage. Second, in our ethical perception, characterized by the experience that I will call "sexuality," the other fundamental question is that of the biological, anatomical-physiological division of the sexes. So, there is opposition between permitted and prohibited localizations of the sexual act, with marriage as the major differentiating criterion, and distinction between homo- and heterosexuality.

In the ethical perception evinced by Artemidorus you can see that the questions are entirely different: not localization, but a hierarchy of isomorphisms. It is not a matter of a division between homo- and heterosexuality, [but] of a whole gradation of isomorphisms and thresholds of heteromorphisms that will render certain acts absolutely blameworthy. So it is an entirely different organization.

To end this chapter on [isomorphism]* I would like to stress the following. In what may be called the modern schema, localization of sexuality and division of the sexes, the presence, mechanism, form, and

---

* M. F. says: heteromorphism.

effects of these principles of localization of the sexual act and division of the sexes quite naturally take the juridical, or quasi-juridical form of the law, which on the one hand separates the permitted sex (the other) and the prohibited sex (one's own), and [on the other hand] separates the permitted partner (the spouse) and prohibits all others. The principle of localization and the principle of the separation of the sexes have the form of law, even if these are not juridical laws, that is to say they are not taken up in an explicit code supported by punishments. Even if these mechanisms do not function within a juridical code, the distribution between permitted and prohibited partner, and the organization that separates from all others the place in which one may make love or the institution within which one may have sexual relations, take the actual form of law. Consequently, it is a juridical form even if there is no judicial mechanism to validate or enforce it. On the other hand, in the Hellenic schema of *aphrodisia*, of which Artemidorus gives us an example, the principle of the valorization of sexual acts does not in any way conform to the form of law. It is a principle of perfection, of appropriateness, a principle of hierarchization around what is considered to be the most perfect form. It is a whole gradation, a whole distribution of possible acts, each of which effectively has its value, or to which a value can be assigned on the basis of these principles, but without there being this juridical form of the law. You will tell me, however, that at several points Artemidorus employs the notion of *nomos* and even organizes a whole part of his analysis around this notion. He says: There are acts that are in accordance with the law and others that are not. But it should not be forgotten that the Greek notion of *nomos* does not correspond solely, exclusively, or even fundamentally to the juridical form of the law as we know it. The *nomos* is also, it is fundamentally, a mode of distribution. To that extent one may actually say that the principle of isomorphism that I have just been analyzing for you is indeed a *nomos*—*nomos* as a principle of distribution of sexual acts and not as juridical law, division between permitted and prohibited. So that's it for the principle of isomorphism that is the first of the organizing principles of this domain of *aphrodisia*, organizing principle of the ethical perception, first principle permitting the formation of judgments, of evaluations that may be brought to bear on sexual acts.

The second principle, after that of isomorphism, is what I will call the principle of activity.* The sexual act is not just related to a social field by Artemidorus, it is also related to a domain of naturalness. Artemidorus does not simply distribute sexual acts in terms of what is or is not in accordance with the *nomos*, he also raises the question of what is in accordance with or contrary to nature, or at any rate of what is outside nature, since *para phusin* means that which is apart from, outside nature, rather than what is exactly contrary to it. In any case, we need to consider this problem for a while. When we try to see what for Artemidorus is [essential] in this naturalness, we find, of course, male activity, that is to say penetration. Sexual penetration is the internal rule of naturalness that will allow separation of sexual acts into natural or not natural. Now this problem of penetration calls for some elucidation. In fact, the penetration to which Artemidorus refers—and which he makes a deciding element regarding the natural-ness of sexual acts—is not thought of by him as a relationship between two terms, between the one who penetrates and the one who is pen-etrated. It is not even exactly a relationship between an active individ-ual and another who would be passive. If it were in fact a relationship between two terms, it would be necessary to give, if not equal, at least a fair degree of attention, even if hierarchized, to both elements in the relationship; it would be necessary to analyze the *aphrodisia* of the one who is passive, penetrated, who receives, and so on, as well as of the other; the two partners would have to be integrated in this field. Now, in fact, for technical reasons (*The Interpretation of Dreams* is, you recall, a family father's book, in which it is a question of helping him to manage his life through his dreams), but [also] more gener-ally, in Artemidorus and the whole tradition he represents, the ethical perception of *aphrodisia* is entirely governed by the point of view of the active individual. Ultimately, in contradiction—I will try to show you how [this contradiction] gets the whole system to function—with what I was telling you about the principle of isomorphism and the impor-tance of social relations, with regard to naturalness, the sexual scene concerns a single character. The principle of isomorphism implies, of course, the plurality or anyway the duality of characters, and their

---

* M. F. adds: here again, I am returning to what I have told you, forgive these repetitions.

social marking within the sexual act. The criterion of naturalness, however, concerns, brings into play, reveals only one of the characters: the
active one. Penetration is not a process that takes place between two
individuals. It is essentially* the activity of *one* subject and the activity *of* the subject. And it is as the subject's activity that it constitutes
the central and natural kernel of all sexual acts (of all the *aphrodisia*).
This must be clarified further to see how this, as it were, unitary, single, non-relational activity of the subject constitutes the very essence of
the sexual act.

[First,] in Artemidorus' text we see very clearly that the principle of activity functions as the sole criterion of the naturalness of the
sexual act in the quite remarkable indifference Artemidorus displays
towards what will later be called—and this will be so important in all
the sexual ethics and morality, [in] the Christian codification of sexual
acts—the natural vessel. You are well aware that the problem of the
natural vessel is very important in the so-called "Christian" conception.† In this Christian conception, in fact, penetration by the male
is, of course, still and always the essential element of the whole analysis of the sexual act. It is indeed always from the male point of view,
from the angle of male privilege, that everything will be analyzed. But
this act of the male entails that it be codified in two ways. On the one
hand, from the point of view of the subject, as an act of the male as
such. But there is also a law that might be called the law of the object.
The act of penetration, characteristic of sexuality or of the sexual act,
must in actual fact be accomplished in a certain way, with a particular
object that is the female sexual organ. This so-called Christian theory
of the natural vessel is therefore a theory with two elements: the naturalness of the sexual act comprises two points of anchorage, on the side
of the subject of course (the male) and on the side of the object (the
woman, the natural vessel—this correlative element is indispensable for
the good naturalness of the sexual act). In this Christian perspective,
sexual acts will be justified or condemned according to whether penetration takes place properly, that is to say according to whether the

---

* M. F. adds: this must be stressed because it is crucial.
† M. F. clarifies: here again, I put in quotation marks and it will be necessary to come back to all
this.

penetrating element is connected with the element to be penetrated. Now what is very clear in the way Artemidorus analyzes dreams and shows how to evaluate them, is that he speaks of course of penetration (it is essential), but he speaks of the correlative element, of the point of outcome, of the place of anchorage, of the place to be penetrated, on only one occasion. He speaks about it on only one occasion and with reference to the place that is abominable for the sexual act, which is the mouth.[3] The mouth is prohibited because of its two functions (eating and speech). But apart from this forbidden place, Artemidorus says strictly nothing about the way in which penetration should take place, whether with a woman or a boy. He is very explicit, he is very detailed, he has the appearance of caring a great deal about the position of the partners: seated, standing, on a bed, crouching, and so on, inasmuch as these positions are in some way the representation, the actualization, the dramatization of social relations. So due to this, the position is important. But on the other hand, [he is not interested in the] place, natural or not, where the penetration takes place, whether this is for women or boys. Now this indifference to the place of penetration is quite intelligible if one fully grasps that in this ethical perception of *aphrodisia* it is not a question of a relationship between two elements. It is the activity of a subject, and of one alone.

A further consequence—[I am extrapolating]—is that if it is true that the naturalness of the sexual act lies in this activity of penetration and not in a relation between penetrating and penetrated, it becomes very difficult to situate the partner precisely in this ethics. There is inevitably a fuzziness, a whole series of uncertainties about the way in which one will appraise, assess the partner's role. First of course there must be partners. But how will they appear? Essentially under three categories [that figure] regularly in the texts. The partner may be a woman, a boy, or a slave. This trilogy is constant, you find it for example in the oath of Hippocrates: the physician must [promise] never to enter a house to which he is called as physician with the aim of making love with a woman, a boy, or slaves.[4] This woman–boy–slave trilogy, these three categories of individual, are the natural correlatives of penetration. The woman, of course, is a natural correlative by virtue of her anatomy, by the arrangement of her body, possibly by her beauty, by the delicacy, the softness of her body, by her inferiority from every point of view (social, intellectual, physical, and so on). The boy is equally the natural object of this activity of penetration, he is its correlative because he too is weak, he too is beautiful, and also because he needs to be formed by someone else. As for the slaves, they are the natural correlative of this activity of penetration quite simply because they are slaves. But natural correlative does not at all mean that they belong to the same field of naturalness as the male individual who penetrates. They are as it were on the outer edges of this naturalness. And they cannot but be on the outer edges since this naturalness is defined by the activity of the subject. The naturalness of the sexual act is not a coupling between activity and passivity. Naturalness is activity. And consequently the passive character is quite naturally the correlative, but the correlative at the outer limit of this naturalness. Let's say again that, as objects or as correlatives of the activity of penetration, they belong to naturalness, but as subjects in themselves they fall outside it.*

This unstable position of the objects of penetration (woman, boy, slave) as correlatives of a natural activity entails of course certain consequences, essentially the mistrust experienced with regard to their pleasure. Actually, these correlative elements of the activity of penetration clearly do not take part in the sexual act strictly speaking, since the latter is defined in its naturalness by activity. How do they take part? By the fact that, voluntarily or not, they give their bodies. They take part also, and this is the delicate point, up to a certain point as subjects, but simply as subjects of pleasure. Now since the whole field of *aphrodisia* is governed, normed by the principle of activity, what is it to have pleasure, and to have pleasure in passivity? Pleasure is an experience of the subject. There is pleasure only because there is a subject, but the naturalness of *aphrodisia* implies that the only subject one can recognize, the only pertinent subject, the only subject who is both subject of *aphrodisia* and of a possible morality, is clearly the one who is active. The woman, the boy—let's leave the slave because, anyway, he poses no problem—belong to this system as objects of penetration. But from the moment that they experience pleasure, that is to say that

---

* Hardly audible passage. One hears only: And one may say that up to a [...] sexual in a conception and in a perception like the latter [...].

they risk manifesting themselves as subjects, then at that point they disturb the system and become problematic elements about which it is necessary to reflect, on which a whole ethics will need to be constructed. A whole ethics, the function of which is essentially to discredit the woman's or the boy's pleasure inasmuch as these pleasures will be what reintroduces them as subjects within a field of *aphrodisia*, in which the only recognizable, valid, and legitimate subject is that of activity. The pleasure of passivity is clearly what puts the whole system at risk of slipping out of control and coming apart, what makes it unstable, or metastable. Hence the extraordinary mistrust of all Greek ethics towards the boy's and the woman's pleasure, the idea that the woman's pleasure is something indefinite, something uncontrollable, that is to say on which the subject can get no hold. It is something that inevitably carries the woman away, as it were, into an unnatural naturalness. The woman's pleasure is an abyss. It is natural that the woman experiences pleasure, but it is woman's nature to leave her own nature, to leave what was foreseen for her by nature, and to lose herself in the worst debauches. The woman is naturally excessive, woman's pleasure is naturally excessive, and due to this it is precisely at the [junction] of nature and the unnatural. The woman is pinned to nature as the correlative of male penetration. She is pinned to nature also because she experiences pleasure in this penetration. But the nature of this pleasure is to drag her outside nature, into every possible excess, which are those of debauchery. The woman's pleasure is the source, in her, of excess. Pleasure is both that which marks the woman as an element in the natural system governed by male activity and that which constantly puts her outside. As for the problem of the boy's pleasure, we will have to come back to that shortly, but we can say that if the woman is marked by the fact that her pleasure is indefinite, and consequently constantly puts her outside, the boy will be a suitable boy only on condition that he does not experience any pleasure in the act of penetration, which it is quite natural, however, for the active subject to impose on him. A well brought-up boy does not experience any pleasure. And I refer you to the book by Dover on this,[5] which deals with Greek ethics of the fifth-sixth century, but it seems that the same type of organizing principle of ethical perception is still valid and at work at the time of Artemidorus.

Finally, the last consequence of this principle of activity is that if the *aphrodisia* are in actual fact governed by the principle of a non-relational activity, this non-relational male activity can and must be subject to the same moral rules, the same ethical rules as those imposed on any activity (on any social, familial activity, on any activity concerning others, men, gods, and so on). That is to say that the man must be moderate in his sexual activity. Since his activity is brought to bear on others in a sovereign fashion, as the subject of an activity that is not relational and could be absolute, without a relational code, the man has to find the restrictive principle of this activity in the sovereignty he must exercise over himself. Sovereign over others, he deploys his activity in a field that is not limited by any physical, anatomical, or other code that would say: "This is what you must do, this is how far you must go." On the other hand, since this activity is an activity solely of the subject himself, it is limited from within by the mastery he must exercise over himself. And faced with a pleasure which has a dangerously unlimited natural tendency characteristic of the woman, the active subject must on the contrary manifest, by himself, a rule of moderation. And the principal effect of this rule of moderation is that the pleasure experienced by the male, unlike that of the woman and the dissolute boy, will not be the source of an impulse leading him to lose control of himself. The pleasure will be as it were inserted within the activity, it will be the just compensation of the male's activity. Pleasure, according to a typical principle of Greek philosophical thought moreover, will be nothing other than the compensation or immediate effect of the activity deployed,[6] of the trouble to which the subject has gone to do what he does, of this expenditure of himself. To that extent, instead of being a source of passivity leading to the boundlessness of the unnatural, this moderate pleasure will remain inserted within a measured activity, an activity measured by the subject himself thanks to the sovereignty that he exercises over himself. And what will happen if in his activity the male allows a pleasure to be unleashed which drags him endlessly towards ever new pleasures? He will, of course, become effeminate. That is to say, the more he deploys his activity without limits, the more that pleasure, which should be the measured correlative of a measured activity, functions rather as the driving principle carrying this activity along and making the subject lose his own sovereignty, the more the male will resemble those who are his natural correlatives (woman or debauched boy) and, like them, he will be an individual carried away by the law of pleasure. Consequently, you understand why, constantly, which is very important to grasp in all the Greek and Latin texts—it will last practically until the heart of the Middle Ages—the individual who chases after women or boys is always defined as effeminate. He is effeminate because he chases women, because he chases boys, possibly because he himself is passive and likes to be passive in the [sexual] relationship. Quite simply this means that, like a woman, the principle of indefinite pleasure and not the principle of measured activity is the driving force of his behavior. In this avidity of sexual pleasure, through the inability to moderate it, to govern it, to be its master, there is a sort of passivization of activity. Some psychoanalysts, I believe—well, I've heard it said—wonder about the homosexuality of Don Juan and so about Don Juan's relation to femininity.[7] But it has been a long time since the Greeks said: Don Juan is feminine because he is a womanizer, and because the law of his pleasure, governing an indefinite activity and not master of itself, is absolutely characteristic of the object and not of the subject of activity. When one is subject of activity it means that one can master it. As soon this activity escapes your control, you become similar [to, or rather] you become the correlative of this activity: you are a woman, or an effeminate being,[8] or a Don Juan.

A few words more, if you will give me five or six minutes. To summarize all this, we can say that the schema that organizes the ethical perception of sexual acts (the *aphrodisia*) comprises a non-relational activity of a subject, an activity that is non-relational in itself but that is exercised in a set of social relations into which it must be fitted if it wants to remain measured, mastered, and to which it must remain as isomorphic as possible. A non-relational activity isomorphic to social relations: this is the heart and the paradox of this conception, of this ethical perception on the basis of which different sexual acts will be assessed, [and] not, once again, formed into a precise hierarchical table in which one will be able to say: "This is better than that, this is prohibited, this is forbidden." Few cultures have ever tried to make such a codification of sexual acts. Christianity tried to do so at a certain moment and in a certain framework, and in a way that was far from being complete moreover. There is nothing similar in Greece or in the Romans. On the other hand, I think it is possible to show, and this is what I would like to do, what the formative mechanisms are on the basis of which, before a sexual act, no matter whether in dream or in life, one will say: "Yes, this is rather good," or: "this is rather bad."

So, when one has this system in which there is a non-relational activity within a field of social relations to which this activity must remain isomorphic as the principle of its own measure, two things can be understood. The first is, of course, that if there is one element about which there is no need to speak, it is marriage. Not that marriage is not important, not that it is not valorized, but it does not raise any problem, since marriage is precisely a socially defined situation, an institutionalized form of social relation that puts husband and wife in a certain type of mutual relationship, and the sexual relations that take place within this relationship are naturally isomorphic to this social form. Marriage does not raise any problem, which does not mean that it is not important in Greek life, which does not mean that it is not valorized.

On the other hand, there is another form that does raise problems (and that is apparently very close to marriage and could be its symmetrical counterpart, like the oriental vase on the other side of the fireplace): this is the man-boy relationship. The hypothesis I would like to put forward is that if the Greeks spoke so much of men-boys relations, it is, of course, because they were tolerated. But if they were simply tolerated by the Greeks, if what characterized this men-boys relation was just the fact that it was recognized, accepted, and so on, I would say that in the end it would be the same for this relation as for marriage, one would not have talked about it so much. It seems to me that it is more the problematic character than the acceptance of the man-boy relationship that should be kept in mind. Why was the man-boy relationship a real headache for the Greeks, a constant incitement to reflect, think, discuss, and speak? Once again, it is not because the boy was a natural correlative of sexual activity. [Otherwise,] it would not raise any problem; one would not talk about it. In fact, what raises problems is that, as correlative of a sexual activity, the boy is in the same position as the woman, that is to say he is only an object. He is not and cannot be a subject. Now, even so, there is something in the boy that means he cannot be assimilated to a slave or a woman. This is that sooner or later he will become a subject. He will become a social subject, a subject in social relations; he will become a sexual subject, a subject in sexual relations. So that the problem arises in the following way. First, one has a man-boy relationship that is morally valid provided that it is isomorphic to social relations, that is to say so long as there is not just a naked sexual relationship between man and boy but [also] a social relationship of pedagogy, exemplification, help, support, and so on, that corresponds to the principle of isomorphism I have talked about. Second, this man-boy relationship will be valid if the one who is the subject is in actual fact a subject, that is to say, if it is, of course, the adult or older man who is active. The opposite situation would be incompatible. So in this relationship the principle of isomorphism and the principle of activity are brought into play: no problem.

Save that, having to become a subject one day, how must the boy be treated, what status can he be given, can one be satisfied with saying only: "For the moment he is not a subject, and then one day his status will change?" It is not possible to say this. Why? Precisely because the social relationship that, in the name of the principle of isomorphism, must frame the purely sexual relationship, or that must frame rather male activity, [to what] domain [does it belong]? To the domain of pedagogical relationships, to the set of actions by which the elder helps the younger to become what he must be, that is to say precisely a social subject. One teaches him to live, one teaches him to become a citizen, one gives him knowledge, one gives him the example of courage, the example of virtue, and so on. One tries to transform him into a subject. [At the same time,] the social relationship has to justify, serve as support, as context for a sexual relationship in which, precisely, the boy is not a subject. So with regard to the boy there is a tension between the two principles, a tension not found in the case of the wife. The social relationship of marriage ensures an isomorphism in sexual relationships, or gives the framework in which sexual relationships have to be isomorphic, but the wife will never become a subject. On the other hand, pedagogy entails a social relationship and social activity around the boy that transforms him into a subject. And at the heart of this pedagogical relationship, sexual activity [is accepted] in which the boy is not a subject. Hence the contradiction, hence the tension between the two, and hence the fact that this relationship cannot be accepted as something immediate. In the final instance, the Greeks did not accept it. They did not accept it because they could not accept it, because, in this system, in the interplay between these two great principles of ethical organization, of ethical perception, [they found these] principles [were] incompatible, aiming at two different objectives.

Hence the need for a whole remedial adjustment to make this acceptable and to make it hold. Hence the need first of all to reveal another element, which is neither that of isomorphism nor that of activity, but that of the erotic. *Erōs* is precisely the feeling, or rather the type of relationship that will be added to the other two in order to make them hold together. What is *erōs*? It is precisely that feeling, that attitude, that way of being that means that, right within sexual activity, one take accounts of the other as becoming a subject. This taking the other into account as he becomes a subject, within this dissymmetric, non-relational sexual activity, is what will make it possible to make the sexual relationship coexist, function with the social relationship of pedagogy to which, in principle, it should be isomorphic. When the sexual activity of the subject is worked from within, when it is directed and governed by the principle of the erotic, it will be quite normal, through its very sacrifice, or through the rules [it] imposes on itself, that this sexual activity is fitted to the form of pedagogy, to the principle with which this man-boy relationship should be isomorphic. *Erōs* will thus be attention to the other, vigilance, devotion, sacrifice, possibly even to the extent of death (the good soldier is the one who agrees to die so that his young friend also becomes a virtuous subject and good soldier). *Erōs* will also be a collection of quite precise obligations entailing a whole art of conducting oneself, or rather a complex art of conducting oneself while conducting the other.

This art of conducting oneself while conducting the other, this *erōs* as attention to, watchfulness over the other, has a very difficult objective: to render what I shall call the "aphrodisiac dissymmetry," which is naturally indifferent to the other, isomorphic to a pedagogical relation that must precisely raise the other as a subject in the social field. Clearly, this objective is very difficult. It involves considerable efforts on the part of both. As regards Greek culture, it also involves an immense work of reflection on the nature of pleasures, *erōs*, and pedagogy. This also entails that one strictly shuns the indefinite relationship and dangers of pleasure. Hence the theme, which we have already encountered, [according to which] the young boy must not have any pleasure, hence the theme that the adult's pleasure, the pleasure of the older person, must be limited by a very strict series of obligations. At the most extreme, *erōs* would entail the older person's renunciation of any sexual activity with the younger person. So that the famous homosexual love, as one says of the Greeks, far from being the locus of tolerance, was the locus in Greek culture of the most visible elaboration of the principle of renunciation of sexual activity. Renunciation of sexual activity, but at the same time of course, increasingly considerable importance accorded to the game of truth, inasmuch as the truth the adult passes on to the boy in the pedagogical relationship will gradually make the boy a subject. This is because it is by loving someone that one teaches him the truth, that one gradually makes him accede to the status of subject. Loving someone to the point of truth, loving him to the point that he becomes himself a subject of knowledge, and consequently as subject of knowledge escapes by right the very relationship established within pedagogy and the status of correlative of a dissymmetrical sexual activity, is what in Platonic language is called "to get a boy pregnant (*engrosser*)."[9] The implication behind the vulgarity of this comparison is that within this complex relationship that is so fraught and difficult to understand, accept, and even think for the Greeks, the boy becomes a subject of the truth, and that when he has given birth to the truth, he will be fully and entirely a subject, a subject in relation to knowledge and a subject in the social field.

We find here the elements with which we now have to work. It is not with regard to marriage, but with regard to this form of sexual relationship between man and boy that we find the three elements that need to be kept in mind. First, the idea that in the *aphrodisia*, and in the way of conducting oneself with regard to the *aphrodisia*, a veritable technology of the self is necessary, that is to say an access of the individual to the status of subject. Secondly, the obligation to tell the truth. So, technology of the self and obligation to tell the truth appear. They appear, of course, in this man-boy relationship, [but] the problem of the technology of the self does not concern the subject of sexual activity, it concerns the other. It is a technology of the self in relation to the other; it is a certain way of getting the other to accede to the status of subject. And the problem of the truth is not at all the problem of the truth grasped by oneself about oneself regarding one's own pleasures. It is the truth that one passes on to the other in the pedagogical relationship.

Clearly, when the *aphrodisia* become linked to an art of conducting oneself in which the technology of the self focuses on oneself and the obligation of the truth no longer consists in passing on the truth to another but in discovering it in oneself, one will be in an entirely different regime from that of the *aphrodisia*. I would like to analyze this transition from a technology of the self in relation to the other to a technology of the self on oneself. I would like to analyze also the transition from an obligation of truth understood as pedagogical duty to an obligation of truth as discovery of the truth in oneself. I will try to show you next week how this double transformation, far from being entirely and exclusively due to Christianity, was already prepared in an ancient philosophy contemporary with Artemidorus' text. That's it, thank you.

1. Artemidorus, *Le Clef des songes*, "Dédicace," p. 16; *Artemidorus' Oneirocritica*, pp. 45-47.
2. See above, lecture of 21 January, p. 72, note 22.
3. Artemidorus, *Le Clef des songes*, I, 79, pp. 91-92; *Artemidorus' Oneirocritica*, p. 147.
4. Hippocrates, "Serment," VI, 631, in *De l'art médical*, trans., Émile Littré and ed., Danielle Gourevitch (Paris: LGF, 1994) p. 83: "In some house that I enter, I will do so for the utility of the sick, keeping myself from any voluntary and corrupting misdeed, above all from the seduction of women and boys, free and slaves"; English translation J. Chadwick and W. N. Mann "The Oath," in G. E. R. Lloyd, ed., *Hippocratic Writings* (London: Penguin Books, 1983) p. 67: "Whenever I go into a house, I will go to help the sick and never with the intention of doing harm or injury. I will not abuse my position to indulge in sexual contacts with the bodies of women or of men, whether they be freemen or slaves."
5. K. J. Dover, *Greek Homosexuality* (Cambridge, MA: Harvard University Press, 1978).
6. One recognizes here the determination of pleasure for Aristotle in his *Nicomachean Ethics*, Book X, 1174 b4.
7. W. Stekel, *Onanie und Homosexualität: die homosexuelle Parapathie* (Berlin and Vienna: Urban & Schwarzenberg, 1923 [1917]); French translation Paul-Émile Morhardt, *Onanisme et Homosexualité. La parapathie homosexuelle* (Paris: Gallimard, coll. "Psychologie," 1951) ch.: "L'homosexualité latente, les masques de l'homosexualité, l'âge critique, Don Juan et Casanova."
8. On the character of the *effeminatus*, see *L'Usage des plaisirs*, pp. 24-26; *The Use of Pleasure*, pp. 18-20.
9. The expression as such appears in the *Phaedrus* to refer rather to a wild and impure relationship; Plato, *Phèdre*, 250e, in *Œuvres complètes*, vol. IV, 3rd Part, trans., Léon Robin (Paris: Les Belles Lettres, 1970) p. 44: "[The one who is not freshly initiated in the truth,] abandoning himself to pleasure, he acts like a four-footed beast, he sets about being covered and begetting [*tetrapodos nomon bainein epikheirei kai paidosporein*], and, becoming familiar with excess, has neither fear nor shame in pursuing an unnatural pleasure"; English translation R. Hackforth, "Phaedrus," in *The Collected Dialogues*, ed., E. Hamilton and H. Cairns (Princeton: Princeton University Press, Bollingen Series LXXI, 1963) p. 497: "(one [not] fresh from the mystery) surrendering to pleasure he essays to go after the fashion of a four-footed beast, and to beget offspring of the flesh, or consorting with wantonness he has no fear nor shame in running after unnatural pleasure." Plutarch freely takes up this passage in a passage of *Dialogue sur l'amour*, trans., Robert Flacelière in *Œuvres morales*, vol. X (Paris: Les Belles Lettres, CUF, 1980) p. 55: "When, out of libertinage and inversion, they [boys] consent, in Plato's terms 'to indulge in being covered and begetting (*bainesthai kai paidoporeisthai*) like quadrupeds' contrary to nature, this is an absolutely 'offensive' and ignoble favor"; English translation W. C. Helmbold, "The Dialogue on Love," in Plutarch, *Moralia*, vol. IX (Cambridge, MA: Harvard University Press, Loeb Classical Library, 425, 1961) p. 323: "(... there is still weakness and effeminacy on the part of those who, contrary to nature, allow themselves in Plato's words 'to be covered and mounted like cattle')—this is a completely ill-favoured favour, indecent, an unlovely affront to Aphrodite." In the *Symposium* (Diotima's speech) Plato will speak more elegantly of "procreation in beauty (*tiktein en tō kalō*, 206c), in order to evoke a purified pedagogical relationship without any carnal relationship." On this ascetic Platonic erotics, see the chapter "Le véritable amour," in *L'Usage des plaisirs*, pp. 251-269; "True Love," *The Use of Pleasure*, pp. 229-246.

five

4 FEBRUARY 1981

Process of valorization and illusion of the code. ᔕ Experience of the flesh and codification. ᔕ The philosophers' new sexual ethics: hyper-valorization of marriage and devalorization of pleasure. ᔕ Comparative advantages and disadvantages of marriage. ᔕ Should a philosopher marry? ᔕ The negative answer of the Cynics and Epicureans. ᔕ The duty of marriage in the Stoics. ᔕ The exception of marriage for the philosopher in the present catastasis, according to Epictetus.

WE CAN SEE THEN, that for the Greeks two fundamental principles control the perception of what they call *aphrodisia*. Anyway, this is what I have tried to extract from Artemidorus' text. It should be understood that these principles are not elements of a code, they are not laws of prohibition, they are not bans. These two principles are ways of evaluating that make it possible to judge according to a graduated scale, to appraise more or less approximately the value or non-value, the greater or lesser value of this or that sexual act, this or that type of physical relationship, physical union, as they may appear in reality as well as in dream, [in] oneiric representation. These two principles of valuation were: first, the principle of isomorphism, which means that a sexual union, a physical relationship is better when it is in accordance with the rules and principles that direct social relations (on the other hand, the more they diverge, they more they risk being judged bad); [second,] the principle of activity, which means that only the position of the active subject is really acceptable, satisfactory, valid, that is to say able to receive a positive value in the sexual relationship.

Singling out these two principles, trying to extract them from Artemidorus' analysis, I was basically looking for two things. First, I was looking for a methodological advantage given that if one adopts the schema of a trans-historical, that is to say universal, general sexuality common to all cultures and civilizations, and if one tries to identify the codes of prohibition that weigh on sexual acts, then it seems to me that one cannot exactly understand what takes place or the way in which these acts are appraised, are evaluated in a civilization like Greek, Hellenistic, or Roman civilization. It seems to me that in this "sexuality-prohibition," "sexuality-repression" grid, one lets slip, or at any rate one cannot account for that whole series of more or less fine and subtle gradations of which Artemidorus has given us an example. Nor can one understand a whole series of paradoxes that stand out either directly in Artemidorus' text, or more generally in the way in which the Greeks evaluate sexual acts, those paradoxes concerning the relations between men and women, between activity and passivity, and those paradoxes [also]* concerning the relations between men and boys—the fact that the man-boy relationship is accepted in itself, and yet the boy's passivity cannot be accepted. It seems to me, that all this is hardly comprehensible if one brings into play the traditional analysis in terms of prohibition and tolerance. It seems to me rather that these paradoxes become much more intelligible on the basis of the point of view I tried to use last week.

A short parenthesis on method. It seems to me that in all analyses of this kind there is a traditional tendency to assume, as a fundamental given, something that would function like a code, that is to say the more or less systematic articulation of areas of behavior that are strictly forbidden [on the basis] of major prohibitions and areas of tolerated behavior. Generally, one assumes a fundamental code and then one [tries] to explain all the rest afterwards, that is to say the gradations, the semi-tolerated, the more or less continuous evaluations, things partly accepted, partly rejected. One tries to explain [them] by

---

* M. F. adds: I have put great emphasis on this because it is clearly one of the delicate points of the problem.

4 February 1981   99

relation to this supposed fundamental code, and one analyzes them as sorts of compromise, adjustments, adaptations, toning down that make the system more or less viable, more or less acceptable, and so on. I would like to do the opposite, and put the nuance before the prohibition, the gradation before the limit, the continuum before the transgression. I would like to show the active, permanent, continuous processes of valorization that organize a graduated perception of things and at certain points of their organization serve as support to some major clear, sharp, and distinct prohibitions. But it seems to me that the major forms of clear, sharp, and distinct prohibitions are in the end only limit cases, the extreme points of something that is a veritable principle of organization of perception and valorization. And these are not negative principles of prohibition, but positive principles of valorization. In sum, and still from the methodological point of view, it will be a matter of getting rid of what may be called the illusion of the code or the juridical mirage that traditionally functions in this kind of analysis. There are a number of reasons for us being victims of an illusion of the code in these analyses, and very precisely an easily identifiable historical reason. This is that we live in fact, we have lived for some centuries (roughly from the period in the Middle Ages between the eighth and the twelfth century) in a society in which a very intense codification of sexual activity has been organized. It is true that from the first penitentials of the seventh and eighth centuries up to the great organization of auricular confession in the twelfth century, and then all that followed this, our society saw a very strong codification of sexual activity, a codification that is no doubt unique in the history of civilizations.[1] This civilization (Christian, Western, European, Modern, as you like, it's not important) is no doubt the only one to have tried to codify sexual relations as it were everywhere and constantly: codification of acts, relationships, thoughts, desires, and temptations, codification in and outside marriage, around the extraordinary and immense family of incests defined in the high Middle Ages,[2] and around the assumed, sought after limit of the unnatural that is never pinned down but ever breaking through, codification made in the form of religious commandments, of civil legislation, and also in the form of medical normativity. The truth is that we have developed a formidable activity of codification around sexuality.

But, on the one hand, is this not an entirely isolated, specific phenomenon peculiar to a certain phase of our history that consequently it would be absolutely illegitimate to want to apply to other forms of society and other periods of our history? On the other hand, is not this codification itself, since it is precisely singular, to be explained on the basis of something else (not explaining the code by the code, not explaining codification of the law by the law, with a more or less capital L)? Shouldn't the question be posed the other way round? Rather than wonder how the individual psyche was able to interiorize pre-existing codes imposed on it from above, would it not be better to wonder what experience was defined, proposed, prescribed to subjects that led them to have a certain experience of themselves on the basis of which precisely the codification of their conduct, of their acts, and of their thoughts became possible, legitimate, and in their eyes almost obvious? In other words, it would be perhaps on the basis of the Christian experience of the flesh—of its very form, of the way in which the subject related to himself through this experience—that we can no doubt understand the logic of the codification of experience presented by Christianity and the way in which this codification was actually accepted. It would be perhaps on the basis of the definition of sexuality, again, not as a trans-historical domain but as a form of experience specific to a society like ours,[3] that we could understand the modern, medical-juridical codification in which we are actually held.

In any case, concerning this Greco-Roman experience of *aphrodisia*, when we turn to what happened during the Greek, Hellenistic, and Roman period, [there is cause] to note the extent to which the elements of code are very few and rudimentary. A light legislative code, which is the prohibition of adultery, to which one refers with a smile, everyone knowing quite clearly that the prohibition is fictive. Prohibition of parent-child [incest], although with some rather different nuances, with once again the father/mother difference that seems quite perceptible.[4] A sort of taboo, horror with regard to oral sex.[5] But apart from these few major points, you have been able to see that most of the elements that allow the value of sexual acts to be differentiated do not correspond to something like the fundamental articulation of a code. And then, if the elements of absolute and strict prohibition are few, dispersed, and seemingly rather incoherent, on the other hand the major formative principles of evaluations are strong, vigorous, and quite clearly legible. This then is the methodological reason: to dissipate the illusion of code, the juridical mirage. This is why, when starting this course, I stressed the forms, the regulating and organizing principles of the ethical perception of *aphrodisia*.

The second reason I have chosen to pay particular attention to Artemidorus' text and the meaning it can be given, is that it serves me obviously as a starting point of the analysis. If one really wants to understand what happened in the ancient world before the spread of Christianity, if one wants to understand what a number of historians more or less explicitly consider to be "the"[*] preparation of Christian morality, if one wants to understand the movement that has sometimes been seen as a slow rise of ancient, pagan morality towards the rigorous and pure ideal of Christianity, I don't think one should look for a transformation of the code or a replacement of one code by another. The philosophers, Stoic or otherwise, the moralists, the spiritual directors, all those who questioned people about their daily life, did not seek to construct a new code. They did, on the other hand, modify the ethical perception. More precisely, within or alongside this general ethical perception,[†] they created, they gave rise to, to a certain extent they invented a new perception that did not entirely replace the other, but was juxtaposed to it and was imposed on certain people, circles, social groups, within a certain class or stratum. It is this new perception, once again, not universal but even so absolutely anchored in the Greco-Roman social field, that Christianity found in front of it, and that it took up.

So, in order that the continuity appears clearly, I would like to indicate schematically, in a quick overview,[‡] the evolution, the transformation that I will study in a bit more detail after. I will say this. It seems to me that the new ethical perception developed by philosophers, moralists, and spiritual directors, that Christianity encountered, that it took up, but that it will transform, but later, may be defined, may

---

[*] M. F. notes: in quotation marks.
[†] M. F. adds: whose general principles I tried to give you last week [see above, lecture of 28 January, pp. 76-77].
[‡] M. F. adds: because it will be necessary to distribute it over several lectures and there will be a fortnight's interruption, which is not my doing, due to problems to do with the room.

be characterized precisely on the basis of the transformations brought about on the two organizing principles I spoke about last week[6] (the principle of isomorphism and the principle of activity). On the one hand, these "philosophers"* questioned the principle of socio-sexual isomorphism. They questioned it, at any rate they profoundly modified it quite simply by what I will call—provisionally, because it will have to be given a more precise content—the hyper-valorization of marriage. This is not in any way to say that marriage was not valorized before this period, these texts, and this philosophical trend in the centuries immediately preceding the spread of Christianity. On the contrary, I have tried to show you how it was very highly valorized within and because of this principle of isomorphism. But it seems to me that these philosophers, these moralists, through the arts of conduct they recommend, give marriage such a place, such a fundamental importance and above all such a specific value—specific to marriage and marriage alone—that marriage will be as it were isolated within social relations. Not certainly that [it will] break with or [be in] opposition to the social relation, but it will have such a specific position and role in the social field that it will not be possible to deduce what is of value in marriage from what has value in social relations. In other words, marriage will become its own criterion of value for itself. As a result, the law and principle of valorization of what takes place in marriage (sexual relations, the physical relations between husband and wife) will not have the general forms of the social relation to which they must be isomorphic. Marriage itself will have to define the law and principle of sexual valorization of the physical acts that take place in it. It will be for marriage to make the law, to give the principle of the valorization of sexual acts in marriage, and not the social field as a whole. This quasi-break, this isolation, this specification anyway of marriage within the social field will bring the general principle of isomorphism into question. As a result, the principle of socio-sexual continuum I have talked to you about, crowned by the conjugal relationship in marriage, will be replaced by a principle of the exclusive localization of the sexual relationship in conjugality. And instead of positing the rule that a sexual relationship must not be foreign to the social relations within

---

* M. F. notes: I say "philosophers," we will come back a little on this, put in quotation marks.

which it is inscribed, one will see this new principle appear, fundamental for the rest of history, that all social relations must now be free from contamination by any kind of sexual relationship. Purification, consequently, of the social field with regard to sexual relations—with the exception of the matrimonial relationship which must alone have the privilege of or responsibility for the sexual relationship. This localization of sexual activity in marriage, [henceforth] considered not so much as the point of convergence of a whole field of social relations but as a specific, singular unit, geographically, as it were, institutionally and morally isolated, is one of the major results, one of the major transformations carried out by and within the arts of conduct.

The second modification I would like to talk to you about in more detail is the modification of what I have called the principle of activity. The principle of activity meant that the only acceptable pleasure is that of the active subject. This pleasure had to be as it were included, built-in to an activity for which it was the compensation and reward. This principle also meant that the other's pleasure (the one who was the correlate of this activity) either counts for nothing or is considered to be a dangerous element, with the risk of the passive subject being passively carried away into the indefinite domain of debauchery, the unnatural, and so on. In the arts of living, the arts of conducting oneself developed by philosophers and moralists of the centuries immediately preceding Christianity, this discontinuity, this dissymmetry— between the pleasure of activity, which is found and is acceptable in the active subject, and the pleasure of passivity, which one doesn't know how to deal with and is either to be eliminated or considered dangerous—will be questioned. Henceforth in these philosophical arts of conduct, any pleasure, even that of the active subject, naturally presents the risk and danger of the subject losing the mastery he exercises on himself. Henceforth, pleasure will be considered as the mark of a dangerous passivity at the heart of every subject, even an active subject. The idea consequently of an entirely localized pleasure, included in (and consequently legitimized by) an activity, is called into question by the [following] philosophical theme: in every pleasure there is a law or a principle of passivity that [makes] any activity guided by pleasure the manifestation and expression of a fundamental passivity. Consequently, rather than the principle of sexual ethics of accepting pleasure only in the form of activity, as was the case in the type of ethical perception I spoke to you about last week,[7] the problem now will be to define conditions of activity such that the passivity intrinsic to pleasure cannot take hold. In other words, one will have to orientate oneself towards an ideal point where, in order to avoid any kind of passivity, the sexual act will be freed from the pleasure that was accepted, tolerated, that was even valorized inasmuch as it was a pleasure of activity. Henceforth, one sees the idea dawning of a sexual act whose value will be linked to the fact that it is detached from pleasure.

So, to summarize this new schema, I will say that in the philosophical arts of living—for which we have some evidence for the first century BCE. and above all for the first and second centuries CE, the period, once again, that interests me—we find a hyper-valorization of marriage that breaks the principle of socio-sexual continuity and defines marriage as the sole place of the legitimate physical relationship. And second, we see a devalorization of pleasure that establishes a continuum of passivity regarding this pleasure and prescribes the endless task of the elision of pleasure from the sexual act, an unrealizable elision perhaps, but still as perfect, as complete as possible. Hence, it is very easy to isolate [the main part] of what may be considered as a new sexual ethic. This valorization of marriage as the sole site of the legitimate sexual relationship leads us to the idea that there can be no other sexual relationship than the conjugal relationship. And the transformation of the principle of activity into criticism of pleasure as passivity leads us to the idea of a hedonically neutralized sexual act. A uniquely conjugal and hedonically neutralized physical, sexual relationship: you see that one approaches, with giant steps, if I dare say so, the model of the elephant I spoke to you about at the beginning. We arrive there therefore, and in a quite logical and intelligible sway, when we try to understand how the arts of conduct, techniques of conduct, technologies of the self begin to shape the field of an ethical perception. It seems to me that this line of intelligibility is much clearer and accounts for many more things than analysis based on the interplay of codes and their replacement. This is a rapid run through of what I would like to do in the next two or three lectures.

So let's study these two points and see if this schema is faithful to reality. First then, let's study the way in which the principle of socio-sexual isomorphism is transformed: how it was questioned, how it was shaped from within as it were by the theme of the hyper-valorization of marriage, and how it was ultimately turned round to become the principle of a discontinuity between marriage as the legitimate locus of the sexual relationship and all other social relationships. Hyper-valorization of marriage: some well-known things, over which I pass very quickly but that I must recall for the consistency of the exposition. In Greek philosophy, or in any case in all those forms of moral reflection on men's daily life, the question of marriage was a commonplace, traces of which are found over a very long period. The question of marriage is found in rhetorical exercises, in diatribes—that is to say those kinds of discussions in public or in a closed circle where the philosopher has to answer questions put to him by the audience. It is found in the precepts of life, in the more or less theoretical discourses like that of Plutarch, for example, on love, and so on. It is found therefore in different forms and over a very long period, from Xenophon's *Oeconomicus*, the second book of the pseudo-Aristotle's *Economics*, up to the Christian authors of the end of the fourth century: Gregory of Nyssa,[8] John Chrysostom[9] will still take up this question of marriage, with the same type of arguments, the same type of considerations as those that can be read four, five, or six centuries earlier. So, through all these authors—from Xenophon to John Chrysostom, by way of Plutarch, Musonius Rufus, Epictetus, the rhetor Libanius,[10] and so on—it is always the same question, in the same form or at any rate with the same themes. I would like first of all very quickly to go back over these traditional themes of the question of marriage, and then to see how, within a certain philosophical tendency, or anyway in the context of certain treatises concerning marriage we see the question transformed.

First of all, the general form of this theme. What is typical in this centuries-old theme of marriage (four, five, six centuries), is that the question constantly raised is not: "What is marriage?" nor even exactly: "How should one conduct oneself in marriage?" The question of the nature of marriage is not raised. Nor is the question of what one should do within it. The question raised is *ei gamēteon*: "Should one marry?" That is to say the question of marriage is posed in terms of a choice between the status of married and unmarried individual. And this question *ei gamēteon* (should one marry?) is quite naturally sub-divided into two, gives rise to two types of discussion, argument. First, what are the comparative advantages and disadvantages of marriage? And then the second, more precise and more particular question: can one and should one marry when one wishes to embrace that very specific, philosophical form of existence? Can one marry and at the same time lead the life of wisdom? These two domains of questions (advantages and disadvantages of marriage; should one or can one marry if one is a philosopher?) are obviously themes that interact with each other but that can be distinguished.

[First of all,] what are the advantages and disadvantages of marriage that may determine us, push us, or incline us to choose or reject the status of married individual? I remind you very schematically of things that you are no doubt familiar with. In this repetitive, rather tedious, centuries-old literature concerning the advantages of marriage in general, and of which a fair amount of evidence has come down to us—here I offer a schema that does not follow the rhetorical developments, often obligatory moreover, in these treatises—things are divided [in the following way]. [First]: is marriage necessary? Second: is marriage useful? Third: is marriage good? Reconstructing this literature from our point of view, we can say that the advantages of marriage can be grouped around four major themes. First, marriage enables one to manage one's household properly. Division of tasks: the tasks of internal management of the household are the woman's responsibility; everything concerning the outside, on the other hand, relations with the world, the social context, and so on, is the man's responsibility. Marriage is indispensable if one wishes to supervise one's domestics, above all when the master of the house is absent. What would happen if the wife were not there to take over from him? So marriage is useful, advantageous for the proper management of one's household. Second, marriage assures the man of rest after the day's bustle when he returns home in the evening, consolation when he has had trouble and difficulties in his public life. Marriage also assures him of physical, medical care when he is sick and when he is old—what happens when the woman is old and sick is less clear. Third, of course, marriage offers the man the possibility of descendants, which is important for the man's name, for the family of which he is both the representative and the relay. The birth of children is also important for the individual's social weight, his social surface in society. What place would the individual occupy, how would he be regarded by his fellow citizens, what respect would he command if he was unmarried and childless? And finally children are useful again for old age when the man will need help and support. Finally, the fourth rubric concerning the advantages of marriage: marriage is not only useful, advantageous for the individual, but also for others. For the city especially, since through his descendants marriage enables the man to provide the city with arms to defend it and people who will enable it to live on across generations. All of this is neither very new nor very interesting.

In this literature the disadvantages of marriage correspond point for point to the advantages I have just mentioned. True, the woman helps the man manage the household. But how much distress, and how many money difficulties and additional cares are due to the woman! True, she must support the man, console him, and care for him. But there are arguments, recriminations over money, deceptions. And then from time to time the woman dies. Children are useful, but they cost money and bring worries. They too may die after one has gone to expense to raise them. And they are more often hard than tender, negligent rather than caring towards their aged parents. And if it is true that it is useful for the city to provide it with children, children often bring shame to the family father in the eyes of his fellow citizens. All of this means that in comparison with the advantages of marriage, you can see how many disadvantages there are. You can see also that all this is not very deep philosophically.

The other side of the question *ei gamēton* [is]: can one marry when one is a philosopher? While mixing the same type of arguments as those I have just [given], this question is nonetheless more interesting. It appears in the following way. Here again I will be schematic. When one wishes to be a philosopher, when one wishes to embrace the life of wisdom, is marriage compatible with an autonomy that is precisely both the goal and form of the philosophical life? Autonomy implies mastery of self, mastery of one's passions, of all the impulses of the soul that may be provoked by external events. Autonomy is also material, social, and moral independence with regard to others, to social demands, and to the very rules of society. And finally, autonomy is the sole care to conduct oneself according to reason, that is to say according to the true judgments one has learned to form. If philosophical existence is thus defined by such autonomy (mastery of self, independence from others; right reasoning, right opinion as the only rule of conduct), can one really be married under these conditions? Would not a married philosopher be contradictory? I refer you to a text from Epictetus whose argument on this point is no more than the repetition of very classical arguments. With regard to the philosopher, he says nonetheless: How can one conceive of a philosopher who has to fulfill a number of duties towards his father-in-law, who has to provide services to other relatives of his wife, and to his wife herself, and who consequently cannot practice his profession, who is forced to become like a home nurse? He needs a pot in which he will heat water for his son in order to bathe him. He will need wool for his wife when she has a child. He will need oil, a cot, a cup, and a whole range of other occupations and distractions. There he is among his children. What leisure does he have chained in this way to his private duties? Must he not procure clothes for his children? Must he not send them to the schoolmaster with their tablets, their stylus, and also to have got ready a bed for them?[11] Such a life is clearly not compatible with the principle of autonomy.

But on the other hand, is non-marriage compatible with a true philosophy? For in the end—and this theme is increasingly stressed when the philosopher's role is seen as that of providing rules of conduct—when one has to give individuals rules of conduct, to form young people for the life they have to lead in society, when one has to tell the truth not only in one's discourse, but in one's life, when one has to be the source of manifestation of truth in speech and by example, to be the manifestation of truth in the middle of society (in one's life, body, and existence, in what one says, in one's manner of being, and so on), can one be unmarried when precisely one is telling those one is educating that they must marry? When one says that it is important for the individual to feel bound to his compatriots, when one says that it is the individual's duty to live in accordance with the general rules of society, to feel brotherhood with other men, can one then live on one's own and so remain indifferent to the needs and demands of human society in general, or of one's homeland in particular, and as a result not take part in the renewal of generations, and so on?

In other words, the problem of marriage for the philosophers is connected to the very important problem of the truth. If, in this general perspective that I am reminding you of and that, again, is centuries old, the philosopher finds himself in a situation, in a critical position with regard to the ethics of marriage, it is inasmuch as he is subject of truth. He is subject of truth in two senses. On the one hand he has to know the truth, his activity is *theōria*. To that extent, anything that may turn him away from the purity of this *theōria*, from the eternity of the object he contemplates, and from the conditions that will enable him precisely to perceive this eternal object in complete clarity, anything consequently that risks blurring his sight and disturbing him in this privileged, fundamental, stable relation to the truth, must be excluded. So marriage must be excluded. From this point of view, as subject of truth, he cannot marry. But on the other hand, insofar as he is subject of truth (that is to say teaches the truth, is master of truth) through his speech, his example, his life, through the match between his speech and his life, marriage is as entailed and necessary for him as for anyone else. The problem of truth-telling and the problem of being married, the problem for one whose status is to manifest the truth, the problem of the status of marriage or of celibacy in relation to this status of manifestation of the truth. We see here, fairly clearly, one of the first major [expressions] of the problem I would like to deal with in all these lectures, [namely] the problem of the relation between the manifestation of the truth as individual duty and ethics, essentially applied to the problem of sexual relationships through marriage or not. [...].*

In normal life the man is active in the social field. It is this activity that authorizes him to be sexually active in marriage also—in marriage insofar as it is the summit of that pyramid of social relations. The philosopher, on the contrary, is someone whose activity in the social field does not consist in being active, but in being a "theoretical individual;" later one will say: a contemplative. When his activity is to be a theoretical individual, that is to say to have a relation to the truth, can he then actually be married? He is in a contradictory position in relation to the ordinary man whose activity authorizes him to be

---

* Gap in the recording. All that is heard is: [...] different from that of ordinary man.

active in marriage. Consequently, there is a discrepancy between the philosopher and the ordinary man with regard to this principle of activity, since the philosopher's activity is theory. You see at the same time a sort of coming together, a sort of symmetry and as it were analogy between the philosopher's position and the young boy's position (which I spoke to you about last week). In fact, you recall that in the field of *aphrodisia* the young boy's position was also critical since, on the one hand, in the field of *aphrodisia* he is the object of a desire, but he has to become an active subject, to learn suitably, correctly, that is to say by way of the truth, to become a social subject. And precisely, because he has to become subject through *paideia*, through the teaching of the truth, he cannot be really, fully, wholly an object of pleasure, as the woman is for example. He cannot identify himself with the role of object of pleasure. He must tear himself away from it or be protected from it, and it is at that point that the ascetic principle arises: the relationship to the young boy must be a love relationship, but not a sexual relationship.

The philosopher's position is fairly close, anyway critical in relation to the *aphrodisia* since it is as a subject of truth that the philosopher cannot be a subject of activity in the social field. Having to become a subject through the truth, the child, the young boy cannot be an object in the field of *aphrodisia*. The philosopher, inasmuch as he is already or wishes to become a subject of truth, cannot be a subject of activity in the field of *aphrodisia*. Consequently, by the very logic of the system, philosopher and young boy appear within this field as critical individuals, as problematic individuals, so problematic that one is obliged to separate them from it, to give them a status apart, to free them as it were, to prohibit their [access]—or recommend non-access—to this field of *aphrodisia* (recommending to the philosopher that he not be married, asking the young boy not to have physical relations with the one he loves).

With this question of the advantages and disadvantages of marriage, with the precise and specific question of the marriage of the philosopher, one sees therefore the problem of truth appearing quite clearly. So now, how precisely will this principle of perception be reworked and transformed in philosophical literature to the point that, from the first and second centuries, the question of the philosopher's sexual activity, of the proximity of philosopher and young boy, and so on, will give way to another question: how actually to conduct oneself philosophically within marriage? And whereas, for centuries, the major critical point will have been, according to a famous text: how *philosophikōs paiderastein*,[12] how "to love boys philosophically?," the problem as it will appear—not, again, in the whole field of imperial Roman society, but at least in a certain group—an entirely new and in truth completely paradoxical problem from the standpoint of the previous point of view, will be: how to be married philosophically, how to have sexual relations with one's wife philosophically?

How did this transformation take place? Once again there are some texts, a tendency, a movement. There is no question of saying that it is a general social transformation, but arts of conduct are developed in which one sees this transformation of the principles of ethical perception. These are above all Stoic texts of the first and second centuries—mainly of Musonius Rufus, the teacher of Epictetus, Epictetus himself, and Hierocles, a Stoic of the second century. But it is somewhat arbitrary to pick out these texts. Firstly because one can see the transformation arise before, and it continues after. And [then] there are many authors, to whom I will refer moreover, like Plutarch, who are not Stoics and in whom the same type of transformation [is carried out]. We will focus first of all now on these Stoic texts. To summarize things somewhat schematically, I think that the hyper-valorization of marriage, which therefore puts the principle of isomorphism in question, takes place in three or four ways—three, at least; next week I will tell you if there is a fourth. First, marriage is defined in the Stoic texts, basically in Epictetus[13] and Hierocles,[14] as what they call *proēgoumenon*, a word that belongs to a relatively technical Stoic terminology. It is not found in Musonius Rufus, but the idea is practically there. What does it mean to say that marriage is *proēgoumenon*? Bonhöffer,[15] the great specialist of the Stoics at the beginning of the century,[*] said that *proēgoumenon* was an absolute duty. This is a rather strong, too emphatic translation that Pohlenz corrects, saying that a *proēgoumenon* act is a *prinzipiell* act.[16] Bréhier translates this saying it is a "*principal* (most important, principal)" act.[17] One could say—this is all something of

---

[*] Adolf Friedrich Bonhöffer, 1858-1919.

a matter of conventions—that the act considered as *proēgoumenon* is a primordial act, of major importance, and so on. What does it mean concretely? In the Hierocles text, saying that marriage is *proēgoumenon*, we see that *proēgoumenon* (primordial) is opposed to something else, which is a *kata peristasin* act, that is to say an act performed according to circumstances. Very precisely, the Hierocles text says this: The life *met gamou* (with marriage) is *proēgoumenos* (primordial). The life *aneu gunaikos* (without a wife) is *kata peristasin*[18] (is of the realm of circumstance, conjuncture). There is the same opposition between *proēgoumenos* and *kata peristasin* in Epictetus.[19] He says that among the different acts some are primordial and whereas others have to be performed only "in case of circumstances." To understand what the Stoics mean by this general opposition between what is *proēgoumenon* and what is *kata peristasin*, and to see how they apply this to marriage, we need to refer to the context and precisely to the forms of thought, the forms of analysis to which they are opposed. Diogenes Laertius,[20] citing texts by Epicurus from *Problems* and *On Nature*, says: The wise man will marry and have children, but only *kata peristasin*, "according to the circumstances of his life." This text of Epicurus has been challenged. Several editors have tried to replace it with a text that would say in part the opposite: The wise man will not marry and will not have children—which actually coincides better with a quotation of Clement of Alexandria in *The Stromata* where it is said that, for Epicurus as for Democritus, marriage is to be rejected, first because of the inconvenience and because it diverts from things *anangkaioterōn* (the most necessary).[21] But in fact, as Bollack in his edition of Epicurus shows, there is no need to replace the Diogenes Laertius text with another, negative text.[22] Epicurus' thought is perfectly clear: marriage is entered into only *kata peristasin biou*, according to the circumstances, it being understood that generally the wise man will not marry. He will not marry except in such circumstances as call [for it], as make it necessary or useful. We find the same position in the Stoics. In *De finibus*, according to Cicero: Some Cynics say that the wise man adopts marriage in certain situations that may arise (*si qui forte casus inciderit*: if the opportunity arises). And he says that this is not the universal opinion of the Cynics and that, for others one must never marry whatever the circumstances. Anyway, if we are to believe Cicero, there would be at least some Cynics who say: Marriage is to be rejected for philosophers, except in certain circumstances.[23] And it is in fact this position of the Cynics that Epictetus reflects in *The Discourses* when he refers to the case of the Cynic Crates. Crates was not very fond of marriage, he was even deeply hostile to marriage to the extent that when his son reached puberty he took him to the house of prostitution, telling him: These are the women, and no others, with whom you must have relations.[24] Nonetheless, this same Crates, who was so hostile to marriage, was himself married. He was married, Epictetus says, due to circumstances, which were that there was a woman who loved him. And this woman was another Crates, that is to say she was as much a philosopher as he.[25] In Book VI, Diogenes Laertius tells us who she was. It was Hipparchia who, Diogenes Laertius says, agreed to share Crates' life. And thus you see something important, because one will see it turn around in the Stoics. Hipparchia was a philosopher herself, she agreed to lead the same life as the philosopher, she accepted therefore the *koinōnia* (the community of existence).[26] So that in the Cynics one sees the idea that marriage is actually something of negative value, [but] that there may be a circumstance or circumstances in which one may marry. At least one of these circumstances is when the wife agrees to *koinōnia*, or when it is possible to establish a *koinōnia* with the wife. In the Stoics, *koinōnia* is not a special circumstance permitting marriage. It is the very essence of marriage—and as a result things are reversed. In any case, this is the context in which the Stoics deploy their analysis.

Let's say that the Epicureans and the Cynics deliver a negative judgment on marriage. The balance is clearly in favor of the disadvantages of marriage, save *peristasis* (save circumstances). It is precisely this idea that the Stoics will overturn and reverse, saying that marriage is—and it is here that they introduce the term—*proēgoumenos*. The Stoics want precisely to reverse the relation between the possibility of marriage and circumstances. Something *proēgoumenos* is something one must do, not unconditionally, whatever the circumstances, and so on, [but] if at all possible, something that ought to be performed, except of course in exceptional circumstances that prevent us. A *proēgoumenos* act is therefore to be performed with the highest priority, in a primordial manner because it is of major importance, major value, except obviously in exceptional circumstances. On the other hand, for an act which is not *proēgoumenos*, one will decide only if circumstances actually encourage or lend themselves to it. This then is how the Stoic position should be understood. Let's look at the texts again. Hierocles, for example, reversing the Cynic or Epicurean proposition term by term, says: The life *aneu gunaikou* (the life without wife, celibate life) must be chosen *kata peristasin* (if circumstances impose it). But unless there is an obstacle, one should marry, the life *meta gamou* (with marriage) is *proēgoumenos*. So what is typical of the Stoic position is this reversal of the relation of circumstance, the turning round of the effect of circumstances.

This reversal of the thesis of marriage by circumstances is obviously a bit more complex in Epictetus. You know that Epictetus does not give exactly the same rules of conduct for good men who lead a public life and for the philosopher in the strict sense, the preaching and militant philosopher, the professional philosopher, whom he calls the Cynic. But, of course, by Cynic he does not mean the Cynic doctrine, but a certain way of being, a way of living, a way of professionalizing philosophy and making it one's own domain of activity. Let's take the case of a good man leading a public life. In this first case there is no problem. Like Hierocles later, Epictetus [maintains that] the good man [should] marry. In the *Discourses*, against the Epicureans who reject marriage and procreation, he says to the good man: "You live in a capital city: you must exercise an office, abstain from the other person's goods, no woman but your wife should seem beautiful to you, no boy handsome."[27] Let's leave the problem of the wives of others and boys. The text proves that this man is considered to be married. It is part of his life, and what is more, a bit further on in this discourse, Epictetus lists the things considered to be *proēgoumena*: *politeuesthai* (being a citizen, leading the citizen's life), *gamein* (marrying), *paidopoieisthai* (having children), *theon sebein* (honoring God), *goneōn epimeleisthai* (honoring parents).[28] All this is part of the list of *proēgoumena* and is not a problem for the good man.

Epictetus' position with regard to the professional philosopher is different and a bit more elaborated, and because of the question of truth I will dwell on it a bit more. The analysis is found in the *Discourses*, Book III, Chapter 22. So, Epictetus calls the professional philosopher, the philosopher who devotes his life to philosophy, "the Cynic." But again, this does not mean the representation of a philosopher who upholds specifically Cynic doctrine. By "cynic," Epictetus means someone who leads the cynic life, who lives as a cynic. What he calls *to kunizein* (living cynically), independently of the precise content of Cynic doctrine, should be the philosopher's condition of life. And this "being cynic," this "living cynically," means: being master of oneself, being scout of the truth.[29] Someone who lives cynically, as scout of the truth, must look critically at reality. He has to interpellate men, reality, and say: Are you conducting yourselves well? Are things really as they should be?[30] Finally, the cynic, the person who lives cynically, that is to say the militant philosopher, is, he says, someone who has no city, no clothes, no house, no wife, no children, no homeland.[31] And why? Well, says Epictetus, because he is the messenger of the gods (or of the god, or the divine, as you like).[32] It is he who serves as intermediary between this world and the universal principle of rationality, order, and wisdom of which he is as it were the representative on earth. On these conditions, if the philosopher who lives cynically, who practices *kunizein*, must be "without marriage," then, says the objector (it is a question of a diatribe), are marriage and children not things that the Cynic must take on *proēgoumenōs* (primordially)?[33] Should not the philosopher, who lectures others, also take on the responsibility of marriage and children? Epictetus' answer is very interesting regarding the position of late Stoicism on marriage. Epictetus says: If we were in a city of wise men (*sophōn polis*) it would of course be necessary to marry. Why would it be necessary to marry, even if one were a philosopher? Quite simply because there would be no need for *kunizein*, one would not need to be the proclaimer, the messenger of the truth, since one would be in the truth. One would not need to interpellate reality and criticize it, since reality would be in line with reason. Consequently, nothing then will prevent (*ouden kōlusei*) the philosopher from marrying and having children [...].*

To show just how marvelous this city of wise men will be, he even adds that the father-in-law will also be a philosopher.[34] So, the

---

* Gap in the recording. All that can be heard is: [...] his wife and his children [...] who had married a woman because she was another him.

philosopher's marriage in the city of wise men is not a problem. You will say to me: even so, is there not a contradiction in Epictetus' text? Either this text means that there will be no cynics. There will be no need of *kunizein*, he says. And then at the same time it seems to say that everyone will be a cynic. Since everyone will have attained the wisdom represented by the Cynic, there will no longer be any need for the Cynic to assert all the negative, critical, provocative aspects of his existence. One will be in a city in which one will be able to be married while being as wise as a Cynic. That is what it means. Must not the philosopher marry *proēgoumenōs* (primordially)? The first element of the answer: In the city of wise men, of course he will have to marry. But in the present state of the world—and Epictetus employs the Stoic term catastasis, which you know will have a very important role in the Christian conception—in the present catastasis, which, Epictetus says, is a battle—and again this is a very important notion for what follows— in this state of things in which one is at war, what is one fighting against? Against evil, against lies, against illusion, and so on. In this catastasis, which is presently a battle, how could the Cynic play his specific role, namely as messenger of the gods and scout of truth, if he let himself be overwhelmed by cares, if he is caught between his wife's nagging and his children's complaints.[35] And here one finds that list of disadvantages of marriage I was just talking to you about. It is inconceivable that someone who is the gods' messenger, who must tell men the truth, and who consequently must exercise a function as guide and as it were father of the whole of humanity should so to speak "localize" that function within a marriage and with regard to children born to his wife. The universal and critical role of the philosopher prevents him from taking on all these functions [connected to] marriage for himself.[36] So you see that Epictetus in no way wants to say that marriage is not something *proēgoumenon*. On the contrary, he says: It is, anyway and whatever happens, something *proēgoumenon*. In the present catastasis, someone who is not a philosopher, someone who wishes to lead a good life, a life in accordance with reason will have to marry. As for the philosopher, who will have to marry and who obviously will marry when the world conforms to reason, he is presently, as it were, forced to suspend his marriage. Why must he suspend it? Because of that circumstance that is the present state of the world, the present catastasis, the battle in which we find ourselves. The bad state of the world is the famous *peristasis,* the circumstance, the conjuncture that in the philosopher's case means he must not marry. Marriage is therefore a primordial act that the public man must perform, except in circumstances peculiar to him. But—and here the text is interesting for the definition of philosophy itself—the philosopher, in contrast to the public man, is someone for whom the entire present world is only a circumstance. Whereas the ordinary man is someone who lives in the world in which certain circumstances might prevent him from marrying, the philosopher is someone for whom the catastasis is a peristasis (*péristase*), for whom the present world is a circumstance. It is due to this circumstance, which he takes on as it were on his own account, and in relation to which he situates himself as critic, proclaimer of the truth, as god's messenger, that he must not marry. We find here in fact a whole core of ideas that will have a particular destiny, first in Origen, but obviously then in all Christianity. The pastor, because he is the pastor, [must] be unmarried and not let himself be caught up in the concerns of daily life. He is the father of everyone, there is his family, there his marriage. In this present catastasis—and before what Origin will call the apocatastasis, that is to say the reversal of everything—celibacy is obviously absolutely compulsory.[37] The difference is that this idea of a catastasis as peristasis, of a state of the world as circumstance from which the philosopher must [distance] himself in order to change it, will be elaborated in the Christian [approach] through a conception of virginity that we will have to talk about later.

I shall stop there [regarding] this Stoic conception of marriage as *proēgoumenos*. In the perfect [and rational] state of the world, marriage will have to be undertaken whatever happens. Unlike in the Epicureans or Cynics, one does not enter into marriage according to circumstances; it is not a circumstantial act. One must marry, save in particular circumstances for the ordinary individual, and [save] in that circumstance that is the state of the world for the philosopher. Which is to say that, anyway, marriage has value in itself; it does not acquire its value from a particular circumstance (Stoic thesis, Cynic thesis). Marriage is not validated from outside, by an effect of the conjuncture. Marriage is of value in itself, even if it means defining afterwards particular or general circumstances that, for the ordinary man or for the philosopher, mean that he may possibly, that he must suspend his marriage until the conditions are established or re-established in which he will be able to realize with his wife the *koinōnia* that is the nature and essence of marriage. These are the themes I will try to develop next week.

1. See M. Foucault, *Mal faire, dire vrai. Fonctions de l'aveu en justice*, ed., F. Brion and B. E. Harcourt (Louvain-Chicago: Presses universitaires de Louvain-University of Chicago Press, 2012) lecture of 13 May 1981 and lecture of 20 May 1981, p. 201; English translation Stephen W. Sawyer, *Wrong-Doing, Truth-Telling. The Function of Avowal in Justice* (Chicago-Louvain: University of Chicago Press-Presses universitaires de Louvain, 2014), p. 201: "the privileging of avowal in penal practices was inscribed, in a general manner, in a sort of broad juridification of Western society and culture in the Middle Ages, a juridification that could be perceived ... in the institutions, the practices, the representations that were part of Christianity." See also *Les Anormaux*, lecture of 19 February 1981 and lecture of 26 February, pp. 171-172; *Abnormal*, pp. 184-185: "How was the confession of sexuality defined before the Council of Trent, that is to say, in the period of 'scholastic' penance between the twelfth and sixteenth centuries? It was essentially organized according to juridical forms. What the penitent was asked about if he spoke spontaneously, were offenses against a number of sexual rules." In a discussion on 23 October 1980, Foucault states: "Sex is something which has nothing to do with law. And conversely. The fact that sexual differentiation, sexual preference, that sexual activity could be a matter of legislation, that, I think, is something which cannot be admitted. Anyway, I would like to know if it is possible to put this principle at the base of a new penal code"; "Débat sur 'Vérité et Subjectivité'" in *L'Origine de l'herméneutique du soi. Conférences prononcés à Dartmouth College, 1980*, ed., H.-P. Fruchaud and D. Lorenzini (Paris: Vrin, "Philosophie du présent/Foucault inédit, 2013) p. 114; "Discussion of 'Truth and Subjectivity,'" in *About the Beginning of the Hermeneutics of the Self. Lectures at Dartmouth College 1980*, trans., G. Burchell (Chicago: University of Chicago Press, forthcoming).
2. See J.-D. de Lannoy and P. Feyereisen, *L'Inceste* (Paris: P.U.F., 1992) p. 8: "The prohibitions [of incest] contained in the Civil Code derive historically from canon law. After the barbarian invasions and during the Middle Ages, persons of the same family living together formed a widely extended group. The Church was concerned with making the fusion of different family groups easier and adopted very restrictive measures on marriage between close relatives, adding prohibitions concerning spiritual kinship, as exist between godfather or godmother and their godchild."
3. *L'Usage des plaisirs*, pp. 43-44; *The Use of Pleasure*, p. 35: "One would have a difficult time finding among the Greeks (or the Romans either, for that matter) anything resembling the notion of 'sexuality' or 'flesh.' I mean a notion that refers to a single entity and allows diverse phenomena to be grouped together, despite the apparently loose connections between them, as if they were of the same nature, derived from the same origin, or brought the same type of causal mechanisms into play: behaviors, but also sensations, images, desires, instincts, passions."
4. On this point, see the differentiated evaluations of incest dreams in Artemidorus, *La Clef des songes*, I, 78-79, pp. 87-92; *Artemidorus' Oneirocritica*, pp. 141-149. See also what C. Vatin says about this in his *Recherches sur le mariage et la condition de la femme mariée à l'époque hellénistique* (Paris: Éd. de Boccard, 1970) p. 4, p. 7, pp. 58-62, pp. 87-90, and p. 178.
5. This "horror" can be read in Artemidorus in *La Clef des songes*, I, 79, pp. 91-92; *Artemidorus' Oneirocritica*, p. 147. It will be the same in Christian authors (Lactantius, *Divine Institutes*, VI, 23).
6. See above, lecture of 28 January, p. 77 sq.
7. See ibid., pp. 76-77.
8. Gregory of Nyssa, *Traité de la virginité*, trans., Michel Aubineau (Paris: Éditions du Cerf, Sources chrétiennes 119, 1966); English translation V.W. Callahan, "On Virginity, in *Gregory of Nyssa: Ascetical Works* (New York: Fathers of the Church, 58, 1967). See in particular Chapter 2 on the comparative advantages/disadvantages of virginity and marriage.
9. See the three *Homélies sur le mariage* de Jean Chrysostome [John Chrysostom], but also the Homily XX on *The Epistle to the Ephesians*.
10. Libanius, "Ei gamēteon," in *Progymnasmata / Libanius' Progymnasmata: Model Exercises in Greek Prose Composition and Rhetoric*, trans., Craig A. Gibson (Atlanta: Society of Biblical Literature, 2008) pp. 511-519.
11. Epictetus, *Entretiens*, III, 22, 70-75, pp. 80-81; Epictetus, *The Discourses*, Book II, pp. 154-157.
12. More precisely Plato employs the expression "*paiderastein meta philosophias*," in *Phaedrus*, 249a.

13. Epictetus, *Entretiens*, III, 22, 67, p. 79: "And are marriage (*gamos*) and children, the young man asks, responsibilities that the Cynic must take on as an essential duty (*proēgoumenon*)?"; Epictetus, *Discourses*, pp. 152-154: "But, said the young man, will marriage and children be undertaken by the Cynic as a matter of prime importance?"

14. Hierocles in Stobaeus, *Florilegium*, 22, ed., A. Meinecke (Leipzig: Teubner, 1860-1863) vol. III, p. 7: "*proēgoumena esti to gamein ... ei ge mē tis peristasis empodōn,*" in *Le Souci de soi*, p. 183; *The Care of the Self*, p. 156, Foucault offers this translation of the Greek: "Marrying is preferable (*proēgoumena*); hence it is an imperative for us provided that no circumstance opposes it." However, the indications of the course suggest translating *proēgoumena* rather as "priority" or even "first duty."

15. A. F. Bonhöffer, *Epictet und dia Stoa. Untersuchungen zur stoischen Philosophie* [1890] (Faksimile-Neudruck Ausgabe Stuttgart-Bad Cannstatt: F. Froman Verlag, 1968) p. 38. See also, *Ethik des stoikers Epictet* (Stuttgart: Verlag von Ferdinand Enke, 1894) p. 210.

16. M. Pohlenz, *Die Stoa. Geschichte einer geistigen Bewegung* (Göttingen: Vandenhoeck un Ruprecht, 1948-1949) vol. I, p. 188: "Das Ziel sei, alles, was in den eigenen Kräften stehe, dauernd und unverbrüchlich zu tun, um das, was primär (*proēgoumenōs*) unserer Natur entspricht, zu erlangen."

17. Epictetus, *Entretiens*, III, 7, 24-25, trans., Émile Bréhier, in *Les Stoïciens*, ed., É. Bréhier (Paris: Gallimard, "Bibliothèque de la Pléiade") p. 977: "In an engraved work, what is best, the silver or the engraver's art? The substance of the hand is flesh, but the important thing (*principal, proēgoumena*) is the work of the hands ... Thus in man, we should not grant our esteem to the matter which is flesh, but to the principal (*principals, proēgoumena*) actions"; Epictetus, *Dialogues*, vol. II, pp. 56-57: "In a piece of plate what is the best thing, the silver or the art? The substance of the hand is mere flesh, but the important thing is the works of the hand. ... So also in the case of man, it is not his material substance that we should honour, his bits of flesh, but the principal things."

18. The Greek text is: "*hōs tō sophō proēgoumenos men estin ho meta gamou bios, ho d'aneau gunaikos kata peristasin*" (Stobaeus, *Florilegium*, p. 7).

19. Epictetus, *Discourses*, III, 14, 7, Fr., trans., Souilhé, p. 50: "There are actions that one performs because they have a value in themselves (*proegoumenōs prattetai*), others according to their relations (*ta de kata peristasin*)"; Eng., pp. 96-97: "Among actions some are performed primarily on their own account, others on occasion ..."

20. Diogenes Laiertius, X, 119, trans., Jean Bollack and ed., Jean Bollack, *La Pensée du plaisir, Épicure: textes moraux, commentaires* (Paris: Minuit, 1975) p. 27: "However, as Epicurus says in the *Problems* and in the book *On Nature*, it is possible that the wise man will take a wife and have children. But he will marry according to the situation in which he finds himself at a certain moment in his life"; English translation R.D. Hicks, *Lives of Ancient Philosophers* (Cambridge, MA: Harvard University Press, Loeb the Classical Library, 185, 1925) vol. II, Book X, "Epicurus," 119, p. 645: "Nor, again, will the wise man marry and rear a family: so Epicurus says in the *Problems* and in the *De Natura*. Occasionally he may marry owing to special circumstances in his life."

21. Clement of Alexandria, *Les Stromates*, II, XXIII, 138, 3-4, ed. and trans., Claude Mondésert (Paris: Éd. du Cerf, "Sources chrétiennes" 38, 1954), p. 139: "Democritus rules out marriage and procreation because of the numerous inconveniences and distractions that result from it and diversion from more necessary occupations (*apo tōn anagkaioterōn*). Epicurus also agrees, and all those who place good in sensual pleasure, tranquillity, and the absence of pain": English translation William Wilson, *The Stromata*, Book Two, XXIII, in Alexander Roberts, James Donaldson, and A. Cleveland Coxe, eds., *The Ante-Nicene Fathers. Vol. II: Fathers of the Second Century* (Grand Rapids, MI: W. B. Eerdmans, 1983) p. 377: "Democritus repudiates marriage and the procreation of children, on account of the many annoyances thence arising, and abstractions from more necessary things. Epicurus agrees, and those who place good in pleasure, and in the absence of trouble and pain."

22. J. Bollack in *La Pensée du plaisir*, p. 40: "The text, as it has come down to us, is not in contradiction with Clement's text."

23. Cicero, *De Finibus/On Ends*, trans., H. Rackham (Cambridge, MA: Harvard University Press, Loeb Classical Library, 40, 1914), Book III, XX, 68, p. 289: "As for the principles and habits of the Cynics, some say that these befit the Wise Man, if circumstances should happen to

indicate this course of action (*si qui ejus modi forte casus inciderit, ut id faciendum sit*); but other Stoics reject the Cynic rule unconditionally."

24. Diogenes Laertius, *Vies et Doctrines des philosophes illustres*, VI, 88, trans., Marie-Odile Goulet-Cazé (Paris: Le Livre de Poche, "La Pochothèque," 1999) p. 754; *Lives of Eminent Philosophers*, vol. II, Book VI, ch. 5, 88, pp. 92-93.

25. Epictetus, *Entretiens*, III, 22, 76: "You are speaking to me about a particular circumstance (*peristasin moi legeis*) in which love was involved and you are citing a wife who was another Crates"; *The Discourses*, vol. II, Book III, ch. 22, 76, pp. 156-157: "You are mentioning a particular instance which arose out of passionate love, and you are assuming a wife who is herself another Crates."

26. Diogenes Laertius, VI, 96, Fr., pp. 760-761; Eng., vol. II, pp. 98-103.

27. Epictetus, *Entretiens*, III, 7, 21, pp. 29-30; *The Discourses*, vol. 2, Book III, ch. ii, 21, pp. 54-55: "You live in an imperial State; it is your duty to hold office, to judge uprightly, to keep your hands off the property of other people; no woman but your wife ought to look handsome to you, no boy handsome."

28. Ibid., III, 7, 26, Fr., p. 30; Eng., pp. 56-57: "citizenship, marriage, begetting children, reverence to God, care of parents."

29. Ibid., Book III, ch. xxii, 26, Fr., p.73: "In reality, the Cynic is indeed a scout of what is favorable to men and what is hostile. And he must first explore accurately and then return to announce the truth"; Eng., pp. 138-139: "For the Cynic is truly a scout, to find out what things are friendly to men and what hostile; and he must first do his scouting accurately, and on returning must tell the truth."

30. Ibid., III, xxii, 26, Fr., p. 73: "He must therefore be able, if the occasion arises, to mount the tragic stage to repeat the words of Socrates: Alas! men, where are you letting yourselves be driven? What evils are you doing?"; Eng., pp. 138-139: "He must, accordingly, be able, if it so chance, to lift up his voice, and, mounting the tragic stage, to speak like Socrates: Alas! men, where are you rushing? What are you doing?"

31. Ibid., III, xxii, 47-48, Fr., p. 77: "Look at me, I am without shelter, without resources, without slaves. I sleep on the ground. I have no wife, no children, no governor's palace, but only the earth, the sky, and an old cloak"; Eng., pp. 146-147: "'Look at me,' he says, 'I am without a home, without a city, without property, without a slave; I sleep on the ground; I have neither wife nor children, no miserable governor's mansion, but only earth, and sky, and one rough cloak."

32. Ibid., III, xxii, 23, Fr., p. 72: "He [the Cynic] must know that he has also been sent as a messenger to men by Zeus"; Eng., pp. 136-139: "he must know that he has been sent by Zeus to men ... as a messenger ..."

33. Ibid., III, xxii, 67, Fr., p. 79: "And are marriage and children, the young man asked, responsibilities that the Cynic must assume as a fundamental duty (*proēgoumenōs*)?"; Eng., pp. 152-155: "But, said the young man, will marriage and children be undertaken by the Cynic as a matter of prime importance?"

34. Ibid., III, xxii, 67-68, Fr., pp. 79-80: "If you give me a city of wise men (*an moi sophōn, ephē, dōs polin*), Epictetus says, it may be that no one will readily take up the profession of the Cynic (*to kunizein*). To what end, in fact, would one embrace this kind of life? Let us assume he does, however: nothing, then, will prevent the Cynic from marrying and having children. His wife, in fact, will be another like himself, as well as his father-in-law, and his children will be raised in the same way"; Eng., pp. 154-155: "If, replied Epictetus, you grant me a city of wise men, it might very well be that no one will lightly adopt the Cynic's profession. For in whose interest would he take on this style of life? If, nevertheless, we assume that he does so act, there will be nothing to prevent him from both marrying and having children; for his wife will be another person like himself, and so will he father-in-law, and his children will be brought up in the same fashion."

35. Ibid., III, xxii, 69-70, Fr., p. 80: "But, in the present state of things, when we find ourselves, so to speak, in the middle of a battle, should not the Cynic remain free from all that might distract him, entirely at the service of God, capable of mixing with men, without being chained by private duties, without being involved in social relations from which he will not be able to withdraw if he wishes to protect his role as respectable man and that he will not be able to keep up without destroying in him the messenger, the scout, the herald of the

gods?"; Eng., pp. 154-155: "But in such an order of things as the present, which is like that of a battle-field, it is a question, perhaps, if the Cynic ought not to be free from distraction, wholly devoted to the service of God, free to go about among men, not tied down by the private duties of men, nor involved in relationships which he cannot violate and still maintain his rôle as a good and excellent man, whereas on the other hand, if he observes them, he will destroy the messenger, the scout, the herald of the gods, that he is."

36. Ibid., III, xxii, 71-72, Fr., p. 80: "Observe, in fact, that he must fulfil certain duties towards his father-in-law, that he has services to provide to the other relatives of his wife, and to his wife herself: finally, he is kept from his profession and reduced to the role of nurse or provider. Not to speak of the rest, he needs a pot in which to heat water for his son in order to bathe him; wool for his wife when she has had a child, as well as oil, a cot, a cup (see how the small furnishings increase); and the other occupations and distractions ... What remains of that famous king who devotes himself unreservedly to public affairs?"; Eng., pp. 154-155: "For see, he must show certain services to his father-in-law, to the rest of his wife's relatives, to his wife herself; finally, he is driven from his profession, to act as a nurse in his own family and to provide for them. To make a long story short, he must get a kettle to heat water for the baby, for washing it in a bath-tub; wool for his wife when she has had a child, oil, a cot, a cup (the vessels get more and more numerous); not to speak of the rest of his business, and his distraction. Where, I beseech you, is left now our king, the man who has leisure for the public interest?"

37. See Origen, *Traité des principes*, trans., Henri Crouzel and Manlio Simonetti (Paris: Éd. du Cerf, "Sources chrétiennes," 1978-1984); English translation G. W. Butterworth, *On First Principles* (Notre Dame, Ind: Christian Classics, 2013), and *Homélies sur le Lévitique*, ed. and trans., Marcel Borrret (Paris: Éd du Cerf, "Sources chrétiennes" 286 and 287, 1981) 2 volumes; English translation Gary Wayne Barkley, *Homilies on Leviticus*, 1-16 (Washington, DC: Catholic University of America Press, 2005). See also H. Crouzel, *Virginité et Mariage selon Origène* (Paris-Bruges: Desclée De Brouwer, 1963).

11 FEBRUARY 1981

[
The kata phusin *character of marriage.* ∽ *Xenophon's* Oeconomicus: *study of the speech of Ischomachus to his young wife.* ∽ *The classical ends of marriage.* ∽ *The naturalness of marriage according to Musonius Rufus.* ∽ *The desire for community.* ∽ *The couple or the herd: the two modes of social being according to Hierocles.* ∽ *The relationship to the spouse or the friend in Aristotle: differential intensities.* ∽ *The form of the conjugal bond: organic unity.*
]

*SO, LAST WEEK I began to study a little the way in which the traditional perception of *aphrodisia* in the Greeks was questioned, not in an overall and general way in the whole of Hellenistic or Roman culture, but in a number of treatises (treatises of conduct, philosophical treatises) that appear as arts of living and are [for] the most part, but not exclusively, inspired more or less directly by the Stoics. I have tried to show you that this questioning of the traditional perception of *aphrodisia* could be approached by seeing how the two major principles that organized and governed this perception (the principle of socio-sexual isomorphism and the principle of the activity of the subject) were progressively

---

* M. F. begins the lecture pointing out that for internal logistical reasons of the Collège de France, he will not be giving a lecture on 18 February: "They need this lecture theater, and I don't think we could occupy another, so as it necessary to honor one of the co-founders of pathological anatomy, we will delay the next lecture to the following week. So: holidays next week."

broken up. I [particularly] tried to study first of all the break-up of the first of these principles, the principle of socio-sexual isomorphism; a break-up due above all to what could be called the insularization (isolation on the one hand and hyper-valorization on the other) of marriage. This insularization, this isolation and hyper-valorization of marriage means and ultimately meant that marriage was no longer, could no longer be considered as the highest point of a system of valorization of sexual acts [according] to the greater or lesser conformity with social relations; through this insularization it became and aimed to become the only legitimate site of sexual activity. I tried to show you last week how one began to see this insularization of marriage, which made it the sole possible place of legitimate sexual relationships, taking shape through the typically Stoic notion of the *proēgoumenon* act, an act that must be considered primordial. So marriage is a primordial act.

To say that, for the Stoics, the act of marriage is a primordial act [signifies], of course, that marriage is behavior, conduct in accordance with nature (*kata phusin*). I would now like focus a bit on this point. In a sense, that marriage is primordial because founded in nature is an entirely traditional conception. But what I would like [to indicate] is the very precise meaning the Stoics, and those more or less directly inspired by them, give to the idea that marriage is founded in nature. The foundation of marriage in nature in the Stoics, and in all those inspired by them in these treatises of conduct in the first two centuries, [is very different from] what traditionally might be said regarding marriage as a natural conduct, a conduct inscribed in nature.

To clearly mark the difference between the foundation of marriage in nature in the Stoics and the justification of marriage as natural conduct in previous authors I will take for reference a text by Xenophon— and so from well before the period I am concerned with, well before those treatises I am talking about now—in the seventh chapter of the *Oeconomicus*. This text is interesting because one finds direct or indirect, voluntary or involuntary, explicit or implicit echoes of it in a literature that stretches over a long time. One finds it in the Stoics and you will see also—if we have time to get there—that at the beginning of the fifth century, Chrysostom, in one of the three Homilies he devoted to marriage when he was Archbishop of Constantinople, takes

up the same themes and the same form.[1] It's a matter in Xenophon's
text of a speech addressed by Ischomachus, a young husband, anyway
recently married, to his wife. We [come across] this genre of speech
to the young wife on several occasions; at any rate, we find it in Saint
Chrysostom, practically eight centuries later, and it is interesting
to compare the two texts. Let us confine ourselves today to speaking
of Xenophon's text.[2] So, Ischomachus addresses his young wife.
Ischomachus says he married his wife when she was fifteen years old,
but that from the first moments of marriage he carefully refrained
from speaking to her, or at least from addressing a speech to her in
which he would have told her in what their marriage consisted,
in what it must consist, and why the devil he married her. She was
only fifteen years old, so he could not deliver this discourse to her.
He waits, he says, until she is domesticated, until she has become able
to talk and converse.[3] At the end of this time—he does not stress the
length of time, he gives no details—when she has been domesticated
and become able to talk he delivers a speech and says to her: Why did
I marry you, why did I choose you, and why, conversely, reciprocally,
did your parents give you to me? Well, says Ischomachus to his young
wife, it is not because you and I had any difficulty in, as the trans-
lators put it, "sleeping," "going to bed together."[4] What should be
understood is not so much a sexual freedom given to the young girl as
well as the man before their marriage. One should understand simply
that there would have been no difficulty, either for her or him, in finding a bed in which to lie, that is to say: in finding someone one could
marry—provided that marriage did not have to be anything other than
[the fact of] going to bed together. On the other hand, the marriage
that they contracted and for which they chose each other, or rather for
which the woman's parents chose the young man and the young man
had chosen the woman, was something else, and much more than [the
fact of] going to bed together. Me and your parents, Ischomachus says,
reflected "on the best partner that we could get for our house and children."[5] "If the divinity gives us any children one day, we will see about
the best means to raise them; it is our common interest to find the best
possible allies and support to provide for us in our old age; for the
moment" there is not a question of children, "it is only this house that
is shared."[6]

So one sees very clearly the two objectives of marriage and the two reasons why he and she were chosen to be married. The two objectives are: on the one hand, children, who must support their parents when they are old; [on the other hand,] the household. And a bit further on, developing the same theme, Ischomachus says again, generalizing the reasons why marriage is a necessity, or anyway an opportunity to be seized: "The gods proceeded to a thorough examination before matching the couple called male and female precisely for the great advantage of their community (*koinōnia*). First of all in view of preventing the disappearance of the animal races, this couple joins together to procreate; then this union enables them, human beings at least, to ensure supports to provide for their old age."[7] These supports to provide for their old age are, of course, children, who are called *gēroboskoi*: feeders of the old, nurses of the old. "Finally, men do not live in the open air like cattle and they need" a shelter, "a roof (*stegos*)."[8] Here again, you see how Xenophon presents the objectives of marriage and the reason for this marriage. The gods have thus matched couples for the advantage of the community (*koinōnia*). How is this community defined? It is defined in three ways. First, it involves procreation, which will allow humankind to develop. Second, it involves precisely the birth of children, who will support the little community they form with their parents and for which they will have to become the support. And then finally, it involves a household.

On the basis of this general theory of the ends of marriage, Xenophon, through the speech he gives to Ischomachus, defines the responsibility of the two sexes in this house, this community, this household that they form. For the man, in particular, it is a matter of working outside, in the fields, in the properties, concerning himself with life with his fellow citizens.[9] And then, for the woman, it will be a matter—her body is formed for this—of feeding the children and staying at home to mind the provisions.[10] The text ends with a completely classical comparison with the bee hive.[11] Humankind is like a bee hive where each is given a certain number of roles, functions, tasks for the greater good of the community. In Xenophon's analysis the man-woman couple thus appears as entirely founded in nature. It was the gods who, after lengthy deliberation, decided to divide humankind into two sexes and couple men and women. But you see that this foundation in nature defines the human couple in terms of objectives and ends that are: humankind in general, the preservation of the city, the birth of children who will support their parents. It is also a question of the management of goods, of their conservation and increase. Let's say that all the ends of marriage are transitive ends in which the couple itself [is not] its [own] end. The couple is justified only insofar as it involves something other than itself (the city, children, goods, and so on).

In the texts I would now like to talk about, written mostly in the first two centuries CE and more or less directly inspired by the Stoics, in these later arts of living, these considerations on marriage found in this period, the couple has, rather, a very particular, absolutely specific nature, irreducible to any other. Marriage gives rise to, as it were, an intransitive couple that is, at least in part, finalized by itself. More precisely, I would like to show you three things today. First, that in this conception of marriage [according to] these arts of living of the first two centuries, the couple has a definite place in the natural order, through the rational plan of nature. In other words, life as a couple is not finalized by something other than itself; life in a couple is in itself one of the ends of nature. Second, in these same texts, the relationship established between husband and wife, the relationship as couple in marriage is absolutely specific, irreducible to any other and heteromorphous to the general field of social relations. Finally, third, in these texts, the function of marriage in relation to the city—for marriage must of course have utility external to the couple, a utility referring to the city, possibly to humankind—is however different from that defined in earlier texts, such as Xenophon's for example.

So, the first thing: the singular place of the dual relation, of the couple relationship in the plan of nature. And here, to compare it as strictly as possible with Xenophon's text, I will refer to a text of Musonius Rufus, a Stoic who lived in the first century CE, and who wrote a number of treatises concerning marriage, the life of the couple, some fragments of which were preserved by Stobaeus.[12] So, in one of these treatises, concerning whether marriage is a handicap (an *empodion*) for the philosophical life, there is the following passage that to some extent—it is difficult to know whether or not explicitly—takes up, refers to, and maybe quotes, implicitly at least, the text from Xenophon I have just been talking about. In this passage Musonius Rufus explains that obviously there is no way that marriage could be an obstacle to philosophical activity for the excellent reason that marriage is in accord with nature. It is *kata phusin*.[13] Moreover, this is what nature herself shows. That marriage was wished by nature is very clearly proven by the fact that nature has precisely divided humankind into two sexes, but that, at the very moment it divided it into two sexes, it implanted, it inserted, it poured into the heart of each of the two sexes a charm, an attraction to the other. It is this charm, this attraction that, permitting physical union, makes the birth of children possible. And when children are born, thus regularly bringing new generations, the species, the race, the *genos* becomes perpetual.[14] This perpetuity of humankind, of the race, which is one of the consequences of this attraction of the sexes for each other, and so of the division of the sexes, shows that men are not just made to live for themselves, [but] that, even in their sexual activity, [they must] care about others. They are bound to each other by this interplay of services and obligations that is passed on through marriage, by and through the series of generations.

Quite naturally [this type] of analysis, which goes from the idea of a natural division of the sexes up to the sequence of generations constituting a humankind in which individuals are linked to each other, ends up at the metaphor of bees, of the bee hive, which is also found in Xenophon.[15] The human species, says Musonius Rufus, should be considered as a bee hive. It is well-known that the isolated bee that does not live in a hive cannot provide for its own needs and dies. Well, the individual human being would also die if separated from the others, if he lived alone, if he was forced to provide for his own needs.[16] You see that we have here, from this idea of division of the sexes up to the metaphor of the hive, something that seems entirely traditional and to be framed exactly in a theme we have already encountered in Xenophon.

However, when we look a bit more closely, I think there are in fact several major differences between Xenophon's text and Musonius Rufus' text. I pass over of course the particular status that, as you know, the Stoics accord humankind. The human bee hive that the Stoics talk about[17] has an entirely different status, value, and force, it entails many other obligations, anyway a different type of obligation than those entailed by Xenophon's bee hive. But what I would like to stress is not this but the passage concerning the attraction of one sex for the other and the kind of mechanism by which nature, after dividing the human species in two (male and female), bound men and women through this particular attraction.[18] In fact, in this passage Musonius Rufus says that after separating the two sexes in the human species, nature implanted in each of them a desire, an avid desire (an *epithumia*, a *pothos*) that he says is strong, vigorous. So a strong desire, a strong appetite of one sex for the other. But is it exactly the appetite of one sex for the other? In fact, if we follow his text, Musonius Rufus does not say that it is the appetite of one sex for the other. He says it is the appetite and the desire—I will quote the terms in Greek, we will explain them afterwards—for *homilia* and for *koinōnia*. [With this] desire for *homilia* and *koinōnia*, Musonius Rufus says, nature shows that it wished that the two sexes *suneinai* and *suzein*.[19] Let's take these words. Desire for *homilia*. *Homilia* means physical relationship and practically always designates the union of two bodies. Desire also for *koinōnia*. *Koinōnia* means something else. It is community, shared life, shared existence. And, having implanted in the heart of men and women a desire for this physical coming together and this shared existence, nature thus showed that it wished human beings *suneinai* (to practice the sexual act, *suneinai* practically always [implies] a reference to the sexual act) and *suzein* (live together). So you see that the will of nature, manifested by the separation of the sexes and then by their coming together, this will of nature [indicates] that it has simultaneously two goals, two objectives. [It wants] the physical relationship (*homilia*), it wants human beings to practice the sexual act (*suneinai*). But also and at the same time, on the basis of one and the same desire, or as the objective of one and the same desire, it wants a shared existence. Human beings are made for *suzein*: living together. They are made to form a *koinōnia*: a community of two, a life together as a couple. So you see that nature has a double objective, or that there are two sides to the bond that attaches man and woman to each other on the basis of and through the difference of the sexes: one concerning the physical coming together and the other concerning shared existence. And continuing this development, Musonius Rufus adds this: Nature, by attracting to each other the two sexes it had separated, wanted that these two sexes *ta pros ton bion allēloin summēkhanasthai* (to arrange the affairs of their life together and reciprocally). The objective of this simultaneous and reciprocal arrangement of the things of life is two things: to have children and to ensure that the race is eternal.[20]

In this great movement from the separation of the sexes and their reciprocal desire up to the formation of humankind, in this movement already described by Xenophon and taken up by Musonius Rufus, you see then that the dual relationship, the couple relationship, has a much more important and consistent place in Musonius Rufus than in Xenophon. In Xenophon, the dual relationship is merely a kind of passage to be gone through to get from the separation of the sexes to the formation of the household and, through the formation of the household, [to] the formation of the city. There is no theoretical halt, no ethical focus on the couple formed by the man and woman, that is necessary, founded in nature, but does not have any intrinsic interest because it is not an end in itself. It is no more than a stage. In Musonius Rufus, on the other hand, the moment of the dual relationship is a strong one that is not defined merely by the existence of a sexual coming together, that is not defined merely by the physical or possible economic complementarity of man and woman. It is a matter in fact of a conjugality of existence that concerns *to pros ton bion* (all the things concerning life). And the *epithumia*, the desire of one sex for the other is just as much the desire of one body for another as the desire of one existence for another. Or rather, it is not just the desire for a coming together between two bodies; it is the desire for a shared existence. Nature wanted that the desire for the coming together of bodies be, in itself and at the same time, a desire for shared existence.

This theme, [perceptible] when looking at the text of Musonius Rufus in more detail, appears even more clearly in a passage from another treatise on marriage, from a bit later, that of Hierocles. In Hierocles' treatise* there is in fact a rather revealing passage on the importance of the dual relationship in the plan of nature. Hierocles uses, directly in this case, some concepts that are concepts, notions of the naturalists. In his *Peri gamou*, his treatise on marriage, he says that nature made human beings *sunagelastikoi* beings, that is to say beings living in a herd, in a group, a multitude.[21] The idea that certain animals, certain living beings are intended by nature for life in a herd is

---

* M. F. adds: which I spoke to you about last week [see above, lecture of 4 February, p. 112].

found in the naturalists. Aristotle employs the word and says it with regard to fish: Fish are made to live in a shoal, to live in a multitude.[22] And the same expression is found in other later texts, like Porphyry for example in *De abstinentia*.[23] So, men belong to that category of animals that live in bands, in a herd, a multitude. But, says Hierocles, nature not only wanted that men live in bands, but that they live, [according to] the expression he uses, *sunduastikoi*, that is to say that they live in a couple, in pairs.[24] They are "syndyastic" animals, and not merely herd animals. The adjective *sunduastikos* is relatively rare. On the other hand, *sunduazein*, "to live as a couple," is a much more frequent expression that is found, for example, in Aristotle in *History of Animals*,[25] in Xenophon,[26] and again in Aristotle, in the *Nicomachean Ethics*.[27] And this expression *sunduazein* (or again the substantive *sunduasmos*: the connection between two elements) designates at the same time, or rather may designate either coupling strictly speaking, the sexual bond, or any union between two elements. The word *sunduastikos*, employed here by Hierocles, refers then to this dual bond that includes the sexual relation as well as establishing a relation between two elements.

When Hierocles says that human beings are *sunagelastikoi* (intended to live in a group), but that they are also intended to live as a couple, you see that he is designating the simultaneity of two forms of sociality inscribed in man by nature. A sociality of the multitude, of a group, a sociality, if you like, that I would call plural—this means that man is intended to live in society, in the city, that he is intended also to be part of the whole of humanity. But at the same time, and as it were by a fork that characterizes [his] nature, man is intended for a sociality of the couple, a dual sociality that is no longer realized in the city, nor in humankind in general, but with his wife. And Hierocles asks what the objectives, the ends are of this quite precise and particular sociality of the couple inscribed in man by nature for the same reasons and at the same time as the plural sociality of the group, but in a specific way that differentiates it from that sociality. In the first place its objective is to produce children, but also to lead a *eustathēs* life to the end, a solid, firm life, a life that holds up well, a well-balanced life.[28] You see then the two dualities that fit together in Hierocles' analysis. First, nature has put two propensities to two different types of sociality: the propensity to plural sociality and the propensity to dual sociality. And this dual sociality has two objectives: on the one hand to produce children, and thereby one rejoins plural sociality since it is through these children that humankind will be able to develop and form itself; and [on the other hand] to lead to the end a well-balanced life. In this the purpose of the couple is precisely placed in the life of the couple, you have a self-finalization of the life of the couple defined by this objective: to live to the end (this is the *diexagōgē*) a solid, firm, well-balanced life. This is shared existence.

To summarize, if we compare the Hierocles text and the Musonius Rufus text I just quoted, you see that Musonius Rufus takes up as it were the framework fixed by Xenophon and tries to describe the movement going from nature's division of the sexes up to humankind, and in this itinerary, this development he gives a particular place to the couple. But he puts it in line between the division of the sexes and the formation of humankind. On the other hand, Hierocles conducts a slightly different analysis. If I may say so, he does not put things in line but in parallel, and he shows how in nature there are two things, two juxtaposed tendencies: one towards the dual and the other towards the plural, one towards the couple [and] the other towards human society in general, whether this is the city or humankind.

Clearly a number of objections could be made to the analysis I am offering of this text by Hierocles, and in particular that of recalling that, after all, Hierocles was not the first to have said that the human beings were intended to live as couples. You find this thesis, you find the expression itself in Aristotle, and you find it precisely twice. First in the *Politics*, and second in the *Nicomachean Ethics*. In the *Politics* (Book I, Chapter 2) Aristotle says that man and woman need to be *sunduazesthai*, to be bound to one another, for one cannot live without the other.[29] And even more explicitly, more emphatically, in the *Nicomachean Ethics* (Book VIII, Chapter 12), there is an important and famous text in which Aristotle says that man is a being more *sunduastikos* than *politikos*, more syndyastic than political; that is to say that he is more inclined to live as a couple than within a political body, within the *polis*.[30] He is made more for the couple than for the city. In saying this, is not Aristotle saying things even more clearly than Hierocles or Musonius Rufus? Do we not then find already in Aristotle exactly the same idea well before these treatises on conduct I am referring to? Comparison between Aristotle and the texts I am talking about is quite revealing and shows clearly, despite everything, the specificity of the type of thought [expressed] in the treatises of life, marriage, and so on, of the first two centuries CE. Let's take in turn the two texts of Aristotle.

In the *Politics*, right at the beginning, the text insists of course on the necessity of the dual relationship. But when Aristotle says in this passage that man is made to live in a couple, he immediately explains what dual relationships are absolutely indispensable to human existence. He says: For man to be able to live, this dual relationship, this *sunduasmos*, which is the man-woman relationship, is necessary, but the dual relationship of master and slave is also necessary.[31] Man is necessarily bound to two dual relationships, he can live only within these two dual relationships, and it is the coexistence of these two dual relationships (man-woman, master-slave) that ends up constituting the family, the household (*oikos*). The first of these dual relationships, the man-woman relationship, enables the individual to reproduce. The relationship between master and slave enables him to produce. And the union of these two relationships makes the household an economic unit and at the same time means that it can be inscribed within a city for which it provides the following generations, the arms to defend it, and so on.[32] So you see that in Aristotle the necessity of a dual relationship is not at all defined by the necessity of a shared existence between man and woman. It is the need, at the starting point of the set of social relations, to deal with the woman in what concerns reproduction and the slave in what concerns production. Hence it is not at all the same type of specificity as in Hierocles and Musonius Rufus.

As for the passage in Chapter 12 of Book VIII of the *Nichomachean Ethics*, here too, notwithstanding the interesting formulation—man is intended to be syndyastic even more than political—I do not think that we find the same type of analysis as in the Stoics, or, more generally, the moralists of the first two centuries. The eighth book of *Nichomachean Ethics* is devoted entirely to the problem of friendship. And Chapter 12 is devoted to a sort of comparative description of bonds of friendship in order to find out which are the strongest. You know that Aristotle distinguishes [between] friendship bonds: those established between comrades, the *philia hetairikē*, the friendship of comradeship, for which he gives the example of the friendship that may link fellow citizens in a political unit like the city, or members of the same tribe, or the members of a ship's crew; these friendships rest on a type of explicit or implicit concord—what he calls *homologia*.[33] And to this friendship of comradeship, fellowship, he contrasts the friendship based on kinship, *suggēnikē* friendship, which, he says, derives in all its forms from a fundamental relationship binding parents to children. Of course, Aristotle says, this second type of friendship (resting on kinship) is obviously much stronger than friendship that entails only relationships of comradeship. And since all of these friendship relationships linked to kinship derive more or less from the parent-child relationship, the parent-child relationship possesses in itself the strongest bond and the most solid and vigorous friendship. He says moreover that it is not just "between" parents and children, but rather from parent to child that the most vigorous friendship is found.[34] For it often happens that children do not display much friendship and gratitude towards their parents, whereas parents see a part of themselves in their children.[35] And since they recognize a part of themselves in their children, it is clear that this bond is the strongest of all. To this parent-child relationship, this friendship of parents for their children, Aristotle compares other types of friendship that might exist between brothers, cousins, [or] shared friendships formed through shared education, and so on. Finally, at the end of the chapter, he comes to the problem of the husband and wife relationship, which it is quite clear is neither comradeship (it is not entirely of the *hetairikē* type), nor a kinship relationship. Between the two, he certainly gives friendship between man and wife an important place and a fairly strong force, but in the phrase we are seeking to understand, saying that man is more syndyastic than political he means simply that between friendship relationships founded on kinship and relationships belonging only to the domain of comradeship, the man and wife relationship is stronger than the purely comrade relationship or the relationship that may be established between fellow citizens. "It is more syndyastic than political" [signifies]: the bond between man and wife will be a stronger bond, a bond that ties the individual more than those that may exist between different citizens of the same city or different members of a tribe.

So you see that for Aristotle the *philia* between spouses is something natural, of course, something strong, but is part of a general typology of friendship. Among all the friendship relationships that can bind individuals to each other, from familial to citizenship relationships, Aristotle [places] the relationships between man and wife mid-way, as it were, with a certain strength, a certain intensity, but he does not give them the fundamental place, the absolutely constitutive place [they have] in the later Stoics, the place that the Stoics, under the Empire, accord to marriage and the life of the couple, to life as a couple, in the plan of nature, [when they say that] nature wanted not only the union of the sexes, but life as a couple.

So specificity of this dual relationship. The second point I would now like to stress is the form itself of this dual relationship. It seems to me that, here again, the texts I refer to give a description, offer an analysis of the relationship internal to the couple that is very different from what could be found previously. Let's take again precisely the text of Aristotle I was just talking about—the one from *Nicomachean Ethics* concerning *philia* (friendship) relationships. Aristotle says that the relationship of *philia* between man and wife is intense, strong, and inclines the man towards the woman and woman to the man [much] more than relationships of comradeship or shared citizenship. And he explains why this relationship between man and woman is particularly intense, more intense than political relationships. If this relationship is so strong, he says, it is first of all because it is useful, because man and wife can continuously provide mutual assistance, that they can divide up the work of the household between them. From this utility in the distribution of work, this mutual assistance, comes a pleasure that is due to the fact that one's activity is useful activity. This pleasure, moreover, may even be strengthened if the two partners are virtuous. To this is further added the fact that the couple have children, and these form an additional bond. Because of all this, Aristotle says, you can see that it is natural that man and woman, that human beings are more syndyastic than political, more inclined to join together in pairs than with the totality of citizens that form a city. In Aristotle's analysis there is basically nothing very foreign to the elements of the traditional discussion I talked about last week. The reason one has an interest in marrying, that it is good to marry, is always situated at the level of the advantages of marriage. Reason inclines towards marriage for reasons of utility. That the marriage relationship is of the same type as political relationships, although stronger than specifically political ties, that it is the same thing in its very nature, is proven at the end of the chapter when Aristotle says that justice must reign between man and woman.[36] The same type of regulation that presides over relations between citizens must preside over relations between man and wife. So, for Aristotle, there is certainly more strength in the man–woman, husband–wife bond than in the bond between fellow citizens, but not a difference of nature. Let's turn to the treatises I would now like to talk about. It seems to me that we finds in them the description of an entirely different form of bond that introduces us to the principle of the heteromorphy of the man–woman relationship in relation to other social relationships.

[The texts of Musonius Rufus are quite explicit on this point.] [In] the treatise on marriage as a possible impediment to the philosopher, Musonius Rufus takes up this theme of the relations, the relationships, the comparison between marriage and friendship. Musonius Rufus does not at all deny that the marriage relationship entails friendship. Up to a certain point he even makes the marriage relationship a form of friendship. But he clarifies straightaway: No friend can be [as] dear to his friend as a wife is to her husband and her husband is to the wife, for the wife is always present in the heart of her husband. She is *katathumios*. And a certain number of things prove that this bond is stronger than any bond of friendship, such as the sorrow of one spouse caused by the other's absence.[37] The husband's absence is for the wife, and the wife's absence for the husband, the cause of a much stronger sorrow than may be felt when a friend leaves. Similarly, the wife's presence is for the husband, and the husband's for the wife, the cause of an infinitely greater joy than in usual relationships of comradeship. As an example of this wholly particular strength of marriage relationships he cites the story of Admetus and Alcestis, which will be one of the commonplaces of all this literature on marriage. As you know, Alcestis accepts death in her husband's place, while the parents of her husband refused to make this sacrifice.[38] Only the wife is capable of sacrificing herself for the life of her husband, which clearly shows that the relationship of friendship between man and woman, husband and wife is not only much stronger than relationships of comradeship, but also than relationships between parents and children.[39] One gets the impression that here again the text of Musonius Rusus corresponds, explicitly or implicitly, to Aristotle's text that I was just talking about. When Aristotle says in the *Nicomachean Ethics* that, in any case, in comparison with the relationship that binds parents to their children, all other relationships are much less strong, Musonius Rufus inverts the analysis and shows on the contrary that marriage relationships are stronger, even stronger than those binding children to their parents. Parents would never sacrifice themselves for their children as Alcestis sacrificed herself for Admetus.

In order to explain the strength possessed by this marriage bond, in contrast with all the relationships of *philia*, it is necessary to get it into one's head that, for the Stoics, the marriage relationship is basically at the extreme limit of the friendship relationship. It is both an effectuation of it and at the same time something different by nature. It is by nature something other than friendship, and it is at the same time the model of all friendship relationships, the point on the basis of which all the other friendship relationships must be ordered.* What is, in fact, the marriage relationship, and in what does it consist in its deepest inner form?

First, unlike friendship relationships, the marriage relationship involves an organic unity. In a text from a bit earlier than the period I am talking about, Antipater puts it very clearly.[40] Antipater, a Stoic from the period of middle Stoicism, says: One often opposes to marriage the fact that there are inevitably difficulties, conflicts, and oppositions once one lives as a couple, that is to say that one is not alone with oneself. Duality, like plurality, is more difficult to manage than unity. But, Antipater says, this argument is wholly insufficient

---

* The manuscript gives the following clarifications:
  "It is necessary to pay attention to this last text for two reasons: the feelings put to work (we know that two feelings were essential in friendship and formed an intense tension: *homonoia* (equality), *eunoia* (benevolence)). Here we have: *kēdosunē* (not just 'being well disposed' but one caring for the other), more than *homonoia*. We are still in the framework of *philia*, in its general form. Plato? The team of the *Phaedrus* which will also end in disaster when one horse goes high and the other low. Doubtless not a direct echo, but the contrast of the two images is important. In any case, if marriage still puts in play a form of *philia*, the latter is no longer at all homogeneous with those relationships of interest found between friends."

and false for the excellent reason that the bond between husband and wife is not a bond between two separate elements. Would one say, for example, that it is better to have one hand rather than two, or a single finger rather than several? It is the same with marriage. Marriage is not the juxtaposition of two different individuals. It allows a single body to be formed, and in this single body there are, of course, a plurality of members. For sure, two legs are better than one. And, Antipater says, if it comes to it, a whole set of legs would be even better than having just one, or even two. Marriage should be thought of in the same way. The duality of the elements in marriage is like the duality of elements within the organism: it perfects it; it is not the source of conflict, a handicap, or an obstacle.

The second series of themes found in the Stoics is that this, as it were, organic unity formed by marriage involves not just bodies and possessions, but also souls. Marriage must be formed of a community of body and soul. This theme [is present] in Antipater and in Musonius Rufus. Musonius Rufus says quite explicitly: Of what other friendship relationship could one say that everything (goods, possessions, and soul) is common as in marriage?[41] And in his treatise *Peri gamou*, Hierocles also says: Everything is common in marriage, including the bodies, and even more than the bodies, the souls.[42]

Third, this organic unity, which involves both bodies and souls, is not to be understood as just the existence of a particularly strong bond between two individuals that ends up uniting them. In reality, it is a matter—and here the texts are important—of a veritable physics of marriage that brings into play a relationship familiar to the Stoics, since they use it in their theoretical treatise concerning physics and the order of the world. They say that marriage is not to be understood as a juxtaposition, of course, but nor even as a mixture of heterogeneous elements that to a certain extent remain heterogeneous. [Marriage falls under] a type of mixture well-known in their physics and which means that the component elements are entirely blended with each other. Antipater, still in this same middle Stoicism, distinguished two types of mixture.[43] What he calls *mixis*: mixture of elements that remain distinct, [as] when one takes two handfuls of pebbles and mixes them together. With this *mixis* he contrasts what, in line with traditional Stoic doctrine, he calls *krasis*: complete fusion of substances that are entirely blended with each other, like, for example, two liquids that one blends (water and wine). This is what the Stoics call *di'holōn krasis*, the complete mixture, total fusion.[44] This idea that marriage is a *krasis*, that there is a matrimonial crasis that constitutes a form of union, a type of bond absolutely irreducible to any other, is found in the Stoics; you find it in someone like Plutarch, hostile to the Stoics on many points, but who takes up a certain number of their themes, particularly with regard to marriage and conjugal morality. An important passage is found in paragraph 34 of the *Praecepta conjugalis* in which Plutarch says that there are three possible types of mixture. You can have the *ek diestōtōn* mixture, that is to say based on separate elements. This is the type of mixture that constitutes an army, for example. In an army, a whole series of individuals remain separate, of course, but the way in which they are joined together and form a unit on the basis of these individuals is a certain type of mixture. Second, you can have a mixture that is produced when, for example, you build a house or a boat. Here also you take separate elements, combine them with each other, arrange them; this is the *ek sunaptomenōn* mixture, in which the elements both remain separate but nonetheless constitute a unity, a physical unity. One cannot remove one of these elements without destroying the unity they produce. After all, one can kill a soldier in an army without thereby destroying the army. Remove the main beam of the house and the house collapses, but the main beam retains its unity. On the other hand, there is a third type of mixture that permits the formation of animals [or] organisms. This is the mixture that produces individuals, *hēnōmena* beings, absolutely unitary. These three mixtures can serve as models for three types of marriage. When one marries solely for the pleasures of the body, the pleasures of the bed, this is a unity, a mixture that is no more solid, no more unitary than that of an army. One joins together and then separates. The second type of marriage, following the second model of mixture, is the marriage one enters into when one's objective is only possessions or even children. Here then, one has the unity whose model is presented by the house or boat, that is to say, in effect, when one marries [for] economic reasons (to increase one's possessions, to mix one's possessions with those of someone else) or when one marries solely to have children and for the continuation of one's name or of the city, one really constitutes a unity, like that of a house or a boat. One cannot in fact take apart this unity. If the unity of the marriage was thus taken apart, the possessions would be separated anew, the family formed with the children would also be broken up. So marriage really is obliged to hold together in this way, as in a house or a boat. But, he says, the true form of marriage, the best possible form of marriage, conforms to the third model, the model found in animate nature when all the elements are combined to form an individual like an animal. This mixture is the *krasis*, the *di'holōn krasis*. Subsequently, the elements combined in this way cannot be separated, the two elements no longer have any autonomy or individuality.[45] In the good marriage on the model of *krasis*, man and wife can no longer be separated from each other because ultimately they form one and the same substance. *Di'holōn krasis* is the type of relationship found in marriage.

This definition of the marriage bond, as a total, substantial mixture, is found in many Stoic texts. You find it in several places in Plutarch, and in the famous *Dialogue on Love* in particular, where love of boys and love of women—or rather love of boys and matrimonial love—are contrasted.[*] [Regarding] matrimonial love, Plutarch, who choses against love of boys, takes up the expression *di'holōn krasis*[46] and says: In marriage, one should consider there to be a complete mixture, a substantial mixture (*di'holōn krasis*), and not something conforming to the model given by Epicurus when, speaking of atoms that remain separate from each other, he defines only their forms of attraction and repulsion. Husband and wife should not be thought of as Epicurean atoms connected to each other by something like an attraction, they should be considered to be like water and wine that blend to form one and the same liquid. When we compare this idea of marriage as substantial crasis that completely, entirely blends two elements to the point that they become inseparable, you see that we have actually passed [to an entirely different] definition of marriage. When you take the text from Xenophon with which I began,[47] and Aristotle's text in the *Nichomachean Ethics* that I also talked about[48] with regard

---

[*] M. F. adds: we will come back to this next time [see below, lecture of 25 February, p. 148].

to relationships of friendship between man and wife, whether it be the naturalness that Xenophon or Aristotle accorded marriage, or the value they accorded to this relationship, or the strength they recognized it as having in comparison with any other relationship, or even the singular place they gave to marriage in the set of social relations, you can see that, first, for Xenophon as for Aristotle, the marriage relationship, even if it is entailed by nature, is not in its intrinsic nature different from any other type of relationship. We are always dealing with a system of advantages and disadvantages, of common interests, of mutual adaptations. We are still and always, as Aristotle put it, in the regime of *dikaios*, of the just and unjust, of a political relationship. Aristotle says so explicitly: The relationship, the power that the man exercises over the woman is a political power. So it is a matter of this regime of the isomorphism of marriage with respect to other social relations. Let's say again that for Xenophon as for Aristotle, marriage is of course good, it is of course natural, but it does not have a specific nature.

On the other hand, in the texts I have talked about and that again are all inspired by Stoicism—[texts of] strict Stoics like Antipater, Hierocles, Musonius Rufus, but also people who are not Stoics like Plutarch—you see that marriage is a type of relationship that is not only natural (*kata phusin*, everyone agrees on this), but has a specific nature. It has a specific nature because the type of relationship put to work within marriage is irreducible to the social relationships found in the city or, of course, in humankind. It is another type of relationship, another form of relationship, another nature of relationship. We are no longer dealing with the regime of the system of advantages; we are already dealing with a physics of the fusion of existences. Marriage comes under a physics of the fusion of existences and no longer simply of a general system of social relations. This marks the break between marriage and other social relations, the beginning of the heteromorphism that will characterize marriage within the system of social relations and that Christianity will exploit in a way we will try to see later, but that needs to be shown to have been already perfectly formulated, marked, outlined in these treatises of conduct I have been talking about.

So it still remains, and I will do it very rapidly in two weeks' time, to talk a little about the role of this marriage relationship in the city and to show what function this elementary cell performs.*

---

* Gap in the recording. All that is heard is: [...] and then I will talk about the other dislocation that governs the perception [...].

The manuscript concludes in the following way:

"Function of life as a couple: this unity constitutes the basic element of the city. Here again, we are very close to very old themes: the idea that one must have children to renew the city; the idea that the city is not made up directly of individuals but of households (cf. the beginning of the *Politics*: 'the first community is formed of families with a view to the satisfaction of needs,' I, 2). The Stoic theses are in fact different. In what concerns households: perfected by the conjugal bond (Antipater). The Aristotelian *oikos* is an economic unit. The Stoic *oikos* is a unit of conjugality."

1. John Chrysostom, *Homélies sur le mariage* in *Œuvres complètes*, vol. IV, *Homélies* (Bar-le-Duc: L. Guérin & Cⁱᵉ éditeurs, 1864).
2. See the analysis of this text in *L'Usage des plaisirs*, the chapter: "La maisonnée d'Ischomaque," pp. 169-183; *The Use of Pleasure*, "Ischomachus' Household," pp. 152-165.
3. Xenophon, *Économique*, VII, 5, ed. and trans., Pierre Chantraine (Paris: Les Belles Lettres, CUF, 1949) p. 59; English translation E. C. Marchant, "The Oeconomicus," in *Xenophon, IV: Memorabilia and Oeconomicus, Symposium and Apology* (Cambridge, MA: Harvard University Press, Loeb Classical Library, 168, 1923) p. 415.
4. Ibid., VII, 11, Fr., Chantraine's translation: "coucher," "se mettre au lit ensemble"; Eng., p. 417: "... we should have had no difficulty in finding someone else to share our beds."
5. Ibid., VII, 11, Fr., pp. 60-61; Eng., p. 417: "I for myself and your parents for you considered who was the best partner for home and children that we could get."
6. Ibid., VII, 18-19, Fr., p. 61; Eng., p. 417: "Now if God grants us children, we will then think out how we shall best train them. For one of the blessings in which we shall share is the acquisition of the very best of allies and the very best of support in old age; but at present we share this in our home."
7. Ibid., VII, 18-19, Fr., p. 62; Eng., pp. 418-421: "For it seems to me, dear, that the gods with great discernment have coupled together male and female, as they are called, chiefly in order that they may form a perfect partnership in mutual service. For, in the first place, that the various species of living creatures may not fail, they are joined in wedlock for the production of children. Secondly, offspring to support them in old age are provided by this union, to human beings, at any rate."
8. Ibid., VII, 19, Fr., p. 62; Eng., p. 421: "Thirdly, human beings live not in the open air, like beasts, but obviously need shelter." On this precise point, see *L'Usage des plaisirs*, pp. 174-175; *The Use of Pleasure*, pp. 157-158.
9. Ibid., VII, 20, 22-23, 30, Fr., pp. 62-64; Eng., pp. 421-423.
10. Ibid., VII, 24-25, Fr., p. 63: "She [the divinity] has given the woman power to nourish the new-born ..., she has charged the woman with responsibility for looking after the provisions"; Eng., p. 421: "God ... created in the woman and ... imposed on her the nourishment of the infants ... And since he imposed on the woman the protection of the stores ..."
11. Ibid., VII, 32-34, Fr., pp. 64-65; Eng., p. 425.
12. These fragments have been collected in the *Reliquiae*, ed., Otto Hense (Leipzig: B. G. Teubner, "Bibliotheca scriptorum Graecorum et Romanorum" 145, 1905. A. J. Festugière has translated them in *Télès et Musonius: Prédications* (Paris: Vrin, "Bibliothèque des textes philosophiques," 1978). Foucault does not use this edition. It is true that it does not provide the Greek text. English translations by Cora E. Lutz in Coral E. Lutz, "Musonius Rufus 'The Roman Socrates,'" in *Yale Classical Studies*, 10 (Cambridge: Cambridge University Press, 1947).
13. *Reliquiae*, XIV, p. 70; *Prédications*, p. 98: "Assuredly, the philosopher is the teacher, I think, and guide of men for all that is suitable for man according to nature (*kata phusin*), and the fact of marrying is manifestly according to nature (*kata phusin*) if anything is"; "Musonius Rufus," XIV, "Is Marriage a Handicap for the Pursuit of Philosophy?" p. 93: "Now the philosopher is indeed the teacher and leader of men in all the things which are appropriate for men according to nature, and marriage, if anything, is manifestly in accord with nature."
14. Ibid., XIV, pp. 71-72; Fr., p. 98: "For why then did the creator of man first divide our race (*genos*) in two, then make two sorts of sexual parts in the human being, one for the woman and the other for the male, then implant in each of the two sexes a violent desire (*epithumian iskhurian*) for intercourse and common life (*tēs th'homilias kai tēs koinōnias*) with the other sex, and mix in the heart of one and the other a powerful passion (*pothon iskhuron*) of one for the other, of the male for the female, of the female for the male? Is it not then visible that he wanted the two to unite (*suneinai*), live in common (*suzēn*), associate to organize life (*ta pros ton bion allēloin summēkhanasthai*), procreate together and nourish children, in order that our race be eternal (*to genos hēmōn aidion ē*)?"; Eng., p. 93: "For, to what other purpose did the creator of mankind first divide our human race into two sexes, male and female, then implant in each a strong desire for association and union with the other, instilling in both a powerful longing each for the other, the male for the female and female for the male? Is it not then plain that he wished the two to be united and live together, so that the race might never die?

15. Xenophon, *Économique*, VII, 32-34, pp. 64-65; *Oeconomicus*, p. 425.
16. *Prédications*, XIV, 5, p. 99: "If on the other hand you concede that human nature principally resembles the bee, which cannot live alone for it dies if isolated, but which conspires and collaborates and contributes its work to its fellows in view of a common work of the animals of its species ... in that case, each must be concerned for his city and make his family its foundation"; "Musonius Rufus," p. 93: "If you will agree that man's nature most closely resembles the bee which cannot live alone (for it dies when left alone), but bends its energies to the one common task of his fellows ... I say, it would be each man's duty to take thought for his own city, and to make of his home a rampart for its protection."
17. See in particular Marcus Aurelius, *Pensées*, VI, 54, in Émile Bréhier, *Les Stoiciens*, p. 1188: "What is not useful for the swarm is not useful for the bee"; English translation Maxwell Staniforth, *Meditations* (Harmondsworth: Penguin Books, 1964) p. 104: "What is no good for the hive is no good for the bee." The idea of the bee as the gregarious animal par excellence comes from Aristotle, *Politics*, I, II, 10, and is widespread in the Hellenistic-Roman period. See the fragment of Philo of Alexandria (attributed to Chrysippus) in *Veterum Stoicorum Fragmenta* (t. II, 733), ed., H. von Arnim (Lipsiae: B. G. Teubner, 1903), and Cicero, *On Duties*, ed., M. T. Griffin and E. M. Atkins (Cambridge: Cambridge University Press, Cambridge Texts in the History of Political Thought, 1991) Book I, 157, p. 61: "Now it is not in order to make honeycombs that swarms of bees gather together, but it is because they are gregarious by nature that they make honeycombs. In the same way, but to a much greater extent, men, living naturally in groups, exercise their ingenuity in action and in reflection." On this point see J.-L. Labarrière, *La Condition animale. Études sur Aristote et les stoiciens* (Louvain-la-Neuve: Peeters, 2005) p. 35 sq.
18. On this subject see *Le Souci de soi*, p. 179; *The Care of the Self*, pp. 151-152.
19. See the quotation above in note 14.
20. See the quotation above in note 14.
21. Stobaeus, *Florilegium*, 22, p. 7; regarding this, see *Le souci de soi*, p. 180; *The Care of the Self*, p. 153.
22. Aristotle, *Histoire des animaux*, IX, 2, 610b: "among fishes, some group together in shoals (*troupes*) and are friends (*hoi men sunagelazontai met'allēllion kai philoi eisin*), others do not group together and are enemies." Aristotle, notes also with regard to "numerous species of fish" that they "live in shoals ([*troupes*] *agelaia*)" and "have a social instinct (*politika*)" (I, 1, 488a, p. 5); *History of Animals*, IX, 951: "Of fishes, some swim in shoals together and are friendly to one another: such as do not so swim are enemies" and pp. 776-777: "many kinds of fishes are gregarious" and "some are social."
23. Porphyry, *De l'abstinence*, III, 11, 1, trans., Jean Bouffartigue and Michel Patillon (Paris: Les Belles Lettres, CUF, 1979) p. 166: "Who does not know what respect for justice towards the other is found in animals which live in society (*ta de sunagelastika*)"; English translation Thomas Taylor, "On Abstinence from Animal Food," in *Select Works of Porphyry* (London: Thomas Hood, 1823) p. 106: "Who likewise is ignorant how much gregarious animals preserve justice towards each other?"
24. Stobaeus, *Florilegium*, 22, 8 (in *Le Souci de soi*, p. 180, Foucault translates "*conjugaux*"; *The Care of the Self*, p. 153, "conjugal").
25. See for example, *History of Animals*, V, 1, 538b-539b.
26. "On Hunting," in Xenophon, *Scripta Minora*, trans., E.C. Marchant (Cambridge, MA: Harvard University Press, Loeb Classical Library, 183, 1925) pp. 388-389: "In winter and summer and autumn the scent lies straight in the main. In spring it is complicated; for though the animal couples (*sunduazetai*) at all times, it does so especially at this season."
27. In the *Nicomachean Ethics* this verb has, however, a more abstract signification: difficult "coupling" or combination of the characteristic features of prodigality (1121a), or of "accidental qualities" (pleasure and utility) in the case of friendship (1157a), or even arithmetical "sum" (1131b).
28. The Greek text is: "*Legō de tēn paidōn genesin kai biou diexagōgēn eustathou*" (Stobaeus, *Florilegium*, p. 8).
29. Aristotle, *Politique*, I, 2, 1252a, ed. and trans., Jules Tricot (Paris: Vrin, 2005) pp. 24-25: "The first necessary union is that of two beings (*anagkē dē prōton sunduazesthai*) who cannot exist without each other: this is the case for the male and female in view of procreation"; English

translation B. Jowett, "Politics," in *The Complete Works of Aristotle*, vol. 2, p. 1986: "In the first place there must be a union of those that *cannot exist without each other; namely, of male and female, that the race may continue.*"

30. Aristote [Aristotle], *L'Éthique à Nicomaque*, 1162a 16, ed. and trans., René Antoine Gauthier and Jean-Yves Jolif (Louvain-la-Neuve: Peeters-Nauwelaerts, 2002) ch. 12, p. 241: "Man, in fact, is naturally inclined to live as a couple, even more than to live in a city (*anthrōpos gar tē phusei sunduastikon mallon ē politikon*)"; English translation W. D. Ross, revised by J. O. Urmson, "Nichomachean Ethics," in *The Complete Works of Aristotle*, vol. 2, p. 1836: "for man is naturally inclined to form couples—even more than to form cities."

31. Aristotle, *Politics*, I, 2, 1252b, Fr., "It is nature that has distinguished the female and the slave"; Eng., 1987: "Now nature has distinguished between the female and the slave."

32. Ibid., Fr., p. 27: "The two communities that we see [man-woman, master-slave] is the family ... On the other hand, the first community formed of several families ... is the village ... Finally, the community formed of several villages is the city"; Eng., p. 147: "Out of these two relationships the first thing to arise is the family ... But when several families are united ... the first society to be formed is the village ... When several villages are united in a single complete community ... the state comes into existence."

33. *Nicomachean Ethics*, 1161b, Fr., p. 239: "In sum, it is shared interest, as we have said, that is the foundation of all friendship (*en koinōnia pasa philia estin*). One might however except from this rule the friendship between members of the same family and childhood friends (*tēn te suggenikēn kai tēn hetairikēn*). It is rather friendships between fellow citizens, members of the same tribe, fellow voyagers, and so on, that resemble community friendships (*koinōnikais*): for it strikes one that they rest on a sort of contract (*kath'homologian*)"; Eng., p. 1835: "Every form of friendship, then, involves association, as has been said. One might, however, mark off from the rest both the friendship of kindred and that of comrades. Those of fellow-citizens, fellow-tribesmen, fellow-voyagers, and the like are more like mere friendships of association; for they seem to rest on a sort of compact." From this quotation one understands that Aristotle wants above all to distinguish on one side familial or childhood friendships and, on the other, community friendship that assumes "a sort of contract" (*homologia*) and remains founded on a "shared interest."

34. Ibid., 1161b, 16, Fr., p. 239: "If it appears that friendship between members of a family is itself divided into several kinds, it is nonetheless clear that it derives entirely from the friendship of parents for their children"; Eng., p. 1835: "The friendship of kinsmen itself, while it seems to be of many kinds, appears to depend in every case on parental friendship."

35. Ibid., 1161b, 18, Fr., p. 239: "Parents cherish their children because they are like a part of themselves, while children cherish their parents because they originate from them"; Eng., p. 1835: "for parents love their children as being a part of themselves, and children their parents as being something originating from them."

36. Ibid., 1162a, 30, Fr., pp. 241-242: "To ask what should the conduct of the husband towards the wife be in their shared life ... manifestly amounts to asking what conduct is just in such a case"; Eng., p. 1836: "How man and wife and in general friend and friend ought mutually to behave seems to be the same question as how it is just for them to behave."

37. *Prédications*, XIV, 7-8, p. 99; *Reliquiae*, XIV, p. 74: "One could not find an association (*koinōnia*) more necessary and more pleasant than that of man and woman. What comrade is as beneficial to his comrade as a desired wife is for the man who has married her (*hōs gunē katathumios tō gegamēkoti*)? What brother for his brother? What son for his parents? Who, when absent, is as missed as a husband is by his wife, a wife by her husband? Whose presence would improve or lighten the sorrow or increase the joy or remedy a misfortune? In what case is everything considered common (*koina einai panta*), body and soul and goods, if not in that of husband and wife? That is why all men consider the love of husband and wife the most precious of all"; "Musonius Rufus," p. 95: "One could find no other form of association more necessary nor more pleasant than that of men and women. For what man is so devoted to his friend as a loving wife is to her husband? What brother to a brother? What son to his parents? Who is so longed for when absent as a husband by his wife, or a wife by her husband? Whose presence would do more to lighten grief or increase joy or remedy misfortune? To whom is everything judged to be common, body, soul, and possessions, except man and wife?

For these reasons all men consider the love of man and wife to be the highest form of love."
See also *Le souci de soi*, p. 188; *The Care of the Self*, p. 160.

38. The story is taken up by Euripides in the tragedy *Alcestis*, trans., Philip Vellacott in Euripides, *Three Plays: Alcestis, Hippolytus, Iphigenia in Taurus* (Harmondsworth: Penguin Books, 1953).

39. *Prédications*, XIV, 8, p. 100; *Reliquiae*, XIV, pp. 74-75; "Musonius Rufus," XIV, p. 95.

40. In Stobaeus, *Florilegium*, 25, pp. 11-15.

41. *Prédications*, XIV, 7, p. 99; *Reliquiae*, XIV, p. 73: "In what case does one judge everything to be common (*koinōnos*), body, soul, possessions, if not that of husband and wife?"; "Musonius Rufus," XIV, p. 95: "To whom is everything judged to be common, body, soul, and possessions, except man and wife?"

42. Stobaeus, *Florilegium*, 24, pp. 8-10.

43. Ibid., 25, pp. 11-15. On the same point see *Le Souci de soi*, p. 190; *The Care of the Self*, p. 162.

44. The Greek text reported by Stobaeus, *Florilegium*, 25, p. 12, is the following: "*hai men gar allai philiai e philostorgiai eoikasi tais tōn ospriōn ē tinōn allōn paraplēsiōn kata tas paratheseis mixesin, hai d'andros kai gunaikos tais di'holōn krasesin hōs oinos hudati kai touto meliti misgetai di'holōn.*" See *Le Souci de soi*, p. 190; *The Care of the Self*, p. 162.

45. Plutarch, *Préceptes de mariage*, 34, 142e-143a, trans., Robert Klaerr in *Œuvres morales*, vol. II (Paris: Les Belles Lettres, 1985) p. 158: "Of bodies, the philosophers say that some are composed of distinct elements (*ek diestōtōn*), like a fleet or army, others of joined parts (*ek sunaptomenōn*), like a house or a ship, of others finally that they form a whole of a single nature (*ta d'hēnōmena kai sumphuē*), as is the case of each living being. It is more or less thus that, in marriage, the union of people who love each other form a whole of a single nature, that of the union of people who marry for the dowry or for children is composed of joined parts, that of people who only sleep together, of distinct elements, and one may think of them that they inhabit the same house together but do not live together. It is necessary rather that, as the physicians say of liquids, there is fusion of all their elements (*di'holōn genesthai tēn krasin*), that bodies, possessions, friends, and relations blend together in the spouses"; English translation F.C. Babbit, Plutarch, "Advice to Bride and Groom," in *Moralia* (Cambridge, MA: Harvard University Press, Loeb Classical Library, 222, 1928) vol. II, pp. 322-325: "Philosophers say of bodies that some are composed of separate elements, as a fleet or an army, and still others form together an intimate union, as is the case with every living creature. In about the same way, the marriage of a couple in love with each other is an intimate union; that of those who marry for dowry or children is of persons joined together; and that of those who merely sleep in the same bed is of separate persons who may be regarded as cohabiting, but not really living together. As the mixing of liquids, according to what men of science say, extends through their entire content, so also in the case of married people there ought to be a mutual amalgamation of their bodies, property, friends, and relations."

46. Plutarch, *Dialogue sur l'amour*, 769e, p. 103: "For what is called 'integral union' is truly the case of spouses who love each other (*hautē gar estin hōs alēthōs hē di'holōn legomenē krasis, hē tōn erōntōn*)"; *The Dialogue on Love*, p. 431: "For this truly is what is called 'integral amalgamation,' that of a married couple who love each other."

47. See above, p. 124 sq.

48. See above, pp. 136-137.

seven

25 FEBRUARY 1981

The new economy of the aphrodisia. ⌣ *Traditional mistrust
of sexual activity: religious restrictions.* ⌣ *Double relationship
of sexuality: symmetry with death, incompatibility with the
truth.* ⌣ *Sexual activity and philosophical life.* ⌣ *The medical
description of the sexual act.* ⌣ *Comparison of the sexual act and
epileptic crisis.* ⌣ *Christian transformation of the death-truth-sex
triangle.* ⌣ *Consequences of the conjugalization of sexual pleasure
in the first two centuries* CE *in philosophical texts; the man-woman
symmetry; objectivation of matrimonial sexuality.*

TODAY I WOULD LIKE to continue the study of some of the texts
I have taken from the first two centuries before Christianity. Rather
than moralist philosophers, I would say: directors of conscience. Or,
inasmuch as the notion of conscience is not applicable to what they
thought, or what they did, or to the domain they referred to, we
should say that a good many philosophers at that time were masters of
life, directors of conduct, directors or masters of existence. Last time[1]
I tried to show you the way in which these masters of life, these directors of conduct—most, but not all of them, Stoics—valorized marriage.
By this should be understood, on the one hand, that they presented
marriage as a general rule of existence, a rule that had to be applied
not only, of course, to common mortals, to common citizens, but even
to philosophers. And, on the other hand, they valorized marriage in

the sense that they presented and analyzed this general rule of marriage as entailing a very particular and specific type of relationship.
The matrimonial relationship was something specific in the order of
nature, specific among the social relations of friendship, specific inasmuch as marriage was supposed to form, create, reveal, give rise to a
singular reality due to the fusion of two beings. This was the famous
notion of *krasis*, of the conjugal crasis that unites husband and wife in
a new and single being. So that is the valorization of marriage as it
appears through these texts.

Now I would like to take up the same genre of texts, with a few variants, and some others also, [in order to] analyze what could be called
the new economy of sexual pleasures, the new economy of *aphrodisia*. In
these texts we find on the one hand an increasingly mistrustful, prudent, economic, restrictive attitude towards sexual pleasures, towards
aphrodisia. But even more than this mistrustful attitude, I would like to
show you above all what could be called the conjugalization of sexual
pleasures, of *aphrodisia*, in a double form. On the one hand, the localization of *aphrodisia*, of the legitimate sexual act within the matrimonial
institution, even more: within the individual relationship itself of husband and wife. And, on the other hand, the codification of these sexual
acts, not only in terms of the juridical or institutional form of marriage,
but in terms of this individual relationship. Consequently there is an
indexation of legitimate sexual pleasure to the specific form of the conjugal relationship. I think we have here a particularly important moment
in the history of the [modes]* of reflection, analysis, and regulation of
sexual activity.

First, then, in this new economy of sexual pleasures, of *aphrodisia*,
I would like to talk about what could be called the new mistrust, or
a certain form of mistrust with regard to these sexual pleasures, an
attitude that becomes, it seems, increasingly marked in the centuries
preceding the development of Christianity. When I say that this attitude
of mistrust is increasingly marked in the Hellenistic and Roman period,
I do not mean at all that, [in a] given [period] of Greek civilization, there was a moment when sexual acts, the *aphrodisia* would have
received, by right, a positive valorization and that they would later

---

* M.F. says: forms.

have been allowed. In fact, in every civilization, in every known society, there are and have always been certain features and characteristics that show what the system of restrictive economy concerning *aphrodisia* is. The problem is how this restrictive economy of *aphrodisia* manifests itself and is organized in Greece. Here, inasmuch as it is a matter of a quick survey of what could be called classical Greece, I will inevitably be very rapid. I would like however to stress a number of aspects. And I will take by turns the three forms within which this economy of *aphrodisia* was reflected and, to a certain extent, regulated and codified. First, within religious practices; second, within philosophical practice and life; and third, within medicine.

First, I will recall the ritual, religious prohibitions that were known in classical, and also archaic Greece, and that tended to establish a relatively strict barrier, limit, or frontier between sexual activity and religious life. You know, for example, the obligations of virginity entailed by the worship of certain goddesses: the virginity of the priestesses of the Artemision,[2] and of the vestals at Rome. You know too the obligation of lengthy abstinences involved in the service of a god. To be quite sure of the abstinence of its servers, they were chosen from children or old women. There was also the obligation of temporary abstinence, either for the duration of the ceremony, or quite simply for the ritual visit to the temple. One could not visit a temple ritually if one had had sexual intercourse immediately beforehand. A number of legislative texts are absolutely explicit. A law of Cyrene,[3] for example, says that one cannot practice sacrifices if one has made love during the day. On the other hand, if one made love during the night one had the right to sacrifice the following day. These themes will persist for a very long time. In Plutarch's *Table-Talk* (Book Three, Question 6), you find the idea that it is better to make love in the evening than in the morning, because if one makes love in the morning one cannot worship the gods, whereas if one makes love in the evening, the night has passed and as it were drawn its curtain between the sexual activity and religious activity.[4] The relation to the gods entails therefore, with different modalities according to cults, regions, and periods, a radical break with the sexual act.

On these things, which are known, and again I am only pointing out from a distance, I would like to make two comments. First, in these obligations or restrictions concerning the *aphrodisia* in relation to ritual and religious acts, one thing appears important to me. This is that the *aphrodisia* occupy an as it were symmetrical position with regard to death and the cadaver. Two great prohibitions protect religious activities and these are, precisely, the sexual act on the one hand, and contact with the cadaver on the other. Having touched a cadaver, having been under the same roof as it, having seen a cadaver face to face, or having had a sexual relation (to which are also attached menstruation, births, abortions), are all equally elements, events, encounters that prevent those who have participated in them from taking part in a religious activity. In a late dictionary from the fifth century CE, that of Hesychius of Alexandria, purification is defined thus: to purify oneself is to cleanse oneself of any contact *apo nekrou kai apo aphrodisiōn* (with a cadaver and the *aphrodisia*).[5] So, symmetrical, analogous positioning of death, of the dead, and the sexual relationship. The second of these religious elements of *aphrodisia* to keep in mind is that the rule of abstinence, the need for purification, bears on the manifestation of the truth in an absolutely particular way. That is to say, truth-telling in the domain of religion, prophetic truth-telling for example, takes place through a pronounced, continuous, definitive sexual abstinence. The god of Delphi or that of Dodona[6] can speak only through the intermediary of a virgin. Abstention from sexual relations must be total for the god's voice to be heard, mediatized by this voice that must be pure of any sexual relationship. What is definitively true for prophecy is true on occasions when it is a matter of hearing or seeing the truth, when one solicits it from a god in a particular circumstance. For example, at the temple of Epidaurus, when one comes to ask the god to send one a dream that will tell one's future, the fate awaiting one, or the illness affecting one, if one wants to receive this dream that tells the truth, if one wants to get the god to speak or appear, or anyway that he send the message that will be able to help and enlighten one, one has to have abstained from sexual relations and to have purified oneself.[7] That the truth can come from the god to man only on the condition that he is purified of any sexual relation is an important idea. It is a traditional idea found not just in Greek religion moreover; it is an idea that will last for a long time—there is the testimony, for example, of Tibullus[8] or Ovid[9] which shows that, here again, someone wanting to get some sign

or indication from the gods regarding his future must abstain from sexual relations. To summarize this, let's say that from the religious point
of view the *aphrodisia* are on the one hand in a relation of proximity, or
symmetry, or analogy with death and, on [the other hand,] incompatible with the manifestation of the truth.

Still in an overview of classical Greece, let's now take the philosophical tradition, more exactly the tradition that is called the *bios
theōrētikos*, the theoretical life. Sexual pleasure appears incompatible
with, or at any rate dangerous for the theoretical life, that is to say
the life dedicated, devoted to the search for, seizure, and contemplation
of the truth. The first manifestation, anyway the most striking, reso-
lute, radical manifestation of this incompatibility is found, of course,
in Pythagoras and the Pythagoreans. To the question: "When should
one make love?" Pythagoras answered: "You will make love when you
wish to harm yourself."[10] In Pythagoreanism, or in the tradition that is
not very well known, not very precise, at least for us, of both Orphism
and Pythagoreanism, it is easy to deduce the incompatibility of contemplation of the truth and sexual activity from a number of themes:
the body seen as a prison in which the soul is confined;[11] the body as
impurity that with its contact soils the soul enclosed in it;[12] the body
as matter destined for death that consequently risks dragging into its
mortality the soul that is destined for immortality. Consequently, to be
freed, to receive the light of the truth, even to be able to receive signs of
that other life through dreams, through the night, the soul must purify
itself by abstaining from any sexual relation.[13] What the Pythagoreans
said about this in their terminology, their metaphysics, or let's say in
their general religiosity, [was] taken up in very different theoretical
contexts. All philosophical life, in any Greek philosopher, entails an
attitude of restriction, of mistrust regarding sexual pleasure.

Of course, the attitudes are different. Not only are the attitudes
different, but the practical consequences are different. For example,
[in contrast with] Socratic *egkrateia*, the mastery of pleasure, you have
Cynic *egkrateia* that consists rather in, as it were, getting rid of one's
pleasure as quickly as possible. There is the story of Crates who did
not even have the time to wait for a prostitute to satisfy his sexual
pleasure. He satisfied it himself with his own hand, for ultimately
he was freed from it more quickly.[14] There is, for example, the total

refusal of Antisthenes who did not want the philosopher to cede to sexual desire in any way and surrender to his pleasure.[15] There is the Epicurean attitude, which is a bit different but no more favorable to sexual activity and pleasure.[16] Epicurus did agree to the philosopher giving way to sexual pleasure, but on a condition, or rather a set of conditions.[17] On the one hand, sexual intercourse [must not be] tied in any way to the impulses, desires, and empty representations of love— so, sexual intercourse without love. And, on the other hand, sex must not provoke any disturbance, any negative consequence, and must be only the pleasure of satisfying a need that is never considered to be necessary, but is nonetheless natural. So anyway, whatever the attitudes and the practical consequences in the different forms and formulae of the theoretical life, we always find a very restrictive attitude towards sexual pleasure.

It is fairly easy to extract the general formulation of this mistrust, the reasons for it that are universally accepted by the philosophers. [First,] sexual pleasure is a violent pleasure that seizes hold of the entire body, shakes it up, and drags the will beyond what it would have wished. Second, sexual pleasure is a fleeting, precarious pleasure that is intense one moment and then converted after into tiredness, exhaustion, sorrow, regret, and so on. In short, sexual pleasure is a pleasure that we would describe in our terms as "paroxysmal" and, to that extent, is opposed to a philosophical life, to a *bios theōrētikos* whose function and aim is precisely the absence of disorder, the permanence of a certain quality of existence characterized by wisdom, self-mastery, and happiness. And finally, the *bios theōrētikos* is characterized by a grasp of the truth that the paroxysmal violence and blinding of the pleasures of love can only prevent. So the theoretical life, which in the Greek conception is opposed to political life by the fact that it is a life of leisure, is opposed to the life of pleasures by *egkrateia*. The three lives—theoretical life, life of pleasures, political life—are distributed in the following way: the theoretical life is opposed to the political life by leisure; *egkrateia* (mastery) opposes the theoretical life to the life of pleasures in which the, if not sole, at least primary, exemplary aim is sexual pleasure. The theoretical life is not a life of inactivity, but of activity, of activity exercised on oneself with the primary function of achieving mastery over all those involuntary impulses of pleasure of which the *aphrodisia* are the most manifest and at the same time most dangerous example.[18] Here again, please forgive the very allusive character of all this, but you see that, without distorting the truth too much, even if we simplify it a great deal, in the Greek conception, from the Pythagoreans up to the period at which I would like to place myself, theoretical life and sexual activity, whatever accommodations there may be between them, are fundamentally incompatible.

Third, now, the medical point of view.[19] In the first lines of Hippocrates' *The Seed* there is a description of the sexual act.[20] The description is carried out in the following way. The Hippocratic author explains that, in the sexual act, the rubbing of the genitals causes warming. This warming of the genitals spreads to the liquids and humors surrounding the genitals. The warming of these liquids produces a foam, as with any warmed and agitated liquid. And it is this foaming liquid, or rather the foam of this liquid that comes out violently in the ejaculation and in this way causes pleasure.[21] This model of pleasure, which is manifestly completely virile, is transposed as such by the author of *The Seed* to feminine pleasure, which also, however dependent upon the first, follows the same mechanics. The womb is agitated in coitus. The humors surrounding the womb are also agitated and warmed, become foamy, and they too spread, bringing about pleasure and releasing that substance, that feminine spermatic matter that, mixed with the masculine spermatic matter, gives rise to the embryo.[22] This then is the schema that Hippocrates' *The Seed* talks about. What is interesting in this general schema of sexual pleasure, and what the Hippocratic author emphasizes, moreover, is that the whole body is affected in this mechanics of pleasure. It is affected as a whole and also essentially. As a whole in the sense that, while it is true that the liquid first warmed by the agitation and rubbing of the genitals is indeed that which surrounds the region of the genitals, the pelvic area, the lower back, and so on, nevertheless, Hippocrates says, all the veins found in this region and that arrive at the genitals through the marrow and kidneys, come in reality from the whole body. And this heat, that sweeps over this region and reaches those regions, does not take long to spread to the whole body. And, starting from the genitals, it is the entire body, as a whole, that through its network of veins begins to be in heat, to reach boiling point. And consequently all the body's humors thus become foamy, so that the sperm that ends up flowing is really produced by the whole body. This explains two things. The first, of course, is that the child resembles its parents, since the different elements, the matter that makes up the embryo comes from the whole body of its parents.[23] [The second is] the great exhaustion of the individual after making love, since his whole body has produced the small quantity of sperm that gives him his pleasure.[24]

So the sexual act concerns the whole body. It also concerns it essentially inasmuch as it is not any parts of the humors and blood that form the spermatic foam in this way, but, Hippocrates says, the richest, liveliest, and strongest parts. And consequently, isolated by the agitation, made foamy, and violently expelled, these are what are detached from the body, so that in the sexual act the body loses what liveliest and strongest in it. It splits, as it were, from itself. In any case, the most essential sources of its existence are wrested from it. This explains how the sperm is able to create another embryo [and] how, by its ejaculation, it exhausts the whole body, but [also] how the individual is taken almost to the threshold of his own death in the sexual pleasure he experiences.

Macrobius claims that Hippocrates likened the sexual act to epilepsy.[25] Actually, this metaphor is not found in the Hippocratic texts. However, if we believe Galen, who cites Sabinus (the citation is found in the third commentaries on the first book of *Epidemics*, paragraph 4), it seems that it was Democritus who compared the sexual act to epilepsy.[26] One thing is certain anyway, which is that this metaphor, the comparison of the sexual act with epilepsy, becomes a very common medical, physiological, and also moral theme in classical and post-classical Antiquity. A physician like Rufus, for example, will say later regarding sexual acts: "One must enjoy venereal pleasures (*aphrodisia*) as infrequently as possible;[27] they are favorable neither to general health nor to the reasonings of the soul, and indeed to the contrary they remove its vigor. First of all the violent movements accompanying coitus belong to the family of spasms, then, the cooling that follows slows down and dulls thought."[28] In the same way, Caelius Aurelianus compares epilepsy and the sexual act term for term and shows that in both cases one finds the same movements, although the contractions are different—the same panting and sweating, rolling of the eyes, flushing of face, then followed by pallor and general weakness of the body.[29] And even where the metaphor of epilepsy is not explicit, in most of the physicians there is at least a constant and implicit recourse to this kinship, [to] the epilepsy-sexual act analogy.[30] Hence the idea, for example, [that] infantile epilepsy is cured during puberty by the first coitus—you find this in Pliny[31] and in Scribonius Largus.[32] There were even physicians who claimed to cure epilepsy with coitus. Aretaeus was indignant, [or] in any case seemed hardly to bear the practice he attributed to some physicians who pushed children to engage in coitus so as to be able to cure their epilepsy.[33] Moreover, the same relation between epilepsy and the sexual act [is implied by] the opposite thesis that consisted in saying that one should be careful not to engage in coitus if one is epileptic precisely because coitus, the venereal act, is often the cause of epilepsy due to their identical nature—you find this thesis in Galen,[34] Celsus,[35] Paulus Aegineta,[36] Alexander von Tralles,[37] and so on. This medical comparison between epilepsy and the sexual act appears significant to me. I pass over, of course, the religious connotations [that are sometimes associated] with epilepsy, according to which epilepsy, like the sexual act, makes impure. In the *Corpus hippocraticum*, VI, 352-356, there is reference to the need for purification of the individual when he has had an epileptic crisis.[38] And Hippocrates again said that the epileptic who feels the onset of a crisis should withdraw from public places and hide his head.[39] According to Apuleius, it was not good to eat at a table with an epileptic.[40]

So, for epilepsy as for the sexual act, we do have a number of ritual and religious prohibitions, but this is not what I want to stress. I have emphasized doctors linking the sexual act with epilepsy for the following reason. In their description epilepsy presents three significant characteristics. First, epilepsy comprises violent and involuntary movements. Second, epilepsy entails loss of consciousness, confusion of thought, loss of memory, and forgetting. And finally, third, epilepsy leads to total exhaustion of the body to the point that after an epileptic crisis it could be thought that the individual was dead. So, involuntary movement, violent tremors, apparent death, and confusion of thought in epilepsy are three elements also found in the sexual act.

Through this metaphor of epilepsy, the sexual act reveals three characteristics already [present] in both religious prohibitions and philosophical thought—[or rather] in the ideal of the philosophical life—namely: first, that the sexual act is by nature a close kin of death; there is something that brings it close to, attaches it to death; second, that the sexual act is incompatible with thought, memory, with any relationship whatever to the truth. An attitude of fundamental mistrust of sexuality manifested itself in classical Greece [through this] closeness to death and exclusion from the truth. At any rate, two elements (closeness to death, exclusion from the truth) are constantly found in religion, philosophical reflection, and medical knowledge. I have laid stress on these elements for two reasons. The first is that, once again, we should not seek behind Christianity, in the myth of a classical Greece, a golden age of sexual freedom that valorized what was later depreciated by Christianity or by that famous and fictional Judeo-Christian tradition that is sometimes invoked. So, in classical Greece, in these different planes of reflection or different codifications of action and life, we [see] a fundamental mistrust of sexual acts. Second, I have emphasized this not just to show what might have been made out without further emphasis: this mistrust, this codification, this regulation, and so on. I have emphasized it because we see clearly the appearance of a sort of sex-death-truth triangle in medicine as well as in philosophy and religion.

⚜

We find this triangle again precisely in the Christian experience of the flesh. But, and this is what I would like to try to show you in the lectures to come: the organization of this triangle in the Greek experience of the *aphrodisia* and in the Christian experience of the flesh is not the same. It is absolutely true that in the Christian experience of the flesh, as in the Greek experience of the *aphrodisia*, the sexual act is regarded as related in a certain way to death. It is absolutely true also that in the Christian experience of the flesh, sexual activity is regarded as incompatible with the relationship to the truth. However, there is a fundamental difference: in reality, there is a whole series of differences, but the most important is [the following]. What differentiates the Christian experience of the flesh from the Greek experience of the *aphrodisia* is not that sex is even more depreciated than in Greek culture, or more profoundly anchored in the domain of death. It is the relationship to truth that is changed. In the Greeks, sexual activity (the *aphrodisia*) rendered the individual incapable, at least temporarily if not definitively, of access to the truth. This is the case for the Greeks and also for the Christians, but with a fundamental modification. If it is true that for Christians, in the Christian experience of the flesh, sexual intercourse excludes access to the truth, and if consequently one must purify oneself of everything that has to do with the sexual act in order to be able to have access to the truth, this purification, necessary for access to the truth, entails for Christians that each establish to himself a certain relationship of truth that will allow him to discover, in himself, anything that may betray the secret presence of a sexual desire, of a relationship to sexuality, of a relationship with anything to do with sex. That is to say that, if one wishes to have access to the truth, and if one wishes to purify oneself in order to have access to this truth, one must, beforehand and as a procedure indispensable to purification, establish a specific relationship of truth that is the specific relationship of truth to what one is. The subject must know what he is. I must know what I am and about sexual desire within me if I wish, first, to purify myself, and second, through this purification, have access to the truth of being. The truth of what I am is necessary to the purification that will allow me to have access to the truth of what is. The manifestation of the truth in myself, the manifestation of the truth that I will bring about by myself, for myself, and in myself is what will enable me to free myself from this tie to sexual desire that has prevented me from having access to the truth. The obligation, consequently, to tell the truth about oneself, to discover the truth of oneself and of one's own impurity, is what, through the intermediary of purification, will finally give me access to the truth.

Consequently, the relatively simple relationship of incompatibility between sex and truth in the Greek experience of the *aphrodisia* is complicated, or at least split. That is to say, in the Christian experience of the flesh it will no longer be a question of asserting that the sexual act, sexual desire is incompatible with the truth. It will be necessary technically to discover the means for recognizing in oneself the very truth of one's desires in order afterwards to have access to that truth promised to one by disappearance of all attachment of one's existence to sexual pleasure and desire. Thus it was Christianity that introduced, as a fundamental question in the relationship to the truth, the question: what about the truth of my concupiscence? Or, as we will say in our terminology, what about the truth, what is the truth of my desire? Thus* it was not Christianity that made the sexual relationship, sexual activity, sexual pleasure, the sexual act fall on the side of the impure. Already, in the clearest way, the Greeks had defined the relation between the sexual act, impurity, and death. It was not Christianity that made sexual activity and the relationship to truth incompatible. The Greeks had already established this. But it was Christianity that as it were split or divided the problem of the sexual relationship to the truth and, in this splitting, revealed the question of the truth of desire as what has to be answered first in order to have access, beyond all sexual desire, to the truth itself. And this is how that relationship of subjectivity and of the truth regarding desire was formed that is so characteristic not only of Christianity but of our whole civilization and way of thinking. Subjectivity, truth, and desire, this is what is formed, in the way I have just told you, when we move from the Greek experience of the *aphrodisia* to the Christian experience of the flesh. But this is the overall movement, the point of arrival that I will try to show you at the end of this series of lectures. [...]†

This movement takes place precisely through the intermediary of these masters of life, these directors of existence who do not belong to Christianity but rather to the (mainly, but not only Stoic) philosophical schools of the Hellenistic period and Imperial period. And this is where we find the problem of the conjugalization of sexual intercourse. Everything I have told you until now has been to show you that even in Classical Greece, which passes so easily for having welcomed and tolerated sexual pleasures, a certain number of elements [testify to] a strongly mistrustful attitude, mistrustful for the reasons I have just given: exclusion of the truth, proximity to death. What the texts of moral philosophy, of the conduct of existence, of the two centuries before Christianity have shown, is first of all a strengthening of this

---

* M.F. adds: and this is the general meaning I would like to give to this year's lectures.
† Gap in the recording: One hears only: [...] from the simple schema of exclusion between sexual relation and access to the truth, to the split schema of the Christian experience [...].

mistrustful attitude. I pass over this very quickly, merely pointing out a very manifest, very clear development in the medical texts of all the precautions that people are called upon to follow when it is a question of defining the regime of sexual acts. The physicians quite openly consider the sexual act as increasingly dangerous and as having to be performed only at the price of both meticulous and numerous conditions. A text of Rufus of Ephesus, for example, says: "The venereal act is a natural act. No natural acts are harmful."[41] But this act, carried out "immoderately, for too long, at the wrong time, may become injurious, mainly for those who are weak with regard to the nervous system, the chest, the kidneys, the side, the groin, or the feet. Here are the indications by which one will recognize the harm" caused by the sexual act. "All human strength diminishes with use; now strength is the natural heat that exists in us. After" sexual acts, "digestion is not good in those who give themselves up to [excessive] coition; they become pale, their sight and hearing deteriorate, none of their senses preserves its strength. They lose their memory, contract a (convulsive) trembling, have articular pains, especially in the side. Some become nephritic, others acquire an illness of the bladder; still others have mouth ulcers, toothache, and feel an inflammation of the throat."[42] All of these things absolutely do not belong, as has been said, to an attitude or movement of thought specific to the eighteenth century or to western, capitalist thought, and so forth. All this [is found] in the physicians of the first and second centuries CE. An increasingly pronounced medical mistrust.[43]

Second, there is strengthening also, of course, of the theme of the incompatibility between sexual activity and philosophical thought. I refer you, these again are just pointers, to Cicero's famous text *Hortensius*—I say famous because the text will be used by Saint Augustine in his *Contra Julianum*—in which Cicero says: "Does not voluptuous pleasure too often bring about the ruin of health ...? The more violent its movements, the more they are inimical to philosophy ... Is not giving way to this voluptuous pleasure, queen of all the others, to make oneself utterly incapable of cultivating one's mind, developing one's reason, and nourishing serious thought? Is not this chasm that night and day constantly strives to produce these violent agitations in all our senses" (you find there the rampant metaphor of epilepsy) "the secret of which is sensual pleasures taken to extremes? What wise man would not prefer that nature had refused us all these sensual pleasures, of any kind?"[44] To be freed from sex, from the pleasure and desire of sex, is what every rational man should wish for precisely to the extent that the use of reason is absolutely incompatible with that kind of activity. And, in the period I am talking about, this line of the total and radical exclusion of sexual activity in and by philosophical activity reveals those completely ascetic philosophers who have wholly renounced sexual activity and whose outline at least you see in the Cynic represented by Epictetus in the *Discourses* and an example of whom you have in Apollonius of Tyana, of whom Philostratus said in Book I, Chapter 13 of his *Life*: "as Pythagoras was praised for having said that one should have sexual intercourse only with one's own wife, he [Apollonius] assured that this precept of Pythagoras concerned others, but that he himself would never marry or have carnal relations ... even in his first youth he was not overcome by that passion: young, and full of physical vigor, he succeeded in overcoming this power of madness and making himself master of it."[45] You see: accentuation of the medical mistrust of sex and also accentuation of the theme that whoever wishes to lead a philosophical life, whoever wishes to lead a full and complete life, must renounce sexual activity.

But it is not especially this that I would like to talk about, it is something else, another [element] that seems to me ultimately much more important, and this is the conjugalization of the sexual act, which has two [major] aspects. First, in the texts I am talking about (these texts of rules of life, arts of conducting oneself, and so on, those of the Stoics, but also of Plutarch), this localization of the sexual act [has the following signification]: in the traditional ethical perception, the perception Artemidorus put to work in his *Oneirocriticism*, we saw that the sexual act had its best possible form and its maximum value when it took place in marriage. So there was not total freedom for the sexual act in this perception. It was actually in marriage that it was better and best. But what we see in the arts of conduct of the first and second centuries is first of all that there could be a legitimate sexual act only within marriage.* Musonius Rufus, for example, says: Men

---

* The manuscript clarifies here: "in the first big philosophical sects that were formed in Greece (Orphism, Pythagoreanism) we find similar themes."

who are not debauched or immoral should regard sexual intercourse as legitimate only in marriage. Any sexual intercourse outside of marriage, whoever the partner, man or woman, boy or girl, and so on, will be condemned for the sole reason that it did not occur in marriage.[46] This is the thesis of Musonius Rufus, Stoic at the beginning of the imperial period, Seneca's teacher. And this is the position taken up by later Stoics, and not only by Stoics but also by Neo-Pythagorean movements, and also by someone like Plutarch.

Second, not only may sexual intercourse take place only in marriage, but more precisely, there will be, and for the first time, a symmetrical condemnation of the two types of adultery. You know very well that in the classical conception* the wife's adultery was considered immoral while the man's was not considered so in the same way. The wife's adultery was immoral, necessarily discredited, inasmuch as, being under the power of her husband and as it were part of the possessions at his legitimate disposal, if the wife gave herself to another she went against that juridico-social structure that formed the general framework of not only marriage but the economy of pleasures. On the other hand, while it is true that the best possible sexual act the husband could indulge in was that with his wife, he could also make use of his other possessions, even though the best of these was his wife. And consequently, he could perfectly well have sex with a professional prostitute, or even with a free woman, and a fortiori of course with a male or female slave. This was absolutely part of everyday and accepted morality—again, I refer you to what was said regarding Musonius Rufus. In this category of texts that try to construct a new schema of sexual behavior, the husband's adultery will be conceived of as symmetrically blameworthy to the wife's. Adultery by the man and adultery by the wife are blameworthy for the same reasons. There is a text [on this point] by Musonius Rufus. [He] is not at all the first [to defend this thesis] since, in the text of pseudo-Aristotle, (the text of *Economics* attributed to Aristotle, but which is not by Aristotle), there is an argument of this kind, but it is the only formulation we have that is so old.[47] So we find the first formulation of this type in pseudo-Aristotle, probably

---

* M.F. adds: I mentioned it when we talked about it [see above, lecture of 21 January, p. 64, and lecture of 28 January, pp. 81-82].

towards the third century [BCE]. Then we do not [encounter] any other formulation establishing a symmetry between the two forms of adultery before that of Musonius Rufus and others also, so from the first and second centuries [CE]. Musonius Rufus' argument is interesting. He says: Many people tolerate a married man having relations, if not with any woman, at least with a slave. We have there the typical situation of the married man whose wife obviously does not have the right to commit adultery. He, of course, has the right to have sex with his wife, but also with his other possessions, so with his slave. Musonius Rufus wonders whether this situation is acceptable. No, he concludes. A married man cannot have [other] relations, not even with a slave belonging to him. And why? Well, [he continues]: Would this same man accept his wife, married [to him], having sex with a slave? Obviously not. If he does not accept it for his wife, he cannot accept it for himself.[48] A very interesting argument. In the first place, because not only a juridical but a moral equality is thus established between man and wife; everything that binds the wife must bind the man in the same way. And we find again here those symmetrical relations, those analogous and intersecting relations of obligation regarding marriage that I talked about last time. The obligations of the [married] man are the same as those of the married woman. But there is even this in the argument of Musonius Rufus: if there is something that was very ugly and even horrible, it was for a married woman to have sex with a slave; it was the very example of the passivity of sexual intercourse, a passivity that entirely reversed the order of social relations. The idea that the pleasure the man takes with a slave is just as blameworthy, as morally unacceptable as the pleasure the married woman takes with a slave actually [indicates] that the man's pleasure is considered in itself as dangerous, as wrong (it is also discredited) as that of the woman. At the moral level, the man's pleasure is of the same type as the woman's pleasure, there is no difference in kind between them. In Musonius Rufus' argument, then, you find both a sort of juridical or juridico-moral elevation of the wife to the level of the man. A juridical elevation of the wife to the level of the man, but conversely a lowering of the man's pleasure to the level of the wife's pleasure: the two pleasures are of no more value than each other. [Man and wife] are at the same level as individuals, but their pleasure is also at the same level. It is the man who establishes the juridical norm of existence; it is the wife who establishes the moral norm of pleasure. This is why the wife is raised and the man thus lowered in this kind of reasoning. So one can say that there is a conjugalization in the sense that sex outside marriage is prohibited—prohibited to the wife, prohibited to the man, and prohibited to the man not only with a woman outside, but even with [a woman] as close to him and falling as much under his power as a slave. This is one of the aspects of the conjugalization of the sexual relationship.

But there is another aspect that is even more important, at least as far as my topic is concerned. This is that through this conjugalization, this localization of the sexual act in marriage, a codification is developed. A whole set of prescriptions appear concerning sexual behavior in marriage. Not only must the sexual act take place in marriage, it must take place in a certain way, taking a certain number of precautions, following a certain number of rules or norms, precisely because it takes place in marriage. What are these regulations, these codifications? The first is that the sexual act in marriage is not to be indexed to pleasure. It is never for its pleasure that one is to have sexual intercourse in marriage. Pleasure, *hēdonē*, is not an end. And it is from this principle that Seneca's phrase,[49] also found in Plutarch,[50] and endlessly repeated by Christianity, gets its meaning: in sexual intercourse with [one's wife] one must never regard her as if she were a mistress. There must be a profound, fundamental difference of kind between the type of sexual relationship one has with a mistress and the type of sexual relationship one has with one's wife.

This exclusion of pleasure as an end in sexual activity takes different forms. The first, of course, which goes without saying, which is traditional, is the appeal to the old purpose of marriage found already in Plato,[51] Aristotle,[52] and all the Greek philosophers: one marries to have children. The *paidopoiia* is the purpose of marriage, it therefore also [becomes] the purpose of the sexual act. Whereas the aim of the sexual act as it was defined previously (inasmuch as it was not localized in marriage) was, by definition, pleasure, here there is [the idea] that the *paidopoiia*, which was the purpose of marriage, becomes the purpose of the sexual act [itself]. The *paidopoiia* (making children) was the reason for marrying—you find this in Plato and in Aristotle.

The sexual act was merely the instrument of this, for in itself the sexual act was performed, when one practiced it, in order to take one's pleasure. Having children was the natural consequence, it was not the aim one pursued, except in the ideal citizens of the Platonic city. Now, with the Stoics, the idea [arises] that any individual having sexual relations with his wife is not to have pleasure in view, but rather the aim that is the very purpose of marriage. In other words: there is superimposition of the purpose of marriage and the purpose of the sexual act, which is the birth of children.

Second, and this will also be very important for the history Christian thought, the production, the birth of children is not the sole aim of the sexual act. It must also be the formation and development of an affective bond between husband and wife. This theme, sketched out in the Stoics, is developed at length by Plutarch in *The Dialogue on Love*, which says: "physical union with a wife is the source of friendship (*philia*), like a sharing in great mysteries. The sensual pleasure (*hēdonē*) is of short duration, but it is like the seed from which mutual respect (*timē*), kindness (*kharis*)" (the goodwill each has for the other), "affection (*agapēsis*) and trust (*pistis*) grow daily."[53] In other words, unlike the Stoics, who completely eliminate pleasure as the aim of sexual intercourse with the wife, Plutarch gives it a place. He gives it an as it were natural and indispensable place, as if despite everything it was not possible not to take some pleasure when one has sex with one's wife. But this pleasure should not be thought of as [an] end. It is just a completely episodic and short-lived moment that should lead to the establishment of a whole network of relations (respect, kindness, affection, trust, and so on) that will constitute the real fabric of the matrimonial union. And it is in this sense that Plutarch gives a really significant and interesting interpretation of an old legislation attributed to Solon. Solon's law—regardless whether this is true or false, in any case Plutarch considered it Solon's legislation—was that every married Athenian male had to have sexual intercourse with his wife at least three times a month.[54] Well, Plutarch says, the reason for this was not the begetting of children, for one could not do this three times a month, but it was necessary thereby to regularly renew and reinvigorate the bonds of affection between husband and wife. So sexual pleasure is either radically eliminated by the Stoics, or subordinated as pure instrument or vehicle for of the objectives of procreation, descendants, and the constitution, reconstitution, and reinvigoration of the bond between husband and wife.

From there a certain number of elements emerge that had never appeared before then in sexual morality or even in conceptions and theories of marriage. This is, if not a complete regulation, at least a whole set of prescriptions concerning the way in which one should have sexual intercourse with one's wife. First, a rule of modesty. Plutarch explicitly criticizes a text in [Herodotus], in which [Herodotus*], recounting the story of Candaules and Gyges,[55] said: Taking off her tunic, every woman loses her modesty. Well, Plutarch says, this is not how it should be between a husband and his wife.[56] When a wife takes off her tunic in front of her husband, or when the husband takes off his wife's tunic, she should straightaway don the even tighter tunic of her modesty. Second, Plutarch says, one should obviously never make love with one's wife during the day, which was actually a fairly common prohibition in Greece. Not only should one not do so during the day, but one should not do so at night with the lamp lit. Why? Because the images of the wife's body could be inscribed in her husband's mind and memory. He might think about them during the day and have a constant attitude of desire towards her that is not appropriate when it is a matter of his wife. And finally, still in Plutarch, there is something that is also very interesting because it will have a long future in Christianity, which is the question of the wife's attitude to her husband's desire. How should she give in to her husband's desire? Should she accept everything he wants? Should she possibly solicit it? What degree of reserve and obligingness should she have towards her husband? Plutarch gives no more indications than these: on the one hand, she should not take the initiative; it is for the husband to take the initiative, she must wait; [on the other hand,] she should not have so surly an attitude as to turn her husband away.

Ten, twelve centuries later, there will be the enormous edifice of the codification of conjugal relations between man and wife in Christian casuistry. But that is something else. One has anyway the kernel of that type of problem in these texts I am talking about. We are very far from

---

* M.F. says: Xenophon.

that complex codification [established in the] Middle Ages, but there is already a whole series of recommendations in these texts that for the first time take the husband-wife sexual relationship as an object of analysis, and they do this in two ways. First, sex between husband and wife is made the expression, element, or a component in a more general relationship that should be one of love, of *erōs*. This *erōs*, which in Greek ethics had hitherto characterized relationships between men and boys, between men and girls, women, but merely as objects of desire, must now have its place, its privileged place, even better: its only legitimate place in marriage. There is an eroticization of the marriage relationship, which I think is the first element to keep in mind.*

Second, a result of this problematization of the sexual relationship between man and wife as a relationship of *erōs* will be that the set of acts, gestures, attitudes, and feelings that form the fabric, the fundamental elements of this relationship, will become an object of analysis, reflection, and codification. In other words, eroticization, on the one hand, and codification, on the other, of sexual relations between man and wife go together. This conjugalization of sexual acts will for the first time lead to analysis of sexual relations between husband and wife in terms of both morality and truth. The question of the truth of what these relations are and must be begins to be raised at this moment. And thus you see the great theme of the mutual exclusion of sexual relations and truth begins to falter, crumble, and shift somewhat. In the classical conception of the *aphrodisia*, where there was sexual intercourse there could be no relationship to the truth. In the traditional conception of pederastic love, the passing on of the philosopher's truth to the child could take place only if the philosopher renounced the sexual act. Where there was sexual intercourse there could not be a relationship of truth; where there was a relationship of truth there could not be a sexual relationship. [Now] you begin to see things shift a little. The sexual act is conjugalized, it is lodged within

---

* The manuscript adds:

  "Hence the interest in a traditional question: the comparison between the two loves, an old question but which is restructured. Previously the love of boys was the focal point of a theoretical elaboration. The problem was how to think of it, differentiating it from the relation to women: *erōs*. Now love of women is the focal point of the questioning, the critical point. And it is in relation to the love of boys that the love of women will be questioned."

the husband-wife relationship. The sexual act, in its legitimacy and validity is indexed to these relations between man and wife, to these legal, juridical, formal relations, but also to the affective relations that should bind man and woman and make the couple a new reality, the sexual act being lodged consequently within this problematic of affective man-wife relations. The sexual act begins to become, not [more] compatible with the exercise of philosophical thought, but an object of analysis for philosophy, an object of analysis for a discourse of truth. The incompatibility of truth/sexual act is maintained at the level of the subject who practices this sexual activity or this philosophical activity. But, on the other hand, on the side of the objects, that is to say in the realm of the domains of objects to be analyzed, the sexual act, in its very nature, and what it has to be in order to be legitimate, begins to become an object of knowledge, an object of truth. And it is thus that one finds here, in the conjugalization of marriage, the seed or the first element of what will be the big question in Christian thought, Saint Augustine's question, but also our question: "What in truth is our desire?"* That's it, thank you.

---

* The manuscript concludes the lecture with the following development:
"It is this question of the subject of desire that will traverse the West from Tertullian to Freud. But it remains to be shown how the subjectivation of the *aphrodisia* as well as the objectivation of the subject of desire is elaborated in Christianity. In this way the subject of desire appears in the West as object of knowledge. One has moved on from the old problematic: how to prevent myself from being carried away by the movement of the desire that incites me and attaches me to pleasure? to this different problematic: how to disclose myself to myself and to those close to me as subject of desire?"

1. See above, lecture of 11 February, p. 137 sq.
2. The Artemision is the Temple of Artemis at Ephesus. It was one of the most important sanctuaries of the goddess of hunting.
3. Foucault is probably referring to the "Sacred Laws" of Cyrene, discovered by Silvio Ferri in 1922. See "La *lex cathartica* di Cirene," *Notiziario archeologico*, IV, 1927, p. 93 sq. See also *Supplementum Epigraphicum Graecum*, IX, no. 72, 1944, p. 34.
4. Plutarch, *Propos de table*, III, 6, 655d in *Œuvres morales*, vol. IX-I, trans., François Furhmann (Paris: Les Belles Lettres, CUF, 1972) p. 135: "We will always refrain from attending the gods' feasts or preparing a sacrifice if shortly before we have performed such an act [sexual intercourse]. It is better therefore to put a sufficient interval between the two to rise purified"; English translation Paul A. Clement, *Table-Talk*, III, 6, in *Moralia*, VIII (Cambridge, MA: Harvard University Press, Loeb Classical Library 422, 1969) p. 257: "I suppose we must, in obedience to our city's law, guard carefully against rushing into a god's sanctuary and beginning the sacrifices when we have been engaged in any sexual activity a short time before. Hence it is well to have night and sleep intervene and after a sufficient interval and period to rise pure again as before."
5. Hesychius Lexicogr, *Lexicon* (A-O) (4085: 002); Kurt Latte, ed., *Hesychii Alexandrini, Lexicon* (Copenhagen: Munksgaard, 1, 1953; 2: 1966) vol. 1-2, alpha. 644.1-2: "*Hagneuin: kathareuein, apo te aphrodisiōn* q. A (Eur. Hipp. 655) *kai apo nekrou.*"
6. Another major oracular site in Epirus, less known but older than Delphi where the priestess of Apollo officiated. It was dedicated to Zeus.
7. The sanctuary of Epidaurus (in Argolis in the Peloponnese), devoted to the god Asclepius, is one of the most important "therapeutic" temples of ancient Greece. The sick person, before giving way to the "incubation" (the sleep within the temple in the course of which the god was supposed to cure him), had to undergo lengthy rituals of purification: baths, inhalation of perfumes and incense, and so on. On these sacred rites of purification in Ancient Greece, see W. Burkert, *La Religion grecque à l'époque archaïque et classique*, trans., Pierre Bonnechere (Paris: Éd. A. & J. Picard, 2011 [1977]) pp. 1145-117. See Plato, *The Laws*, 783e-784b.
8. Tibullus, *Élégies*, Book II, I, ed. and trans., Max Ponchot (Paris: Les Belles Lettres, CUF, 1968) p. 85: "And you, far from here, I order you, get away from the altars, you to whom, last night, Venus brought pleasure; chastity is pleasing to the gods"; English translation A. M. Juster, *Elegies* (Oxford: Oxford University Press, Oxford World's Classics, 2012) p. 61: "I also that ask all of *you* depart; let those / whom Venus thrilled last night avoid the altar. / The gods are pleased by abstinence."
9. Ovid, *Les Fastes*, II, v. 328-330, ed. and trans., Robert Schilling (Paris: Les Belles Lettres, 1992) vol. I, p. 41: "This was how they dined and drifted into sleep, resting separately on beds placed side by side: the reason was that they were preparing a sacrifice in honor of the inventor of the vine (*et positis juxta secubere toris causa: repertori vitis pia sacra parabant quae facerent pura*)"; English translation by Sir James George Frazer, Ovid, *Fasti*, (Cambridge, MA: Harvard University Press, Loeb Classical Library, 253, 1931) p. 81: "In such array they feasted, in such array they resigned themselves to slumber, and lay down apart on beds set side by side; the reason was that they were preparing to celebrate in all purity, when day should dawn, a festival in honour of the discoverer of the vine."
10. Diogenes Laertius, *Vies et Doctrines des philosophes illustres*, VIII, 9, p. 948; *Lives of Eminent Philosophers*, vol. II, p. 329: "Asked once when a man should consort with a woman, he replied: 'When you want to lose what strength you have.'"
11. The comparison between the body (*sōma*) and the tomb (*sēma*) is recounted by Plato in *Cratylus*, 400c (on this passage see *L'Herméneutique du sujet*, p. 175; *The Hermeneutics of the Subject*, p. 182).
12. On this point, see the lecture of 10 March 1971, on defilement and Orphism, *Leçons sur la volonté de savoir. Cours au Collège de France, 1970-1971*, ed., D. Defert (Paris: Gallimard-Seuil, "Hautes Études," 2011) pp. 161-174; English translation Graham Burchell, *Lectures on the Will to Know. Lectures at the Collège de France 1970-1971*, English series editor Arnold I. Davidson (Basingstoke: Palgrave Macmillan, 2013) pp. 167-182. Daniel Defert, in his notes to the lectures, refers to L. Moulinier, "Le Pur et l'Impur dans la pensée et la sensibilité des Grecs jusqu'à la fin du IVe siècle av. J.-C." copy of thesis (Paris: Sorbonnne, 1950).

13. On "the purifying preparation for the dream," see *L'Herméneutique du sujet*, p. 48; *The Hermeneutics of the Subject*, p. 48.
14. "One day, it is said, he had agreed with a courtesan that she come to his house; as she was late, he rid himself of his sperm by rubbing his sex with his hand, and after this he sent away the courtesan who had just arrived, saying that his hand had arrived ahead of the marriage song" (Fr., 197 of Diogenes in the edition of Giannantoni, 1990, vol. II, reported by Galen, in *De loc. affect*, VI, 15) cited by M.-O. Goulet-Cazé in "Le cynisme ancien et la sexualité," *Clio. Femmes, genre, histoire*, no. 22, 2005, pp. 17-35.
15. Diogenes Laertius, *Vies et Doctrines des philosophes illustres philosophes*, VI, 3, p. 682: "He [Antisthenes] constantly repeated: 'May I be mad rather than experience pleasure'"; *Lives of Eminent Philosophers*, vol. II, p. 5: "He used repeatedly to say, 'I'd rather be mad than feel pleasure.'"
16. See Galen's declaration in his *Peri aphrodisiōn*: "According to Epicurus coitus is never favorable to health," in *Œuvres d'Oribase*, ed. and trans., Charles Daremberg (Paris: Imprimerie nationale, vol. I, 1851) p. 536.
17. Epicurus, *Lettre à Ménécée*, 1132, in *Les Épicuriens*, trans., Daniel Delattre, Joëlle Delattre-Biencourt, and José Kany-Turpin (Paris: Gallimard, "Bibliothèque de la Pléiade," 2010) p. 48: "It is neither the endless succession of binges and parties of pleasure, nor the enjoyment of boys and women, nor that of fish and all the other dishes ...; it is rather a sober reasoning, which seeks the exact knowledge of the reasons for each choice and each refusal and which repels opinions that allow the greatest disturbance to seize hold of souls"; English translation Russel M. Geer, *Letter to Menoeceus* in Epicurus, *Letters, Principal Doctrines, and Vatican Sayings* (Indianapolis: Bobbs-Merrill Educational Publishing, "The Library of Liberal Arts," 1964) p. 57: "Neither continual drinking and dancing, nor sexual love, nor the enjoyment of fish and whatever else the luxurious table offers brings about the pleasant life; rather, it is produced by the reason which is sober, which examines the motive for every choice and rejection, and which drives away all those opinions through which the greatest tumult lays hold of the mind"; Epicurus, *Sentences vaticanes*, 18 and 51, in *Les Épicuriens*, p. 64 and p. 69: "The suppression of [sensual] gazes, of intimate relations, and of common life frees from amorous passion"; "... with regard to sexual pleasure: if you do not overturn the laws, do not attack the good morals in force, if you do not cause grief to one of your neighbors, do not exhaust your flesh and use up the things necessary for life, exercise then, as you wish, your choice. However, it is impossible not to be restrained by at least one of these conditions. For sexual pleasure is never profitable: it is already fine if it does not cause any harm"; *Vatican Sayings*, in *Letters, Principal Doctrines, and Vatican Sayings*, p. 67 and pp. 69-70: "If sight, association, and intercourse are removed, the passion of love is ended"; "... Follow your inclination [toward sexual passion] as you will provided only that you neither violate the laws, disturb well-established customs, harm any one of your neighbors, injure your own body, nor waste your possessions. That you be not checked by some one of these provisos is impossible; for a man never gets any good from sexual passion, and he is fortunate if he does not receive harm." See too, Lucretius, *De la nature*, Book IV, 1058-1287, ed. and trans., Alfred Ernouth (Paris: Les Belles Lettres, CUF, 1985) vol. II, pp.42-50; English translation A.E. Stallings, *The Nature of Things* (London: Penguin Books, 2007) pp. 138-145; Ibid., 1073-1075, Fr., p. 43: "To avoid love is not to deprive oneself of the enjoyments of Venus, it is rather to take their advantages without penalty"; Eng., p. 138: "Nor does the man who gives the slip to Love lack for its fruit; Rather he enjoys it without penalty or pain."
18. In the lecture of 25 March, Foucault will develop this theme of the "three lives," studying its presentation by Pythagoras in a saying reported by Diogenes Laertius.
19. For another presentation, see *L'Usage des plaisirs*, pp. 143-146; *The Use of Pleasure*, pp. 127-130.
20. Hippocrates, *The Seed*, in G. Lloyd, ed., *Hippocratic Writings* (London: Penguin Classics, 1983). On this text see also *L'Usage des plaisirs*, pp. 142-147; *The Use of Pleasure*, pp. 126-130.
21. Hippocrates, *The Seed*, p. 317: "*All things are governed by law.* The sperm of the human male comes from all the fluid in the body: it consists of the most potent part of this fluid, which is secreted from the rest. The evidence that it is the most potent part which is secreted is the fact that even though the actual amount we emit in intercourse is very small, we are weakened by its loss. What happens is this: there are veins and nerves which extend from every part of the body to the penis. When as the result of gentle friction these vessels grow warm

and become congested, they experience a kind of irritation, and in consequence a feeling of pleasure and warmth arises over the whole body. Friction on the penis and the movement of the whole man cause the fluid in the body to grow warm: becoming diffuse and agitated by the movement it produces a foam, in the same way as all the other fluids produce foam when they are agitated. But in the case of the human being what is secreted as foam is the most potent and the richest part of the fluid. This fluid is diffused from the brain into the loins and the whole body, but in particular into the spinal marrow."

22. Ibid., IV-V, pp. 319-320.

23. Ibid., VIII, pp. 321-322.

24. Ibid., I, 1, Eng., p. 317.

25. Macrobius, *Saturnales*, II, 8, in *Œuvres complètes*, ed., M. [Désiré] Nisard (Paris: Firmin-Didot, "Collection des auteurs latins avec la traduction en français," 1875): "Hippocrates, that man of divine knowledge, thought that the venereal action was a sort of frightful illness. These are his words: 'Coitus is a little epilepsy (*tēn sunousian einai mikran epilēpsian*)'"; English translation Robert A. Kaster, *Saturnalia*, Books I-II (Cambridge, MA: Harvard University Press, Loeb Classical Library, 510, 2011) pp. 386-387: "Hippocrates, a man of godlike understanding, thought that sexual intercourse has something in common with the utterly repulsive illness we call the 'comitial disease'"; the same statement is found in Aulus Gellius, *Attic Nights*, XIX, 2.

26. This citation of Sabinus by Galen forms the fragment 68 B 32 of Democritus; H. Diels and W. Kranz, *Die Fragmente der Vorsokratiker* (Berlin: Weidmann, 1954). See English translation of Democritus by Jonathan Barnes, *Early Greek Philosophy* (Harmondsworth: Penguin Books, 1987) p. 271: "*Coition Is Mild Madness; for a Man Rushes Out of a Man.*"

27. The Greek text has: *aphrodisiōn ap spainiaitatē hē khrēsis estō.*

28. Rufus of Ephesus, *Œuvres. Fragments extraits d'Aétius*, 75, in *Œuvres d'Oribase*, vol. I, Book VI, trans., Daremberg, pp. 370-371. It seems to be this edition of Daremberg that Foucault used to assemble the main theses on the relation between epilepsy and sexuality in Antiquity. On this point see *Le Souci de soi*, pp. 132-133; *The Care of the Self*, pp. 109-110.

29. Caelius Aurelianus, *On Acute Diseases and on Chronic Diseases*, ed. and trans., Israel E. Drabkin (Chicago: The University of Chicago Press, 1950). In *Le Souci de soi*, p. 135; *The Care of the Self*, p. 113, Foucault refers to *On Chronic Diseases*, I, 4, p. 314.

30. On this point, see the work of O. Temkin, which is still authoritative, *The Falling Sickness. A History of Epilepsy from the Greeks to the Beginnings of Modern Neurology* (Baltimore and London: John Hopkins University Press, 1994) pp. 31-32. See also the excellent article by P. Chiron, "Les représentations de l'épilepsie dans l'Antiquité gréco-latine," in *Épilepsie, connaissance du cerveau et société*, ed., Jean-Paul Amann et al. (Quebec: Presses de l'Université Laval, "Bioéthique critique," 2006) as well as Foucault's reference in *L'Usage des plaisirs*, p. 142; *The Use of Pleasure*, pp. 126-127.

31. Pliny the Elder, *Natural History*, trans., W.H.S. Jones (Cambridge, MA: Harvard University Press, Loeb Classical Library 418, 1963) Book XXVIII, X, p. 33: "Many kinds of illness are cleared up by the first sexual intercourse, or by the first menstruation, if they do not they become chronic, especially epilepsy."

32. Scribonius Largus, *Compositiones*, 18, ed. Georghius Helmreich (Lipsiae: In aedibus B. G. Teubneri, 1887).

33. Aretaeus of Cappadocia, *Traité des signes, des causes et de la cure des maladies aigües et chroniques*, Book I: *De la cure des maladies chroniques*, I, 4: "De la cure de l'épilepsie," trans., M. L. Renaud (Paris: E. Lagny, 1834); English translation F. Adams, "On the Cure of Chronic Diseases," in *The Extant Works of Aratæus, the Cappadocian* (London: Sydenham Society, 1856).

34. Galen, *Des lieux affectés*, V, 6, cited by C. Daremberg in *Œuvres d'Oribase*, p. 668, and O. Temkin, *The Falling Sickness*, p. 32. For a complete presentation of Galen's doctrine of epilepsy, see O. Temkin, ibid., pp. 60-64.

35. It is true that Celsus (III, 23) recommends continence to prevent epileptic crises. Having said that, one also finds in him the idea (mentioned by Foucault himself a few sentences earlier) of a possible disappearance of epilepsy with the first sexual relations. See Aulus Cornelius Celsus, *On Medicine*, trans., W. G. Spencer (Cambridge, MA: Harvard University Press, Loeb Classical Library, 292, 1935) vol. I, pp. 332-335: "That malady which is called comitialis, or the greater, is one of the best known. The man suddenly falls down and foam issues out of

his mouth; after an interval he returns to himself, and actually gets up by himself. This kind affects men oftener than women. And usually it persists even until the day of death without danger to life; nevertheless occasionally, whilst still recent, it is fatal to the man. And often if remedies have been ineffectual, in boys the commencement of puberty, in girls of menstruation, has removed it."

36. Paulus Aegineta, *Libri,* III, 13, ed., J.L. Heilberg (Leipzig-Berlin: B. G. Teubner, 1921-1924, 2 volumes), cited by O. Temkin, *The Falling Sickness,* p. 32 (on the designation of the uterus as cause of epilepsy in pregnant women), and in C. Daremberg in *Œuvres d'Oribase,* p. 668.

37. Alexander von Tralles, I, 15, Original-Text und Übersetzung nebst einer einleitenden Abhandlung. Ein Beitrag zur *Geschichte der Medizin,* ed., Theodor Puschmann, I. Bd., Vienna, 1878; reprinted Amsterdam, 1963.

38. Hippocrates, *La Maladie Sacré,* I, 12 (VI 363) ed. and trans., Jacques Jouanna (Paris: Les Belles Lettres, CUF, 2003) pp. 8-9: "They then have recourse to purifications and incantations, thus committing a very sacrilegious and impious action, in my view at least. They purify in effect those who are prey to illness with blood and other similar things, as if it were a question of people bearing a defilement, or pursued by a vengeful demon, or bewitched by humans, or authors of a sacrilegious act"; English translation J. Chadwick and W.N. Mann, "The Sacred Disease," in G. Lloyd, ed., *Hippocratic Writings,* p. 240: "In using purifications and spells they perform what I consider a most irreligious and impious act, for, in treating sufferers from this disease by purification with blood and like things, they behave as if the sufferers were ritually unclean, the victims of divine vengeance or of human magic or had done something sacrilegious."

39. Ibid., XII, 1, Fr., p. 22; Eng., p. 247.

40. Apuleius, *Apologie,* XLIV, ed. and trans., Paul Valette (Paris: Les Belles Lettres, CUF, 1960, p. 54): "The slaves, his comrades ... are here for the most part. All can tell you why they spit when they see Thallus, why no one wishes to eat with him from the same dish, or drink from the same cup"; English translation H.E. Butler, Apuleius, *The Defence,* 44, ⟨classics.mit.edu/Apuleius/apol.html⟩: "Many of his fellow servants, whose appearance as witnesses you have demanded, are present in court. They can all tell you why it is they spit upon Thallus, and why no one ventures to eat from the same dish with him or to drink from the same cup."

41. Rufus of Ephesus, *Œuvres. Fragments extraits d'Aétius,* 60, in *Œuvres d'Oribase,* vol. I, Book VI, trans. Daremberg, p. 318. The Greek text has: "*phusikon men ergon hē sunousia esti. Ouden de tōn phusikon blapheron.*"

42. Ibid.

43. The objects and forms of this increased mistrust are set out in *Le Souci de soi,* pp. 126-169; *The Care of the Self,* pp. 99-144.

44. Saint Augustine, *Contre Julien, défenseur du pélagianism,* Bok IV, 72, trans., Abbé Burleraux, in *Œuvres complètes de saint Augustin,* trans., under the direction of M. Raulx (Bar-le-Duc: L. Guérin & Cⁱᵉ, 1869); English translation Mathew A. Schumacher, *Against Julian* (New York: Fathers of the Church, 1957) p. 229: "What injury to health ... is not evoked and elicited by pleasure? Where its action is the most intense, it is the most inimical to philosophy. The pleasure of the body is not in accord with great thought. Who can pay attention or follow a reasoning or think anything at all when under the influence of intense pleasure? The whirlpool of this pleasure is so great that it strives day and night, without the slightest intermission, so to arouse our senses that they be drawn into the depths. What fine mind would not prefer that nature had given us no pleasures at all?"

45. Philostratus, *Vie d'Apollonios de Tyane,* Book I, 13, in *Romans grecs & latins,* ed. and trans., Pierre Grimal (Paris: Gallimard, "Bibliothèque de la Pléiade," 1958) pp. 1041-1042; English translation Christopher P. Jones, Philostratus, *The Life of Apollonius of Tyana,* Books I-IV (Cambridge, MA: Harvard University Press, Loeb Classical Library, 16, 205) pp. 60-61: "Now Pythagoras was praised for saying that a man should not approach any woman except his wife, but according to Apollonius Pythagoras had prescribed that for others, but he himself was not going to marry or even have sexual intercourse. In this he surpassed the famous saying of Sophocles, who claimed that he had escaped from a raging, wild master when he reached old age. Thanks to his virtue and self-mastery, Apollonius was not subject to it even as an adolescent, but despite his youth and physical strength he overcame and 'mastered' its rage."

46. Musonius Rufus, XII, 2-3, *Prédications*, p. 95; *Reliquiae*, pp. 63-64: "Those who are not debauched or immoral must regard as just only those sexual relations that take place in marriage and with the aim of procreation, because only these are legitimate ... No temperate man would agree to sleep with a courtesan, or with a free woman outside of marriage, or, by Zeus, with his slave"; *Musonius Rufus*, p. 87: "Men who are not wantons or immoral are bound to consider sexual intercourse justified only when it occurs in marriage and is indulged in for the purpose of begetting children ... So no one with any self-control would think of having relations with a courtesan or a free woman apart from marriage, no, nor even with his own maid-servant." See also *Le Souci de soi*, p. 197; *The Care of the Self*, pp. 168-169.

47. Aristotle, *Economics*, Book III, ch. 2-3, trans., G.C. Armstrong, in *The Complete Works of Aristotle*, vol. 2, pp. 2148-2150. On this precise point, see *Le Souci de soi*, p. 203; *The Care of the Self*, p. 174, and for a more general vision of this text attributed to Aristotle, *L'Usage des plaisirs*, pp. 193-200; *The Use of Pleasure*, pp. 174-181.

48. Musonius Rufus, *Prédications*, XII, 8, p. 95; *Reliquiae*, XII, p. 66: "If it seems to someone not to be shameful or improper for a master to go to bed with his slave, especially if she is a widow, let him reflect on what he would think if a mistress went to bed with her slave. Would it not seem intolerable, not only if a wife with a lawful husband allowed herself to have relations with her slave, but even if she were to do so and was not married? However, I do not think one will judge men worse than women, nor less able to discipline their desires"; *Musonius Rufus*, pp. 87-89: "if it seems neither shameful nor out of place for a master to have relations with his own slave, particularly if she happens to be unmarried, let him consider how he would like it if his wife had relations with a male slave. Would it not seem completely intolerable not only if the woman who had a lawful husband had relations with a slave, but even if a women without a husband should have? And yet surely one will not expect men to be less moral than women, nor less capable of disciplining their desires."

49. Foucault is alluding here to a passage of Saint Jerome's *Against Jovinian* (I, 49), in which Jerome freely evokes a now lost work by Seneca on marriage (mentioning also, at the beginning of the paragraph, those by Aristotle and Plutarch): "Any love for another's wife is scandalous; so is too much love for one's own wife. A wise man should love his wife with discernment, not passion, and consequently control his desires and not let himself be drawn into copulation. Nothing is more impure than to love one's wife as if she were a mistress." Translated [into French] by P. Ariès in his article, "L'amour dans le mariage," *Communications*, no. 35: *Sexualités occidentales. Contribution à la sociologie de la sexualité*, 1982, pp. 116-122; English translation by Anthony Forster, "Love in Married Life," in Philippe Ariès and André Béjin, eds., *Western Sexuality. Practice and Precept in Past and Present Times* (Oxford: Basil Blackwell, 1985) p. 134: "Any love for another's wife is scandalous; so is too much love for one's own wife. A prudent man should love his wife with discretion, and so control his desire and not be led into copulation. Nothing is more impure than to love one's wife as if she were a mistress." This volume also contains what Foucault presents as "an extract from the third volume of the *History of Sexuality*" (*The Confessions of the Flesh*), "Le combat de la chasteté"; "The Battle for Chastity," and an article by Paul Veyne, "L'homosexualité à Rome"; "Homosexuality in ancient Rome."

50. Plutarch, *Préceptes de mariage*, 16-17, 140b-c, pp. 151-152; *Advice to Bride and Groom*, pp. 308-311.

51. Plato, *The Laws*, VI, 783e-784b.

52. Aristotle, *Politique*, I, 4, 1252a, p. 100: "The first union (*sunduazesthai*) necessary is that of two beings who are unable to exist without each other : this is the case for the male and female in view of procreation (*geneseōs eneken*)"; *Politics*, p. 1986: "In the first place there must be a union of those who cannot exist without each other; namely, of male and female, that the race may continue."

53. Plutarch, *Dialogue sur l'amour*, 769a, p. 100; *The Dialogue on Love*, pp. 426-427: "physical union is the beginning of friendship, a sharing, as it were, in great mysteries. Pleasure is short; but the respect and kindness and mutual affection and loyalty that daily spring from it ..."

54. Ibid., Fr., p. 101; Eng., p. 427.

55. The account is in Herodotus, Book I, 7-14.

56. Plutarch, *Préceptes de mariage*, 10, p. 49: "Herodotus was wrong to say that a woman sheds her modesty at the same time as she sheds her tunic; rather, the one who is wise then dons modesty in exchange, and for spouses the greatest modesty is the token of the greatest reciprocal love"; *Advice to Bride and Groom*, pp. 305-306: "Herodotus was not right in saying that a woman lays aside her modesty with her undergarment. On the contrary, a virtuous woman puts on modesty in its stead, and husband and wife bring into their mutual relations the greatest modesty as a token of the greatest love."

This page intentionally left blank

eight

4 MARCH 1981

The three great transformations of sexual ethics in the first centuries CE. ᨆ *A reference text: Plutarch's* Erōtikos. ᨆ *Specificity of Christian experience.* ᨆ *Plan of* The Dialogue on Love. ᨆ *The comic situation.* ᨆ *The young boy's place: central and passive position.* ᨆ *The portrait of Ismenodora as pederast woman.* ᨆ *The break with the classical principles of the ethics of* aphrodisia. ᨆ *The transfer of the benefits of the pederastic relationship to within marriage.* ᨆ *The prohibition of love of boys: unnatural and without pleasure.* ᨆ *The condition of acceptability of pederasty: the doctrine of the two loves.* ᨆ *Plutarch's establishment of a single chain of love.* ᨆ *The final discredit of love of boys.* ᨆ *The wife's agreeable consent to her husband.*

IN SCANNING THE REGIMES of existence and the arts of living of the first two centuries CE, I have tried to bring out three transformations that are significant for the subject I would like to consider. These three transformations are the following: first, a shift that consists in the question of sexual pleasure, of the *aphrodisia* and their regime, being raised less and less with regard to the love of boys—which had been the sensitive, difficult question of the classical period—and increasingly in relation to the problem of man-woman love, and more precisely of the love between husband and wife. So much for the shift. Second, it seems to me that in these arts of living we can also detect a tendency for those *aphrodisia*, those pleasures that are the only legitimate pleasures to be localized within marriage. And finally, third, we can note an attempt to articulate, to connect the regime of *aphrodisia*, to index the rules of their use to the individual, personal relationship of the couple within marriage. A text bears a quite specific, clear, and revealing testimony to this triple transformation, and it poses clearly the problem of sexual ethics in Greco-Roman Antiquity right on the eve of the spread of Christianity. This text, which both represents this triple transformation and takes it to its clearest, most formulated point, is Plutarch's *Erōtikos* (the dialogue on love), and this is the text that I would like to talk about.*

So, let's talk about this *Erōtikos* by Plutarch, this *Dialogue on Love*.[1] It seems to me worth the effort dwelling on this text for a while. Actually, in its form as a dialogue on love it refers to very old forms. The theme it evokes—comparison between the two loves: the love of boys and the love of women—is also very old, very classical. Moreover, the text is peppered with explicit citations or references to great classical texts, Plato's *Phaedrus* and *Symposium*.[2] So, it brings to bear a clear, explicit, deliberate gaze on the whole of the Greek classical culture of love and the *aphrodisia*. And at the same time it utilizes a whole series of references to contemporary texts, that is to say texts falling under those arts of living more or less inspired by Stoicism that can be found in the first centuries CE. So there is a double direction in this text: a gaze turned to the past and at the same time a quite explicit reference to [contemporary] conceptions of marriage. Thanks precisely to this double direction, this double reference, through this text we can see very clearly the pivoting of the whole ethics of *aphrodisia* in Greek or Roman culture, from the classical period up to the period immediately preceding the spread of Christianity. To grasp this pivoting in the ethics of *aphrodisia* seems important to me for a number of methodological reasons concerning the way in which this problem has to be posed.

First, grasping this pivoting enables us to understand that the sexual morality of paganism should in no way be treated as a single

---

* M. F. stops to say:
   I get the impression that there is a horrible whistling [...] yes? So we will try to lower it [...] How is it, better? My impression is yes. Stop me if it is really unbearable, because I don't hear very well the whistling that you can hear.

whole, a unity opposite which another unity, just as coherent, would
have to be set up: the Christian morality of sex. There is not *one* sexual
morality of paganism, just as no doubt there is not *one* sexual morality of Christianity. Second, it seems to me that grasping this pivoting
internal to Greek and Roman ethics also enables us not to attribute
to Christianity, and to Christianity alone, a whole series of elements,
principles, and rules of morality that are often attributed to it but
which, if one actually looks at the historical processes, were formed
much before, and precisely in Greek, Hellenistic, and Roman thought.
Finally, it seems to me that grasping this pivoting is also interesting
in order to understand clearly how Christianity, after having integrated precisely this new morality that we have seen taking shape
right within paganism, and after having as it were played the game of
this morality for a certain time, had then, later in its history, from the
fourth and fifth centuries, developed something new, which is precisely
the experience of the flesh. In other words, the historical schema that
I would like to illustrate a little this year is this: to show how, within
paganism, the classical ethics of *aphrodisia* was transformed, how it carried out what could be called a sort of conjugalization of the regime
of *aphrodisia*; how this conjugalized ethics of sex was then transferred
into Christianity; and how finally, in a third stage, within the history of Christianity, there was a whole re-elaboration of this morality and the emergence of the experience of the flesh. In other words,
the problem is to show how what is called conjugal sexual morality
does not belong to the very essence of Christianity, both because one
sees it being formed well before Christianity and because the relation
of Christianity to sex [is not constructed] at all [through] a conjugal
morality, [but through] something else, [which is] the experience of
the flesh formed a bit later, after the first developments of Christianity,
on the basis of the development of monasticism, of the asceticism of
the fourth and fifth centuries. This then is why[*] I would like to dwell
a bit on Plutarch's *Erōtikos*. The theme of the *Erōtikos* is the comparison between the two loves. A classical theme. A large number of
texts, which unfortunately are lost but of which we have traces, whose

---

[*] M. F. adds: since we are roughly half way through the lectures, to close what I would like to tell
you this year regarding the Greek and Roman ethics of sex.

existence we know about, dealt with this theme and this subject for
centuries.

Plutarch's text comes as an argument in three parts. First of all, a
part mainly reserved for a discussion between partisans and adversaries
of the love of boys, or partisans and adversaries of the love of women,
with reference to an anecdote about the marriage of an attractive young
boy called Bacchon. So a small comic scene. Then comes a lengthy
eulogy of love that is celebrated as an all-powerful god, or at least as a
god more powerful than the others. So, after the comic scene, we have a
sort of development in the style of Plato, in the Platonic style and with
quite explicit references to Plato. Finally, in a much more modern-
ist style compared to [contemporary] texts, the third part of the text
is devoted to the famous *topos* of the advantages and disadvantages of
marriage and what the morality of marriage should be. Here the style
is much closer to Antipater, Musonius Rufus, or, later, to Hierocles,
about whom I have spoken. I will not follow the twists and turns and
developments of this dialogue that, not without a few difficulties,
moreover, tries for the subtleties of one of Plato's dialogues. I will try
to study it a bit systematically and [then] take an overview [of it].

First, to start with I would like to turn my attention to the kind
of comic situation of Bacchon's marriage that is developed throughout
the text but especially in the first part. When looked at closely, this
comic situation is actually quite interesting. The situation is this: an
attractive young man called Bacchon has been courted both by men,
one of whom, Pisias, was his most fervent lover, and by a woman called
Ismenodora, who wanted to marry him, and wanted to marry him
so much that she finally took him from under the nose of the good
Pisias, who was obviously very vexed. She has taken him and will
finally marry him. The dialogue's problem is this: should one save the
young man from the hands of this ardent person and prevent him from
marrying? An entirely comic situation, but one that should be taken
seriously for a number of reasons.

The first remarkable element in this is the way the alternative
between the two loves is presented. The structure of the choice itself,
"love of women/love of men," is strange. In fact in the traditional
debate, in the texts that raise the question, the problem is generally
posed in the following way: is it best for someone to love young boys or

girls or women? That is to say the choice is between loving young boys and loving women, with, hesitating between them, the man, that is to say the male adult who is already old enough to desire or to be of marrying age, and old enough also to pursue young boys in turn. So, in the middle is the man, and then: love of women [or] love of boys. Here you see the situation is reversed in relation to this classical schema, since the central character as it were, the one placed in the middle between the two alternatives, is the young boy, with a woman on one side and a man who is pursuing him on the other—the dialogue's question then being what is the best course to follow: should one let him marry or should one rather restore him to the folds or the sphere of influence of the men who love and pursue him?

In fact, the dialogue is in a way a bit skewed in relation to this schema of the boy between the love of a woman and the love of a man. [First,] because, although it is the young man who is at the crossroads, although in a sense it is his problem, [that he is the one who] has to choose one or the other, we do not see the young man in the dialogue. He is spoken about, but he does not take part in the dialogue, he does not express his views, he does not decide, his position is in fact one of passivity. He is courted unilaterally by his lovers, and then the woman who pursues him well and truly takes him away and marries him straight off in the same exuberant impulse. So he is at the same time at the center, [but] neutral, passive. [Second,] the dialogue is skewed because despite having this schema "young man between man and woman," the theme discussed will never be: is it better for a young boy to have a lover or a mistress? Without taking account, as it were, of the real situation—which is both subjacent and ineffective in the dialogue, not effective, not operative, not an organizer of the dialogue—the problem will be: in what does the love of boys consist when one compares it with the love that may, that must take place in marriage? So the boy is at the center, but never in a situation of choosing. I think it is nonetheless important to emphasize this strange situation because the mechanics of the discussion do not start from the ideal philosophical situation—is it better for a man to love boys or women?—[but] from the boy's position between man and woman, a boy who, the text says, is still wearing the chlamys,[3] that is to say a youth of eighteen years, who is at the turning point, the point between the age when he was still an object of pleasure for men and [the age] at which he will become an active subject—active in the city, the army, and in relation to women. The youth is placed [and] described here, grasped at that famous moment when he starts to grow a beard and, in principle, ceases to be a desirable and legitimate object for men, the moment when he will become an active subject in the world of *aphrodisia*.

The second remarkable element in this comic situation, in which a woman carries off a youth under the nose of [the latter's] lover, is that this woman, Ismenodora, is presented, and this is an essential element of this little intrigue, as distinctly older than the youth—from crosschecking the text she must be a bit more than thirty years old, while the youth is eighteen. The woman is older, richer, from a better family, and furthermore is already a widow. This is a traditional comic situation—the somewhat mature woman who throws herself at a youth is often found in Greek literature—[but] while being comic, while causing Bacchon, who is not very happy being pursued in this way by the older, richer person, some shame, it is [nonetheless] a relatively frequent real situation due to the well-known scarcity of women in Greek and Hellenistic cities, that was due, even more than to death in childbirth, to the fact that daughters were killed at birth, or more simply abandoned, left to their own fate, which was usually fatal. With this scarcity of women in Greece, marriage to an older woman, a rich widow, was not rare, however it was a situation that provoked some laughter.

Now what is very interesting in Plutarch's text is that this familiar and somewhat laughable situation is immediately corrected by a number of important elements. [First,] this already widowed, rich, thirty year-old woman, who throws herself at an attractive youth, is presented as someone who has many qualities, who is very virtuous, wise, and also full of experience, thanks to her first marriage. She is a woman who is generally respected and held in good opinion in the city. Second, apart from her intrinsic qualities, this woman is presented as amorous, as actively amorous as it were: she pursues the youth, she hunts him. Of course, she cannot follow him into the gymnasium, but she lies in wait and tries to attract him when he returns from the gymnasium and the palaestra. And why does she love him? Because, the text says, she had been charged by the youth's family to marry him,

to find a wife for him, and consequently to help and guide him in
his life. Getting to know him in this way, since she had to guide and
help him, she was able to appreciate his qualities. Not only was she
able to appreciate his qualities, but everywhere she heard many good
things said about this youth. I quote the text: Seeing several worthy
men were seeking the love of this young Bacchon, she came to love him
too. Moreover, her intentions in loving him were honorable. There was
nothing shameful, nothing ugly (nothing *agennes*) in her mind when
she pursued the youth. The noble and serious aim she was pursuing
was marriage.[4]

Take all these different features that characterize the woman: she
is older, richer, and has a better social status; she is virtuous, experienced, and is respected by everyone; she is amorous, she chases the
boy, runs after him insofar as she can; she loves him because he has
a good reputation, because everyone speaks well of him, and because
men love him and she enters into rivalry with them; and then, her
aims are honorable, there is nothing ugly, nothing *agennes* in her projects. All of this is the very definition of pederastic love, the very definition of the ideal lover for a youth. If he wants to have an ideal lover,
the boy should have someone older, wealthier, and better situated,
someone who loves him and pursues him, but with honorable intentions, and who, by his virtues and qualities, can be an example and
guide for him. So Ismenodora occupies very exactly a pederastic position in relation to Bacchon. She is the pederast woman, the respectful and respectable pederast, with two variations, however, that one
may or may not regard as important: to start with, she is a woman;
and [then,] what she wants, the noble aim she seeks, is not the boy's
aretē, his virtue, it is marriage. But apart from this, apart from this
endpoint and starting point, everything in her designates the lover of
boys, the pederast.

What effect can this somewhat paradoxical situation have on how
the discussion unfolds, on its stake and its tactics? I think the woman's
pederastic position should be emphasized for two reasons. On the one
hand, you see that by making the woman this active character pursuing the youth, this older, wealthier, and so on, character is in a way
a challenge to the most traditional ethics of *aphrodisia* that Plutarch
defines. In effect those principles of isomorphism and activity that we

noted as fundamental features of the ethical perception of *aphrodisia*[*]—if we accept that these do indeed organize the morality of *aphrodisia*, then you can see the degree to which the situation of Ismenodora is not only paradoxical but scandalous. She is a woman, she wants to marry, to have a husband, and what role does she play? She plays the role of active individual, of the male, and in the home she will be the one who provides wealth and reputation, she will be the one who brings virtue, who will play the role of initiator and pedagogue. All this is absolutely the opposite of a morality that would have it that the relations between the two marriage partners and sexuality are isomorphic to social relations and the social field in general—[so] that it must be the man who is wealthy, active, older, who acts as guide, and so on. Now this is exactly the reversal, it is the rigorously scandalous situation. But on the basis of this scandalous situation, and precisely because it is scandalous, you can clearly understand that when Plutarch has succeeded in demonstrating that the legitimacy of this marriage that is so foreign to the fundamental rules of the ethics of *aphrodisia* and marriage, then a fortiori all marriage will be legitimate, founded, and justified. This marriage is [therefore] justified [even though it] binds characters to each other [who reverse the poles] of that, as it were, regular and normal structure of isomorphism and activity. By this one shows that if marriage has the requisite qualities—that is to say if it binds two characters who are in actual fact virtuous—it may be as it were independent of and unaffected by those rules of isomorphism and activity I have been talking about. Without it being spelled out, what Plutarch's text shows is the increasing imperviousness of the ethics of marriage to the prevailing rules of isomorphism and activity.

This is one of the effects produced by this comic situation in which it is the woman who has the initiative. In this comedy schema, not only is the woman active—as I was telling you, she reverses the schema of isomorphism and activity that regulated the morality of *aphrodisia*—but she is quite precisely in a pederastic situation. She combines all the qualities of the good lover of a noble youth. As a result, by the very fact of this situation, Plutarch is able to transfer as it were [en]

[*] M. F. adds: you recall what I told you regarding Artemidorus' *Interpretation of Dreams* [see above, lecture of 21 January, p. 52 sq].

bloc all the positive values, all the beneficial effects traditionally attributed to the love of boys, to the man-woman relationship, and quite precisely to the matrimonial relationship.

The dialectical benefits of this discussion, of this situation within which the discussion between the different characters unfolds, [will thus be the following]: first, to reveal marriage as impervious to the principles of isomorphism and activity that in general rule the social field and that hitherto had been the major directing principles of sexual morality—marriage will thus appear as isolated, autonomous in relation to this, and only the two partners' virtue [will be] important for defining the morality and the value of a marriage and the sexual relations [that ensue]; second, to be able to show that marriage itself can bring about the same transformations, can provide the same beneficial effects, can have the same positive consequences as that old pederastic relationship in which the Greeks saw one of the instruments, if not one of the conditions of the young man's formation and his initiation, his passage to the position of active subject within the social body. That is what I would like to say about the signification of this comedic situation within which Plutarch's dialogue unfolds.

Now, starting from this situation, how does the transfer take place? How is it that we see marriage inherit all the virtues that hitherto, in the discourse favorable to the love of boys at least, were attributed to pederasty? I think the text shows three things. First, it carries out a general revision of the concept of love. In terms that are obviously not suited, that are somewhat anachronistic in relation to Plutarch's text, we could say that the text sets up the definition of the complete chain of the one and only love. Second, the text conducts a critique of the love of boys in relation to this normative chain of the complete, integral, sole love. Third, the text establishes, it recognizes that the conjugal relationship must—and that anyway it can, is able to, liable to—be identified with this chain of love and constitute the privileged place where it can be realized. So: definition of the complete chain of love; critique of the love of boys; recognition of the new isomorphism of the conjugal relationship and chain of love. These are the three points.

But before showing you how these three elements are established in Plutarch's text, I think it necessary to start from a prior point, which will be the fourth or rather the first point in the exposition,

[that is]: the way in which the partisans of the love of boys present
their own conception in Plutarch's text. This passage at the beginning of the text is interesting. It shows very clearly, coherently, and
fairly exactly the conception of love that was immanent to pederasty,
to the love of boys, in philosophical reflection, in moral reflection. In
any case the partisans of the love of Bacchon, the friends of Pisias,
and in particular Protogenes, his spokesman, develop the reasons for
which, according to them, the love of boys is preferable to the other
love. The thesis, [or] the assumption, rather, on which they establish
the preferable character of the love of boys to the love of women is the
heterogeneity of the two loves. This heterogeneity is marked in three
ways. First, a heterogeneity that is a natural heterogeneity as it were,
or more exactly that is the following: one of the two loves is natural,
and the other is not natural. The partisans of the love of boys say: But
of course there is a love that is in accordance with nature, of course
there is a natural love. And which is it? It goes without saying, it is
the love of man for woman that is natural. Nature, Protogenes says,
has put in man a drive, an *orexis*, a *hormē* for the woman. These words
are important. *Orexis*, *hormē* are traditional words but especially valorized by the Stoics to designate a natural impulse of the individual
towards something.[5] So there is a natural impulse (*orexis*, *hormē*).
Now, Protogenes says, this natural drive towards the woman is so
natural that it is found not only between man and woman, but [even]
between a fly and the milk on which it settles. It is also found in
the bee as impulse towards the honey. That is to say this impulse of
the man towards the woman, this *hormē*, and this *orexis* cannot really
be regarded as loves, precisely because they are natural impulses and
are found even in non-rational beings, even in animals, even in those
for which it would be absolutely impossible to conceive there being
something like *erōs*.[6] So the naturalness of the desire of man for
woman is, in the mouth of Protogenes and, I think, in the traditional conception of the love of boys, [incompatible with] *erōs*. If the
impulse that carries a man towards a boy may be called *erōs*, it is precisely because it is not natural. We touch here on one of the points that
makes thinking about pederasty, reflecting on pederasty at once so difficult and stimulating for the Greeks. [If] love of boys cannot pass as
natural, [this is by virtue] of the forms it has to take and the reasons

4 March 1981   185

that justify it, [namely:] the friendship, the benevolence of one for the other, the need for pedagogy, the need [for the elder] to serve as guide, to give a good example, and so on. This impulse toward love of boys, since it belongs to the domain of *philia*, of friendship, can in no way be assimilated to a natural impulse.[7]

But you know too that *phusis*, nature, without having the constraining, regulatory, normative, quasi-legislative value that it will be accorded later, is nonetheless a principle that is referred to in order to justify certain things. In other words, for the Greeks, there is a contrast between natural and unnatural, between what is in accordance with nature (*kata phusin*) and what is contrary or foreign to nature (*para phusin*). It is certainly not as binary as it will be later—to be unnatural is not at all to be monstrous for the Greeks—[but] nevertheless a certain dangerous domain opens up with what is *para phusin*, a dangerous domain inasmuch as the reference points, the internal rules indicated by nature are missing. So, there is something that is difficult to think in this conception, according to which love of boys can be defined only as external to nature, in contrast with the love of women that is internal to it. And authors do not always take the same position on this. Some [of them] say: Yes, absolutely, the love of boys is *para phusin*, external to nature. They say so explicitly and they see in this a reason to positively value this love. You find some indications [of this kind] in Plato,[8] and in Lucian also. In a very interesting passage of *Affairs of the Heart*, Lucian says: Listen, all the same, love of boys is terribly good. It is even so much better than love of women that you will never see two animals of the same sex loving each other. Have you ever seen a pederast lion?[9] Consequently, unnaturalness (the *para phusin* side) is very clearly marked as a positive form [...] [In fact,] the need to distinguish the love of boys in relation to nature and the difficulty of conceiving in absolutely positive terms something that is *para phusin*, foreign to nature, makes his position much more nuanced, much more muffled. And Protogenes says: You women, if you desire us it is, of course, through an effect of nature, and consequently is neither very valorous nor very important. But Protogenes will never say [directly]: When we love boys we are doing something *para phusin*. Whatever the nuances, the somewhat distorted way in which Protogenes speaks, it is the case, nonetheless, etc. that in this whole current of thought the first great mark of heterogeneity, the first major difference between the love of boys and the love of women is therefore the insertion of men's love of women in nature and the non-natural, foreign to nature, borderline character [of] relations between men and boys.

The second difference Protogenes emphasizes between the love of women and the love of boys, and which derives moreover from the first, is this: being attracted to women, because and insofar as it is natural, has an end, a clear objective, indicated, marked, veritably offered by nature. The objective, aim, and endpoint of this attraction is pleasure and enjoyment: *hēdonē* and *apolausis*.[10] The fact that nature had arranged a *hormē* (an impulse, a drive) for women, that the aim of this *hormē* is to obtain pleasure and enjoyment, [enables] the following series to be established: nature, an impulse implanted by nature; an impulse directed towards a pleasure; [an impulse] that, at this point, taking the form of an impulse towards pleasure, may be called desire, that is to say *epithumia*. Nature, impulse, pleasure, desire. [*Phusis,*] *hormē, epithumia, hēdonē.* The attachment to boys is developed in a completely different series, since it does not start from nature and is not orientated towards pleasure at the end—for what [this] love seeks is not the pleasure of the one who pursues the boy. Love of the boy is justified, is in line with its true nature and essence, only if, I have stressed this already, it seeks virtue rather than pleasure. Whose virtue? The virtue of the subject who loves, but also and especially the virtue of the one who is loved. Consequently, to go from the love of the boy to the boy's virtue, the lover will have to devote himself to and develop a whole activity regarding and around the boy of watching over, supervision, care, concern, attachment, and so on, which will possibly take shape in pedagogy. In this conception the care of the other, fundamental in the love of boys, is what is called *epimeleia.*[11] *Epimeleisthai* is care for oneself, care for the other.[12] So we won't have the series nature-impulse-desire-pleasure. It will be a series that does not start from nature but takes the form of *philia.* There is no *hormē,* there is *philia* (friendship). There is no *epithumia* (desire), but *epimeleia* (care, concern for the other). And there is no *hēdonē* (pleasure) for the subject, but *aretē* (virtue) for both and especially for [the beloved]. So you see that the two series are entirely heterogeneous and do not comprise the same elements.

Finally, the third difference between the two loves is that attraction to women—so the movement that goes from nature to pleasure by
way of the drive and desire—is taken to excess. There is no internal
limit and, [becoming] excessive, desire undergoes that metamorphosis,
that mutation I have already spoken to you about: become excessive,
the love of women is feminized. By loving women too much, by taking too much pleasure in women, one becomes effeminate. Why is it
that one is made effeminate, why does one effeminize oneself? Precisely
because pleasure becomes the law of the individual's behavior who
loses his activity and become passive in relation to his own pleasure.
As Plutarch says in this text, which I think confirms this analysis: By
virtue of frequenting women, of living in their midst, one becomes
moist, *hugros*, and one becomes housebound, *oikoiros*. That is to say, one
takes on the same qualities as the woman's body, which one knows
is moist, and one takes on the same form of feminine existence that
revolves around the *oikos*, the household.* [...] And here, in a passage
that is very interesting and clear—while not exactly establishing its
mediation, he juxtaposes, but one senses that for him the things are
the same—with regard to this love of women, he says: It is very bad to
give oneself up to it, since one becomes moist and housebound; [this
love] is as ugly as that which a slave may bring to a free boy.[13] In other
words: to love women is to passivize oneself, just as one passivizes
oneself when one agrees to be loved by a slave. On the other hand, if
the love of women has no internal regulation and if the *hēdonē* towards
which it is directed, by a sort of internal law of excess, becomes passiv-
ized, this is indeed because the relationship with boys has its internal
regulation because it presupposes precisely that one renounces *hēdonē*,
that one renounces pleasure in order to be able to achieve the final aim,
which is virtue.

All this is not new. It is not precisely intended to be new in
Plutarch's text since all he does is take up and make very explicit
[a shared conception. But] here, then, a little methodological reminder:
I do not mean that this whole conception was really present behind the
practice of pederasty itself and that [it] gives the truth of Greek pederasty. I mean that these were the conditions that allowed Greek pederasty

* Gap in the recording. One hears only: housebound and moist.

to be thought; [that] this was the only way the Greeks had to make this practice of pederasty, their practice of pederasty, acceptable in a rational and theoretical discourse. That is the postulate. To be able to account for this pederasty, to be able to accept it in their discourse, thought, and reflected consciousness, there had to be two loves. The conditions of reflexivity of pederasty were this. Later, much later, a time will come when to make the relationship between man and man, men's love of boys acceptable, one will be led to say: But you know it really is one and same love in both cases. [In] Plutarch,* we have evidence that for the Greeks pederasty could be justified only through the very specific definition of the difference between one love and the other.

The reintegration of pederasty at the time of Proust, anyway at the turn of the nineteenth and twentieth centuries, will take place rather through a completely different operation that consists in showing that there is indeed only one and the same love. But if one came to want to justify pederasty by asserting that it is one and the same love that is present in both, it is because meanwhile the idea had been constructed, that you may or may not call chimerical, that there is in fact only one and the same love, and that it is this love that is found, with different forms, in both one and the other relationship. Now this idea that there is one and the same love is found very exactly in Plutarch. I am absolutely not saying that it was Plutarch who constructed it, but at any rate it seems to me that one finds in him one of the first expressions, and anyway the clearest expressions for the time of the existence of one and the same love, the existence of a single complete chain of the sole love. And in the history, once again, not of sexual behavior, of sexual practices, but of reflection on sexual practices, in the history of the way in which sexual practices emerge within a cultural or normative consciousness, this is an absolutely important moment. How is it that, unlike what the partisans of the love of boys say, Plutarch shows that there is in fact a single love? How does he demonstrate it?

---

* M. F. says:
  "but this will hardly exist before the nineteenth century, and in Proust, let's say anyway that from Plutarch to Proust, or if you like with Plutarch, we have evidence ..."
  One understands then that Foucault traces back to Plutarch the conception of a single love, but that in *The Dialogue on Love* he finds evidence of an earlier conception of two loves formulated by the defenders of pederasty.

This is the great watershed in the history of reflection on love. It is not with Christianity, it is before, and Plutarch is evidence for this. Plutarch establishes the existence of a great complete and single chain of love with a number of arguments. The first—the theme, the starting point, the beginning [are] absolutely classical—[consists in] unmasking what could be called, with a somewhat anachronistic vocabulary, pederastic hypocrisy. In effect, the partisan of the love of women and Plutarch's representative—who is, specifically, Daphnaeus, his father, since the dialogue assumes that Plutarch himself does not speak, but his father, responding to the theses of Protogenes[14]—says to the pederasts: You claim that your love is distinguished from the love of women because you would not have sexual intercourse (no *sunousia*, no *koinōnia*, no physical union) [with] the boys you love, and you say that thanks to this, thanks to this elision of pleasure and sexual union, you are able to establish a bond of a completely different kind, which would be *philia* (friendship). Now, Plutarch says, you know perfectly well that this is not true and that if you pursue boys it is in actual fact to arrive at this *koinōnia*, this *sunousia*.[15] So don't be hypocrites. Anyway since, in actual fact, regardless and whatever you say, you seek a physical pleasure with the boys, either you maintain the theoretical thesis that where there is a physical relationship there can be no friendship (*philia*)—and at this point, since, in actual fact, you practice these physical relations, acknowledge that your love is not able to produce friendship, is incompatible with friendship, and consequently do not say that your love is distinguished from ours by the fact of being capable of *philia*; or then you acknowledge that you have sex with the young people you pursue, and you admit that these sexual relations are not incompatible with friendship, that actually you have friendship for your lovers, your beloved, [while having] sexual intercourse. But when you admit that your friendship for the boys does not exclude the sexual relations you have with them, why then do you refuse this coexistence in us who love women? And why do you claim that love with women, which in actual fact aims for pleasure, for that *koinōnia*, that *sunousia*, that sexual relation, is foreign to *philia*, that it excludes *philia* since there is *koinōnia*, *sunousia*, sexual intercourse? Since in fact you sleep with your loved ones and feel friendship for them, do not cut us off from friendship with the women with whom we sleep. There is then at least a common point between our two loves, which is that the sexual relationship is not incompatible with friendship. Do not make a division between what includes friendship while excluding sex and what, including sex, excludes friendship.

The second way of bringing the two loves closer and outlining the form of a single chain is Plutarch's designation of the same origin and formation of love. Plutarch says: Don't deceive yourself, all the mechanisms one may imagine to show how one becomes in love with someone are inevitably the same whether it is a case of a girl or a boy. I will simply quote you the texts since they are very clear: "the causes that are said to be at the origin of love are not specific to one sex, but common to both."[16] And here, by turns, Plutarch, although he does not cite the names, refers explicitly to two great, two extreme doctrines: Epicureanism and Platonism. And he says this: "One claims that corpuscles formed in the image of the loved object leave it, insinuate themselves in the body of the lover and pass through it, stimulating the mass of atoms in a way to set it in movement, to make it slide at the same time as its corpuscles and in this way produce sperm."[17] This is plainly the reproduction, the summary of the Epicurean conception. So love would be produced because there would be corpuscles coming from the image of the body. These corpuscles insinuate themselves into the lover's body, agitate it, and this agitation gives rise to sperm. If, Plutarch says, this is the Epicurean mechanism, is it not obvious that these famous corpuscles that are supposed to enter the body of the one who loves and provoke in him the agitation from which the spermatic flow arises, may emanate just as well from a girl as from a boy? If the Epicurean mechanism is valid for boys, it must also be valid for girls. The same thing can be said for the Platonic explanation. Here again, Plutarch does not cite Plato explicitly, but [his] presence is visible. Listen to the text: "those beautiful and holy recollections that return us to the true, divine, and Olympian beauty of the beyond and give wings to our soul."[18] If these recollections that draw our soul towards contemplation of the beautiful may be provoked by boys and youths, what prevents them from also being provoked by young girls? So, first argument: *philia* is not incompatible with sexual relations. Second argument: the causes are the same.

The third argument is the great eulogy of love that I have spoken to you about. The general considerations on the nature of *erōs* and

the relations between *erōs* and Aphrodite clearly show that one cannot separate them. In fact, Plutarch says, if *erōs*—the much vaunted love
that you lover of boys claim to reserve for your form of attachment—
does not go as far as *koinōnia*, as *sunousia*, as real physical union, it will
inevitably remain incomplete. *Erōs*, if we call [thus] an impulse of love
of someone that does not go so far as *koinōnia*, [will have to be] completed by Aphrodite, if we put physical union under the sign of *koinōnia*.
Erōs will remain in suspense. Without Aphrodite, Plutarch says, *Erōs*
is like drunkenness without wine.[19] And by this rather strange and
bizarre expression he no doubt means that *erōs* without Aphrodite
is like a sort of drunkenness brought about without one even having
had the pleasure of consumption that is drinking wine. A drunkenness
unaccompanied by the pleasure of drinking wine would be a vain and
empty drunkenness that remains in suspense, like an *Erōs* not completed by the enjoyment of Aphrodite. Or again he says: It would be
something *akarpon*, without fruit. And in a much more coded, clearer,
strong philosophical way, he says: *erōs* without Aphrodite is *atelēs*.[20]
It is without purpose, it has no end, it has therefore not fulfilled its
essence, its nature, it has not met that end for which it is made. It is an
incomplete act. In this conception, physical pleasure is indeed inserted
and marked as the end, the completion of all love. So there is no *erōs*
without this *telos* (this end, this purpose) that is physical pleasure.
And conversely, if we put sexual union under the sign of Aphrodite,
Aphrodite alone, without *Erōs*, would be no more than a tiny moment
of pleasure. If it comes to it, it is what could be experienced with prostitutes.[21] It is the pleasure of a moment that immediately fades. But if
one wishes to prolong and continue the pleasure of Aphrodite, if one
wishes Aphrodite to be a mode of relationship one has with someone,
let's say with a woman since the love of women is placed under the sign
of Aphrodite, then between individuals, between man and woman, a
certain type of continuous and durable relationship must develop that
is to be put under the sign of *erōs*. The duration of Aphrodite is assured
by *Erōs*, just as the end, the completion of *erōs* is assured by Aphrodite.
The whole long, pseudo-, quasi-Platonic praise of *erōs* in Plutarch's dialogue is really a way of reversing the Platonic thesis, since for Plato the
praise of *erōs* was precisely a way of isolating the love of boys and showing how it was specific, and what it was for him, its philosophical value,

to the exclusion of the relationship with women. On the other hand, we have a praise of *erōs* here that will permit the repatriation, as it were, of relationships with women within *erōs*. Far from being the privilege of some, far from being the privilege of those who love boys, *erōs* presides over the whole of nature. It is as if the Platonic praise of *erōs* is inverted. A long, single chain of love, identical for girls and boys, must therefore be accepted. This single and identical chain of love has the same origin, the same causes, and the same activating elements. You see too that in its unfolding, this chain includes of course desire, *epithumia*, but includes also, at the same time, without any dissociation, friendship, *philia*. And finally this chain must end on its natural conclusion: enjoyment (*jouissance*). So there is only one and the same complete chain of love. I think that this is an absolutely fundamental thesis that, historically, will have considerable effects.

Now, and this is the third point I would like to emphasize in the account of Plutarch's dialogue—so after, [first,] the contrast between the love of boys and the love of girls; second, against this thesis, the organization, the definition of the single chain of love—Plutarch's text is even so very perverse. After having established that the love of girls and the love of boys is the same, after having, as it were, merged together the qualities, marks, and properties of love of girls and love of boys, after having established the uniqueness of the chain, you can see a little how Plutarch situates love of boys in relation to this integrative chain, this complete and unique chain of love that has to establish the rule. At the same time as saying that there is only a single love of both, he makes use of this single chain of love as rule, scale, and instrument of measure to see whether the love of boys actually corresponds to it. And, of course, he shows that the love of boys does not correspond. He constructs a single chain of love combining the two loves, and then he withdraws the love of boys and shows the extent to which the latter is inadequate to this chain that he has just assembled. Why inadequate, deficient? For two reasons. One is what could be called the mythico-historical origin that he gives to love of boys. He takes up a classical theme, but his use of it is interesting. He says: Nature combined first of all the love of men and of women. Such then was the situation to start with and that endured for a long time. And then, at a given moment, the love of boys was introduced. Referring to an idea that is often found before him [and] that will often be taken up again after, [he says that] it was when the bad habit was introduced of stripping boys entirely in the gymnasium and palaestra that the ugly habit began of loving and pursuing them and that this nasty epidemic spread everywhere.[22] We have here evidence that the idea of a sexual epidemic is extremely old and that it absolutely does not date from the eighteenth century as is sometimes said. Practically, since Greek Antiquity the great denunciation of sexual faults has been made [in the following terms]: it is a novelty that is spreading, an epidemic and dangerous novelty. It is historically true that the nudity of boys in the stadia was something relatively late in Greek societies. Relying on this fact, Plutarch thus defines this love as coming after, as second historically. And as always in these analyses, these mythico-historical genealogies, what is second is of less value. But Plutarch says, which is interesting: Coming after in this way, this love was ashamed of itself, it had to conceal itself, it dared not show itself as it was.[23] And that is why you lovers of boys are so hypocritical that you claim not to have any sexual desire, any *epithumia* for them, that you are not seeking *koinōnia* and *sunousia*. The hypocrisy of the love of boys is thus historically grounded. The need to conceal itself, since it is a bastard love, a love that comes after, a love that crept in following a definite historical conjuncture.

Love of boys is thus deficient by its mythico-historical origin. But it is deficient for another, absolutely fundamental reason. It is fundamental, and here in Plutarch's *Erōtikos* there are two quite decisive passages, which are relatively brief but on which it is necessary to linger for a bit. In the first of these passages, just before the great praise of *erōs*, one of the partisans of the love of women, a certain Daphnaeus (who is not Plutarch's father) says that there is after all an advantage on the side of the love of women. This advantage is that this love of women, while being a love that entails pleasure, physical community, is capable of leading to friendship* through the intermediary of what the text calls, what it designates as *kharis*.[24] *Kharis*, that famous word that is translated as "grace." Of course, it does not mean grace here. The [French] translator says: "*complaisance* [kindness, obligingness]." What actually is involved? The word is important enough for the interlocutor, to whom Plutarch

---

* M. F. adds: this is the them I was just talking about [see above p. 185].

gives this remark, to define what he means by this. He says: *Kharis* is *hupeixis*, the woman's acquiescence, consent to the man. *Hupeikhō* actually means: to give way, make concessions, consent.[25] The *kharis* women have for men is therefore the fact of giving way, but of doing so willingly, granting something in a positive gesture. And here again—which proves how important the notion was for him and how difficult to pick out—Plutarch gives a number of quotations, one in particular from Pindar and the other from Sappho. In the Pindar text—it is in the second Pythian at verse 42, the text is a little distorted by Plutarch, but no matter[26]—it is said that, Zeus having substituted the goddess Nephele (*Nuée, cloud*) for Hera when Ixion wanted to throw himself on her, seize her, and join with her, Nephele had received Ixion's embraces unwillingly, she had received the embrace *aneu karitōn*,[27] precisely without *kharis*, without that willing acquiescence, that agreeable consent that is characteristic of *kharis*. And in a verse from Sappho, it is a matter of a young girl who is still *akharis*, that is to say not yet able to grant her favors with pleasure—in any case with obligingness, for the word pleasure precisely does not come into this.[28] What is the importance of this notion of *kharis*? Why is it so crucial?

I remind you of the problem of the object of pleasure [in] the Greek ethics of pleasures. It seems to me[*] that the Greeks [perceive] the ethics of pleasures solely from the point of view of the one who is active, from the man's point of view, the man who takes his pleasure. The other's pleasure does not enter into it and one does not have to talk about it. Hence the woman's pleasure has to be problematized from either the medical or the moral point of view. However, in terms of the principle of isomorphism, inasmuch as by social status and function the woman has to be object of pleasure, it is precisely part of her role to consent to this, to experiences not directly a pleasure from it but a sort of gratefulness, to accept it with good grace. This notion of *kharis* could be translated as "obligingness" or perhaps as "agreeable consent." *Kharis* as agreeable consent is an absolutely regular and perfectly acceptable way for a woman to identify as subject with her role as object of pleasure. *Kharis* is a way of willingly playing, of very readily agreeing to play the role of object of pleasure. *Kharis* is therefore the bond, the only

---

[*] M. F. adds: I come back to what I said [see above, p. 192].

acceptable bond, that will be established between the man's pleasure, which defines the fundamental element of the sexual ethics, and that hidden pleasure, that unspoken pleasure, that pleasure out of sight that is the woman's pleasure. *Kharis* is the woman recognizing and accepting herself in a field defined entirely by male activity. Of course, commenting on Plutarch's text I have maybe pushed it beyond its limits. In any case, this is what Plutarch's text seems to me to [express] when it says: In love of women, in love with women, there is a *kharis*, and this *kharis* is a source of friendship since, through this obligingness, this agreeable consent, the woman has the possibility of accepting the role one makes her play, and therefore the possibility of experiencing *philia* for the man—and the man, in acknowledgment of the woman's obligingness towards him, is able to conceive *philia* for her. So *kharis* is very precisely the hinge that allows pleasure and friendship to be connected to each other. It is *kharis* that means that the woman has friendship for the man who takes his pleasure in her. The woman's *kharis* is what enables the man to acknowledge that the woman has friendship for him, and consequently to conceive that it is possible to have friendship for someone who, after all, is an object of pleasure. There is the place of *kharis*.

And now, why is it that the love of boys [cannot] give rise to *kharis*? I refer you to what I told you last time in a previous lecture:[29] the boy is precisely someone who cannot permit himself to identify himself as the object of pleasure he is since, as boy, he is someone who will become a free citizen. Since one day he will be an active subject, since he will have to be a soldier, a politician, a family father, since he will have to be an orator also, he cannot have been someone who was an object of pleasure. And what happens when he is an object of pleasure? It can only be against his consent since, if he were to consent, he would accept being that for which he was not made, that is to say object of pleasure. Consequently, there is what may be called the boy's paradox, which is this: he is an object of pleasure, but cannot consent to being an object of pleasure.[30] And when, in actual fact, he is seized by a man, when he is possessed by a man, he [is faced with the following alternative]: if he consents he becomes a pure and simple prostitute, someone at any rate for whom one cannot have any respect; if he resists, as soon as he resists he will not be able to have any feelings other than hostility towards the person who forced him.[31] In both cases, *philia*, as highly valued social bond, does not have a place. Either the boy consented: he is no more than a prostitute; one can have no *philia* for him. Or the boy was taken with violence. As a result he can have no *philia* for the one who possessed him. So there can be no *kharis* in love between man and boy.

We have here a fundamental point for understanding what will now be the exclusion of the man-boy relationship from the single great chain of love that Plutarch had defined. Why, how did this single great chain of love, in which therefore friendship and desire, pleasure and virtue, etcetera were connected to each other, hold together? Precisely by the fact that the woman experienced *kharis*, that there was this *kharis*, this obligingness, this agreeable consent. This is what assured the synthesis of *philia* and *epithumia*, of friendship and desire, of pleasure and virtue. The boy cannot [experience] *kharis*. There cannot be that agreeable consent in the relationship between boy and man. Consequently the chain no longer holds and there can no longer be *epithumia* and *philia*. In their case there cannot be both *hēdonē* and *aretē*. They have to choose one or the other, and if virtue is chosen, love will be *atelēs*, it will not arrive at its completion. On the other hand, if love goes so far as *koinōnia*, then, as Plutarch says referring to Plato, it will be a love of the quadrupeds.[32]

So it should not be thought that the Greeks were people who welcomed love of boys in the great figure of a love unconcerned about the sex to which they might address themselves. You see that in the classical conception of *aphrodisia*, Greek philosophical and moral reflection welcomed the love of boys only on condition of clearly separating love of boys and love of women from each other. For the Greeks—even and especially for those of the classical period—love of boys could be admitted, accepted, thought, and reflected only as another love, a profoundly different love, the theory of which it was necessary to reflect on and produce, the theory therefore that goes entirely outside nature, desire, and pleasure. And when, in this Hellenistic and Roman civilization of which Plutarch is the witness, a great single and complete love was formed, when one succeeded in constructing a conception in which friendship and desire, pleasure and virtue could be joined together, [then] the love of boys could not fail to be ruled out. The love of boys became *aneu kharitos*, a disgraced love, a love that, from that point and throughout Christian civilization, lacked grace (*kharis*).

I am going to stop here because from this point things become easy. So you understand that on the basis of this constitution of the great chain of the single love, from which the love of boys was excluded because literally disgraced, it is easy for Plutarch to show that, opposite it, the love between spouses is precisely the only one to realize the great chain of the single and complete love. Plutarch's text runs by itself, it's enough to quote it: "Physical union with a wife is a source of friendship, liked shared participation in great mysteries. Pleasure (*hēdonē*) is of short duration, but it is like the seed from which day by day springs ..."[33] What? *Timē* (respect), *kharis* (agreeable consent), *agapēsis* (both love and friendship), and *pistis* (trust).[34] That is to say the relationship between man and wife is thus charged with all the elements (*timē, philia, pistis*, and so on) that characterized the relationships between men, and that were those by which previously the love of boys could be distinguished from the love of women.

On this basis, since conjugal love between man and woman thus inherits all these virtues, we understand that Bacchon's marriage could be blessed by the gods, despite the paradoxical situation of the rich, powerful, widowed matron having abducted this appetizing youth against everyone's will. Not only is such a marriage blessed by the gods, but Plutarch explains how it will be able to serve as a pedagogical framework as it were, benefiting from what could be called positive pederastic effects, from the fact that all the tasks of education previously attributed to man-boy relations will be combined in the affective bonds of man and woman. This old, rich, experienced woman will be able to guide her young husband.[35] All her wisdom, all the lead she has over her husband, as it were, in the realm of reflection, knowledge, and experience will enable her to serve, without this causing any scandal, as guide for her slightly too young husband, until the day when their roles will be gently and imperceptibly reversed and finally it will be the husband who will supplant the woman, finally having acquired the experience and authority that will enable him to do so. Experience and authority that Plutarch emphasizes straightaway are necessarily moderate precisely because the husband will have been under the moral tutelage of his wife for a long time. Those imperious, authoritarian husbands who treat their wives badly, those husbands who do not recognize the necessary community of the two partners, the couple, those husbands who treat their wives only as objects, who are not capable of a genuine *sumbiosis* ([shared] life), belong of course to the first morality of *aphrodisia*. On the contrary, now that a wife will be able to teach her husband to behave, he will of course not behave badly with her and will be able to moderate his authority over her, since in the end it was she who taught him this authority. And this is how the *sumbiosis* that appeared so improbable in the case of an older woman with a young husband will actually be realized so much better in a situation like this. And as a result the dialogue can come to an end. The garlands are prepared for the feast at which the unity of the couple will be assured. Meanwhile, the dialogue's discussants prepare to take part in the feast and to mock the graying old man who had pursued Bacchon and now no longer has any role to play.

1. See Plutarch, *Dialogue sur l'amour*; *The Dialogue on Love*.
2. See ibid., 751e, 758d-e, 764a, Fr., p. 55, p. 73, p. 87; Eng., pp. 322-323, pp. 360-363, pp. 396-397.
3. Ibid., 755a, Fr., p. 63; Eng., pp. 340-341.
4. Ibid., 749d, Fr., p. 50: "She [Ismenodora] heard and said many good things about him, and she saw that many men of merit sought Bacchon's love; she too came to love him, but there was nothing dishonorable (*agennes*) in her intentions: she wanted to marry him publicly and live with him"; Eng., pp. 310-311: "what with hearing, what with saying many kind things about him and observing the throng of noble lovers who courted him, she was carried so far as to fall in love with him herself. Her intentions were far from dishonourable: she desired to marry him and be his companion for life."
5. See the first statement by Epictetus in his *Encheiridion* or *Manual*: "Depend on ourselves the judgment of value, the impulse to action (*hormē*), desire (*orexis*) or aversion, in a word everything that is our own work" (P. Hadot's translation); English translation W. A. Oldfather, Epictetus, *Echeiridion* in *The Discourses*, vol. II, pp. 482-483: "Under our control are conception, choice, desire, aversion, and, in a word, everything that is our own doing." On these concepts, see A. J. Voelke, *L'Idée de volonté dans le stoïcisme* (Paris: PUF, 1973) and P. Hadot, *The Inner Citadel. The Meditations of Marcus Aurelius*, trans., Michael Chase (Cambridge, MA: Harvard University Press, 1998), especially the chapters "The Discipline of Desire, or *Amor Fati*" and "The Discipline of Action, or Action in the Service of Mankind."
6. Plutarch, *Dialogue sur l'amour*, 750c-d, p. 52: "As for me, I do not call love (*eran*) the feeling you experience for women or young girls, just as we do not say that flies feel love of milk or bees for honey, or that the breeders and cooks love the calves and fowl they fatten up out of the light! Nature excites in us an appetite (*orexin*) for bread and food that is moderate and sufficient, but the excessive and passionate desire for food is called greed or gluttony. In the same way nature has placed in us the need for the mutual pleasure (*ap'allēlōn hēdonēs*) man and woman give each other, but the desire (*hormē*) that carries us away, when it becomes violent, passionate, unbridled is certainly not worthy to be called love": *The Dialogue on Love*, pp. 314-315: "I deny that it is love that you have felt for women and girls—any more than flies feel love of milk or bees for honey or than caterers and cooks have tender emotions for the calves and fowls they fatten in the dark. In a normal state one's desire for bread and meat is moderate, yet sufficient; but abnormal indulgence of this desire creates the vicious habit called gluttony and gormandizing. In just the same way there normally exists in men and women a need for the pleasure derived from each other; but when the impulse that drives us to this goal is so vigorous and powerful that it becomes torrential and almost out of control, it is a mistake to give the name Love to it."
7. Ibid., 750d, Fr., p. 52: "Love which attaches itself to a young and well-endowed soul leads to virtue (*eis aretēn*) by way of friendship (*dia philias*); Eng., pp. 314-317: "Love, in fact, it is that attaches himself to a young and talented soul and through friendship brings it to a state of virtue"; ibid., 750e, Fr., p. 52: "Love (*erōs*), when it loses the hope of inspiring affection, ceases to beset with its cares the fleeting brilliance of youth in bloom, if the latter does not produce the proper fruit of its character, proof of friendship and virtue (*philian kai aretēn*); Eng., pp. 316-317: "Love, if he loses the hope of inspiring friendship, has no wish to remain cultivating a deficient plant which has come to its prime, if the plant cannot yield the proper fruit of character to produce friendship and virtue."
8. On the love of boys, see Plato, *The Symposium*, 181a sq. and 192a-b. The expression *para phusin* is used in the *Phaedrus*, but with a negative connotation (relating more to the position of passivity than to the homosexual relationship as such). See Plato, *Phaedrus*, 250e, trans., Hugh Tredennick in Plato, *The Collected Dialogues*, ed., Edith Hamilton and Huntington Cairns (Princeton, NJ: Princeton University Press, Bollingen Series LXXI, 1963) p. 497: "Now he whose vision of the mystery is long past, or whose purity has been sullied, cannot pass swiftly hence to see beauty's self yonder, when he beholds that which is called beautiful here; wherefore he looks upon it with no reverence, and surrendering to pleasure he essays to go after the fashion of the four-footed beast, and to beget offspring of the flesh, or consorting with wantonness he has no fear nor shame in running after the unnatural (*para phusin*) pleasure." See also M. Foucault, *L'Usage des plaisirs*, ch. 4, "Érotique," pp. 205-248; *The Use of Pleasure*, "Erotics," pp. 187-225.

9. Lucian of Somosata [Lucien de Samostate], *Les Amours*, 22, in *Dialogues des courtisanes*, followed by *Amours* and *Taxaris*, trans., P. Maréchaux (Paris: Arléa, 1998) p. 85: "If each remained faithful to the laws that Providence has prescribed for us, we would content ourselves with the intercourse of women and our life would be purified of all infamy. Look at the animals: they cannot corrupt anything by a vicious disposition and the legislation of nature remains pure. Lions do not burn for other lions"; English translation M. D. MacLeod, "Affairs of the Heart (Amores)," in *Lucian Volume VIII* (Cambridge, MA: Harvard University Press, Loeb Classical Library, 432, 1967) pp. 184-185: "If each man abided by the ordinances prescribed for us by Providence, we should be satisfied with intercourse with women and life would be uncorrupted by anything shameful. Certainly, among animals incapable of debasing anything through depravity of disposition the laws of nature are preserved undefiled." According to this quotation, Lucian's dialogue is not as favorable to the love of boys (indeed quite the contrary) as Foucault allows us to assume here; on this dialogue see *Le Souci de soi*, pp. 243-261; *The Care of the Self*, pp. 211-227.

10. Plutarch, *Dialogue sur l'amour*, 750d, p. 52: "These desires (*epithumiai*) experienced for women, even in the best of cases, permit only the enjoyment of sensual pleasures and fleeting pleasures of the body (*hēdonēn periesti karpousthai kai apolausin hōras kai sōmatos*)"; *The Dialogue on Love*, pp. 316-317: "but the appetite for women we are speaking of, however well it turns out, has for net gain only an accrual of pleasure in the enjoyment of a ripe physical beauty"; ibid., Fr., p. 52: "The end of desire (*epithumia*) is only pleasure (*hēdonē*) and enjoyment (*apolausis*)"; Eng., pp. 316-317: "The object of desire is, in fact, pleasure and enjoyment."

11. Ibid., 751a, Fr., p. 53: "You will see [the love of boys (*paidikos erōs*)] always simple and without indulgence, frequenting the meetings of philosophers or, sometimes, the gymnasia and palaestrae, in search of young people, in order to exhort to virtue (*pros aretēn*), with strong and noble voice, those who are worthy of its care (*tois axiois epimeleias*)"; Eng., pp. 318-319: "Love, the love of boys ... is simple and unspoiled. You will see it in the schools of philosophy, or perhaps in the gymnasia and palaestrae, searching for young men whom it cheers on with a clear and noble cry to the pursuit of virtue when they are found worthy of its attention."

12. On the notion of *epimeleisthai* (caring for, taking care of oneself and others, caring about), see the lectures *L'Herméneutique du sujet; The Hermeneutics of the Subject*. Associating the figure of the master and the lover of boys, Foucault asserts: "The master is the person who cares about the subject's care for himself, and who finds in his love of his disciple the possibility of caring for the disciple's care of himself" (Fr., p. 58; Eng., p. 59).

13. Plutarch, *Dialogue sur l'amour*, 751b, pp. 53-54: "But this other love, enervating and housebound (*hugron douton kai oikouron*), attached to the dresses and beds of women, always seeking sensual pleasure and pleasures unworthy of a man, without friendship (*aphilois*), without enthusiasm, should be proscribed. As Solon did: he prohibited the love of boys and gymnastics to slaves, while allowing them intercourse with women; in fact, as much as friendship (*philia*) is beautiful and shared, the love of a slave for boys could be neither noble nor honorable, for this love is a carnal union (*sunousia*), like the love of women"; *The Dialogue on Love*, pp. 318-319: "But that other lax and housebound love, that spends its time in the bosoms and beds of women, ever pursuing a soft life, enervated amid pleasure devoid of manliness and friendship and inspiration—it should be proscribed, as in fact Solon did proscribe it. He forbade slaves to make love to boys or to have a rubdown, but he did not restrict their intercourse with women. For friendship is a beautiful and courteous relationship, but mere pleasure is base and unworthy of a free man. For this reason also it is not gentlemanly or urbane to make love to slave boys: such a love is mere copulation, like the love of women."

14. In fact, in this dialogue Plutarch stages his own son (Autobulus) as narrator and reporter of a conversation between his father (Plutarch) and several protagonists (including Daphnaeus, son of Archidamus, who shares with Plutarch the same suspicion towards the love of boys). Foucault will correct this mistake himself later in the lecture.

15. Plutarch, *Dialogue sur l'amour*, 751b-c and 752a pp. 54-55; *The Dialogue on Love*, pp. 320-321 and pp. 324-325.

16. Ibid., 766e, Fr., p. 95; Eng., p. 413: "the causes that they give for the generation of love are peculiar to neither sex and common to both."

17. Ibid., Eng., "For is it really the case that visual shapes emanating from boys can, but the same from women cannot, enter into the body of the lover where, coursing through him, they

stimulate and tickle the whole mass and, by gliding along with the other configurations of atoms, produce seed?"

18. Ibid., Eng., "... those beautiful and sacred passions which we call recollections of the divine, the true, the Olympian beauty of the other world, by which the soul is made winged ..."

19. Ibid., 752b, Fr., p. 56: "If, therefore, there is a Love (*Erōs*) without Aphrodite (*khōris Aphroditēs*), it is like drunkenness without wine"; Eng., pp. 324-325: "But if, on the other hand, there is an Eros without Aphroditē, then it is like drunkenness without wine."

20. Ibid., Fr., "[a Love without Aphrodite] can only be an effort without fruit and without fullness (*akarpon autou kai atelēs*)"; Eng., "No fruit, no fulfilment comes of the passion."

21. Ibid., 759e, Fr., p. 76: "The work of Aphrodite, when Love is absent, is bought for a drachma"; Eng., p. 369: "the work of Aphroditē, if Love is not present, can be bought for a drachma."

22. Ibid., 751f-752a, Fr., pp. 55-56; Eng., pp. 322-325.

23. Ibid., 752a, Fr., p. 56: "If the love of boys denies sensual pleasure, it is because it is ashamed and fears punishment": Eng., pp. 324-325: "Boy-love denies pleasure; that is because it is ashamed and afraid."

24. Ibid., 751c-d, Fr., p. 54: "If the unnatural union of males does not destroy loving tenderness or harm it, with all the more reason the love of men and women in accordance with nature leads to friendship by way of obligingness/kindness (*eis philian dia kharitos*)"; Eng., pp. 320-321: "if union contrary to nature with males does not destroy or curtail a lover's tenderness, it stands to reason that the love between men and women, being normal and natural, will be conducive to friendship developing in due course from favour [*kharitos*]."

25. Ibid., 751d, Fr., pp. 54-55: "For the Ancients, Protogenes, called the woman's acquiescence to the man's desire 'obligingness (*complaisance*)'"; Eng., pp. 320-321: "For you see, Protogenes, a woman's yielding to a man was called by the ancients 'favour.'"

26. See the editor's note in the Belles Lettres edition, p. 135: "The expression is read in fact in Pindar, *Pyth.* 2, 42, but it is applied to the cloud that Zeus substitutes for Hera when Ixion wishes to embrace this goddess." This is what Foucault precisely recounts also, whereas Plutarch in his dialogue writes, 751d, Fr., p. 56: "Pindar says that Hera conceived Hephaestus 'without obligingness (*complaisance*)'"; Eng., p. 321: "Pindar declared that Hephaestus was born from Hera 'without favour.'" [W.C. Helmbold notes that this quotation is "perhaps a confusion of *Pyth.* ii, 42 with Hesiod, *Theogony*, 927; G.B.]

27. Ibid.

28. Ibid., Fr., "Sappho addresses a young girl, who is still not of marriageable age, in these words: 'In my eyes you were only a young girl/To whom obligingness was still foreign (*smikra moi pais emmen ephaineo kakharis*)'"; Eng., "And Sappho addressed a young girl not yet ripe for marriage: 'You seemed to me a small child without favour.'"

29. See above, the end of the lecture of 28 January and the beginning of that of 4 February.

30. On this point, see Foucault's arguments in *L'Usage des plaisirs*, the chapter "L'honneur d'un garçon," pp. 225-236; *The Use of Pleasure*, "A Boy's Honor," pp. 204-225.

31. Plutarch, *Dialogue sul l'amour*, 768f, Fr., p. 100: "Thus the young men who willingly agree to serve as instruments of such debauchery are ranked in the category of the most degraded, and we not grant them the least parcel our trust, respect, or friendship ... As for young men who, without being naturally perverse, have been induced to give way and surrender their bodies, there are no men who inspire in them more distance and hatred than those who have abused them, and they take a terrible vengeance on them when given the opportunity"; Eng., pp. 424-425: "That is why we class those that enjoy the passive part as belonging to the lowest depth of vice and allow them not the least degree of confidence or respect or friendship ... Young men not naturally vicious, who have been lured or forced into yielding and letting themselves be abused, forever after mistrust and hate no one on earth more than the men who so served them and, if opportunity offers, they take a terrible vengeance."

32. Ibid., 751e, Fr., p. 55: "Boys, when their favors are taken by violence, are victims of an assault, and when they consent, through libertinage and inversion, according to Plato's words 'to allow themselves be covered and impregnated like quadrupeds' contrary to nature, this is an absolutely 'offensive' obligingness, indecent and ignoble"; Eng., pp. 322-323: "But to consort with males (whether without consent, in which case it involves violence and brigandage; or if with consent, there is still weakness and effeminacy on the part of those who, contrary to

202 SUBJECTIVITY AND TRUTH

nature, allow themselves in Plato's words 'to be covered and mounted like cattle')—this is completely ill-favoured favour, indecent, an unlovely affront to Aphrodite."

33. Ibid., 769a, Fr., p. 100; Eng., pp. 426-427: "... in the case of lawful wives, physical union is the beginning of friendship, a sharing, as it were, in great mysteries; but the respect and kindness and mutual affection and loyalty that daily spring from it ..."

34. Ibid., Fr., "... mutual respect, obligingness, affection, and trust (timē kai kharis kai agapēsis allēlōn kai pistis); Eng., see previous note.

35. Ibid., 754d, Fr., p. 62; Eng., p. 339.

nine

11 MARCH 1981

[
The new ethics of marriage. ⌐ *Evolution of matrimonial practices:
the historians' point of view.* ⌐ *Institutional publicization, social
extension, transformation of the relationship between spouses.* ⌐
The evidence of writers: the poems of Statius and Pliny's letters.
⌐ *Games of truth and reality of practices.*
]

STARTING FROM SOME TEXTS of the first two centuries CE, we
have picked out some quite important things. First, of course, the
general development of what is ultimately a highly restrictive system
of sexual acts and pleasures, what the Greeks called *aphrodisia*. Second,
the definition of a regime that localizes the legitimate use of *aphrodisia*
in marriage and only in marriage. Third, the formation of an internal
ethics of marriage in which there is a very strong connection between
the traditional roles of husband and wife (the economic and social
role of the two partners in relation to each other), but also affective
relations (community of thought, tenderness, affection, and so on, all
that the Stoics called the *krase* of marriage), and then also sexual relations, which should be governed by rules of decency, forms of intensity,
as well as serve as the starting point and forms of expression of these
affective ties.

On the basis of this new marriage ethics, linking traditional roles,
affective ties, and sexual relations, a general and canonical form of love

was formed* through the intermediary of what the Greeks called *kharis*. Through the intermediary of *kharis*, that sort of obligingness, consent, and reciprocal pleasure each takes in the other's pleasure, we see the definition of a general love that enables *erōs* and *aphrodisia* to be linked, a synthesis that was so difficult in classical Greek thought. And this definition of a love that links *erōs* and *aphrodisia* through the intermediary of *kharis* has the effect of accommodating, tolerating, and accepting as complete, standard, and regular only love between man and woman—to tell the truth: between husband and wife—and of excluding love between man and boy, which is defined as incomplete and insufficient because of the inability to bring about the *erōs-aphrodisia* synthesis (the synthesis of love and sexual pleasures) due to the inevitable absence of *kharis* (of obligingness). This transformation in the representation, in the definition of the way of conducting oneself in relation to *aphrodisia* as well as to marriage, raises two questions.

First, what is the connection, the relation between these techniques† and the reality within which they appear and are defined? A question of synchrony. And then the second question, which I will broach another time, is the fate of these techniques, these regimes of life, these modes of existence in Christianity. So the first question that I would like to talk about today: where does this new regime of *aphrodisia*, now so profoundly linked to marriage and the man/woman relationship, come from? What is it that can give rise to this type of thought, this mode of prescription? I will formulate the question in an intentionally ambiguous and naive way: what relationship can this new regime of *aphrodisia* have with what is called reality, what relationship can it have with the actual behavior of individuals? You will see that as the investigation makes some headway we will clearly need to develop this question and see how it is ultimately an inadequate and false question. But still, I think we may now turn somewhat hastily to the historians and question them on what it is possible to know regarding the social and daily life of Greek and Roman populations,

---

* M. F. adds: this is what I tried to show you last week, at least in someone like Plutarch (see above, lecture of 4 March p. 193 sq.].
† M. F. adds: that maybe I have tried to set out in too much detail.

or of Mediterranean populations, in the Hellenistic and also imperial Roman periods.[1]

What can be known on this subject is very fragmentary, certainly geographically, since apart from some particularly or relatively well illuminated areas like Rome or Greece, and Egypt also, we do not know a great deal about the rest. The historians' ray of light illuminates what are ultimately only some very limited geographical areas and social strata. This is inevitably the behavior of elites, of the elites who speak and of the elites to which one speaks, of the elites who leave a trace of some part of their existence, and it is only about these classes that we can get some useful information. So the information that historians can use is inevitably fragmentary. Very rapidly, let us try to see what they have been able to deduce from these fragmentary indications regarding what one might suppose the general and overall evolution of both sexual behavior and marriage to have been. Without pushing the results they have obtained too far, it seems to me that we can note three things: modifications in the matrimonial institution; modification also in its practice and the spread of this practice; modification finally in the type of relationship established between husband and wife in the matrimonial institution.

The first modification identified by historians concerns the matrimonial institution itself. One must keep in mind that in classical Greece, and also in Rome until the end of the Republic, marriage was a private act that was a matter for the family and only the family, that is to say a matter of family authority, of the rules commonly accepted in the family or in that type of family, and finally of the economic and social strategies peculiar to these families. Marriage was a transaction between two heads of the family: the father of the future wife, and the future husband, seen as prospective head of the family. The act of marriage took place as a private transaction between these two family heads, or between these two families; consequently the public authority had nothing to do with it. It had so little to do with it moreover that marriage, and here I refer to Paul Veyne's studies on the subject, was not considered a juridical act.[2] It was a private act that no doubt had legal effects on the individuals, but that was not in itself a juridical act. Now things changed. They changed in the last centuries BCE.

One sees them changing—here* I am only passing on to you a somewhat synthetic point of view on a whole series of works by historians³—in Egypt for example.⁴ Between the third century BCE and the first centuries CE, during the whole intermediate period of the Romanization of Egypt, marriage became a sort of public institution. Public institutionalization [then], publicization of marriage that takes place through the important intermediary of religion. Of course, marriage as a private act had a religious dimension inasmuch as one of the reasons it was necessary to marry and have legitimate heirs was the continuation of the cult of ancestors. But it was the family religion itself that required marriage, and it was within that familial, as it were religious strategy of the family itself, that marriage got its meaning. What we see emerging in the Greek communities of pre-Roman Egypt is a public institutionalization of marriage, that is to say it now becomes an episode of private life inscribed and sanctioned in religious ceremonies that form part of public rituals. It is the public religion that sanctions the marriage. And through this relay we see marriage become increasingly a public institution that ultimately will also be sanctioned in civil institutions.

Let us take the case of Rome a bit later, that is to say in the Augustan period, when marriage was the object of a whole series of important laws. These laws concern, for example, depravity, that is to say rape, and also adultery. These Augustan laws are traditionally interpreted in terms of moralization. It was thought to have been a matter of Augustus moralizing a Roman society that had become slack in its ethics, in its observance of the fundamental moral principles. In fact, things should be seen a bit differently, and especially if one takes for example the famous law against adultery (*lex de adulteriis*).⁵ The law *De adulteriis* condemns any married woman who has a relationship with a man [whoever he may be, other than her husband]. On the other hand, the man, married or not makes no difference, is condemned by the *lex de adulteriis* only if he has sex with a married woman. So it is still this problem of the married woman that is at the center of all this legislation on adultery. The married woman cannot have a relationship with any [other] man and the man cannot have a

---

* M. F. adds: here again, I am merely pointing out some indications, the clearest, most decisive, I am not going into details ...

relationship with a married woman. Now, when we were talking about the moral perception of *aphrodisia* in classical Greek thought, we saw precisely these same rules at work. That is to say that the condemnation of adultery on the basis of the status of the married woman, and on the basis of this status only, was absolutely taken for granted by what could be called the social morality of the period. When the Augustan law was formulated as it was, it did nothing other than take up what was in actual fact part of the current morality, with the practical effects that this might have inasmuch as families, up to a point at least, could themselves sanction offences against their own morality. So it would seem that the Augustan law regarding adultery does not mark a new moralization so much as a certain way of giving legal sanction to a set of ethical principles that were already perfectly recognized by morality, and by current morality. The Augustan law transforms into an offence subject to sanctions imposed by the public authority an offence recognized for a long time by morality and according to the same principles of this morality. So it seems that much more than a moralization of behavior—although to some extent this perspective, this intention [was subjacent]—the function, the effect anyway of this Augustan legislation was to publicize the private transaction of marriage and to publicize by law the moral principles that governed its practice and custom.

This institutional transformation that increasingly made marriage a public act is linked to a number of modifications. It is, as it were, only their expression, or also their relay and instrument. By this [I mean] first and foremost of course, the spread of matrimonial practice. You will understand why publicization is the correlative of the spread of matrimonial practice, that the two phenomena are linked. In fact, as marriage was a private act between families, it is quite understandable that it was ultimately a fairly limited practice. People did not really marry much, as marriage was essentially seen, perceived, and organized as a private act between families. This is because, as a private act between families, it corresponded to private objectives, tactics, and strategies. It was a matter of passing property from one family to another, or rather from generation to generation. A name also had to be perpetuated. A caste had to be perpetuated in relation to other castes of society. Family cults had to be maintained, as I was just telling you.

And it is easy to understand why this kind of objective, this kind of matrimonial strategy had meaning only for those with property to pass on, those for whom social alliances of power had meaning and effects. To marry, one had to have an interest in marrying. The interest in marrying was linked to family strategies and these only had meaning, only [offered] possibilities of course to a very limited number of people. Marriage was an elite practice. It could only be a very elitist practice with strong meaning and value for no more than a small stratum of citizens, the highest, most powerful, most influential stratum certainly, but at the same time the most limited of the social body. And then, when marriage becomes a public type of institution, when it ceases to be a private institution, you can see why it separates off from familial strategies. Conversely, you can see that when marriage separates off from these strategies, then as a result it no longer seeks its sanction from a private act between families. Those who marry will look for the security and guarantee of their marriage from, and seek to inscribe it in, a public institution rather than in an institutional game between families.

The spread of marriage as a practice and the publicization of the institutional form of marriage are two correlative phenomena. Marriage becomes an increasingly less elitist practice. It becomes a practice increasingly open to common mortals. One does not need to have family objectives, very precise social objectives in order to treat oneself to the luxury of a marriage. This double movement of publicization of the marriage institution and spread of the practice of marriage is linked, of course, to a number of phenomena, to economic, political, and social processes and so on. Let's say that the development of the monarchies in the Hellenistic period, also the development of autocratic power at Rome, or [again] the diminished role of families in the cities, in political life, makes the intra- and inter-family strategies less and less important. We will add to this, no longer taking things from the side of the higher strata of society but from the mass of the population, a whole urban development notable in Hellenistic Greece, [but] also in Rome, [that] brings a rural population into the towns. It brings to Rome a provincial population [for which] marriage was much more traditional, for in the countryside it had a much clearer economic function, even for poor people. And this urbanized, poor population will find much more favorable conditions of existence in the matrimonial structure than by remaining single. So, again going by the documents that historians can use, we can accept that there was an [increase in] the real practice of marriage [and], together with this, its institution-alization as a public act.

What is much more important for us is that a modification of the relationship between husband and wife is linked to this spread of matrimonial practice. This modification, such as we can observe it, appears in two ways. First, it seems that matrimonial practice increasingly involved the free consent of the man and woman, of the two partners. From a juridical and institutional point of view, the well-known act of *ekdosis* in Greek marriage—that is to say the act by which the father gave the woman to the future spouse by way of a system of exchange, the act that marked both the father's authority over his daughter and the husband's future authority—disappears, or at any rate tends to disappear. And, according to the contracts that it has been possible to preserve, above all for purely material reasons, contracts [coming from] Greek communities in Egypt, we see that, increasingly, the woman who has received her dowry at the moment of marriage retains possession of it [and] it remains at her free disposal in the marriage. When the marriage was dissolved, the contract provided—for the contracts always provided for what was going to happen when there might be or had to be divorce—that the young woman would keep her dowry possessions. It even happened that she could keep what will be called the acquests of marriage; which gives the woman a much more favorable juridical and economic position than in the past. In the same order of ideas concerning the autonomization of the two partners, and especially of the woman, we also see very interesting court decisions, still in Egyptian documents from the Roman period, still in Hellenistic communities, but in the Roman period in Egyptian territory. For example, the old right of the father of the young wife to dissolve the marriage in an authoritarian way if it did not suit him or no longer suited him disappears and court decisions take into account the woman's wishes and grant divorce only if she herself shares her father's views and also wishes to dissolve her own marriage. So, in this [ever] more public legal act, husband and wife increasingly appear as two partners whose own wills constitute the fundamental element.[6]

What I would like to focus on obviously a bit more[*] is that within these marriage acts, which clearly show that the wishes of both partners are taken into account, we see emerging the clear and precise codification of relations between husband and wife concerning their daily life, and even their sexual life. The marriage contracts that have been preserved from Egyptian documents are actually very detailed. Some have been found that cover the period going from the third century, that is to say from the Hellenistic period, to the Roman period. Comparison between documents taken from the beginning and end of the period show quite a transformation, an interesting evolution. In the marriage contracts of the third century BCE, the codification of relations between husband and wife is really relatively precise, but much less so than later, however. In these contracts of the third century we find, for example, the definition of the husband's and wife's obligations. The obligations of husband and wife are obviously dissymmetrical, they are not the same [but] contain nevertheless some complementarities. In the contract the wife's obedience to her husband is, of course, explicitly required. Second, the wife is forbidden to leave the house, either at night or during the day, without her husband's authorization and consent. Third, still in the terms of the contract, any sexual relation of the wife with a man [whoever he may be, other than her husband] is excluded. The obligation not to ruin the household is prescribed, that is to say to manage or take part in the management of the household in such a way that the economic situation of the couple, of the family is not compromised. And finally, a general prohibition of[†] dishonoring the husband. This clearly means, as well as [the prohibition of any sexual relationship outside marriage,] the general prescription to behave herself in according to her status as a married woman. What are the obligations and prescriptions on the man's side? The contracts make clear that the man must [first] maintain his wife, [that is to say] assure her material conditions of life. Second, he must not have a concubine in the house. Third, he must not mistreat his wife. And finally [he must not] have children from a relationship outside the marriage. For the domains that concern us we have then

---

[*] M. F. adds: because we come back here finally to the subject I want to deal with.
[†] M. F. says: of not ...

these two prohibitions: no concubine in the house, no children from a relationship outside the marriage. The second prohibition has meaning only insofar as these children would be, if not legally recognized, at least effectively maintained by the father. In sum, the father must not have a family outside the one he is in the process of forming, and no concubine in the house, that is to say a status of de facto bigamy. These are the obligations of man and wife in the third century [BCE].

In the later contracts we note that things have changed, not so much on the side of the wife's obligations, which are more or less similar, but on the side of the husband's obligations, which are really much more precise. First, the modes of material, economic maintenance that the husband must assure the wife are specified in terms of the quantity of the dowry and his own wealth, etcetera. Economic details. And then, on another side, in the marriage contracts of the period from the beginning of the Roman Empire, we see that the husband undertakes not to have any mistress in general. Of course, this does not mean the husband is forbidden all sexual relations outside marriage, but that in any case he cannot have a permanent sexual tie with a woman other than his wife. He must not have, and this clause obviously never appeared in the contracts of the earlier period, a *paidikon*. He cannot have a sweetheart. He cannot have a boy with whom he would have constant and permanent relations. And finally, he must not have a home [other than his own,] that is to say a place where he might keep a sweetheart, a mistress, and so on. You can see that in this kind of contract the husband is still not forbidden sexual relations [outside] marriage. This is of course forbidden for the wife, not for the husband. But these relations are passed over in silence, and so tolerated, only to the extent that they are not permanent or are merely sexual acts engaged in casually. What absolutely must not appear in conjugal life, what absolutely must not weigh in the slightest on conjugal life are those no doubt episodic but relatively permanent characters with whom one may be linked sexually: the mistress, the boy, and so on. There is certainly no obligation of sexual fidelity as we now understand that, and in the sense moreover that the Stoics understood it in the arts of living I spoke to you about in previous lectures. So it is not a question of sexual fidelity. However, it is a matter of fidelity of existence in which the sexual dimension has an important place. In one's existence, including the sexual relationship, one must be faithful to one's wife in the sense that one must not have a double of the [wife] in the form of a boy, a mistress, and so on. So sexual activity is not confined to marriage alone, [but on condition that it does not involve a] permanent sexual bond. It is impossible to have a sexual bond and a form of shared existence with someone other than one's wife. The new element in the formation of this new morality is the idea that sexual activity and shared existence belong to each other. What is important is not sexual fidelity strictly speaking.[7]

From all this and other documents of the same type—here again I am doing absolutely no more than referring to some studies by historians—we can see the appearance of what Paul Veyne calls a "new idea" in this period, that is to say: the couple.[8] Or again, in somewhat different terminology, it could be said that in a complex social field in which individuals were [joined together] by ties of family and caste, by adherence to the place where one lives, by various forms of comradeship, marriage in this period was certainly not an invention, a discovery. When Paul Veyne speaks of the couple, he is not talking about marriage. Marriage is not a discovery, but the practice of marriage is generalized and at the same time becomes an increasingly intense, specific bond irreducible to all other social ties, and there is a very precise relationship between the marriage bond and the sexual bond, the sexual relationship. This is what actually constitutes the novelty of the couple.

We could find other signs that the family is becoming a new reality in this period. Apart from these institutional signs studied by historians, the literature of the period could also give us some signs and evidence. I will dwell on some of these texts because they make it possible to advance a little in the analysis of this phenomenon, and especially to situate better what I shall be able to explain to you, that is to say the role of these arts of living in relation to these social processes I have been talking about. The literary texts I shall rely on concerning the emergence of this new reality of the couple, of the married couple, are, on the one hand, the letters of Pliny, which are well-known, and on the other some poems by Statius from a slightly later period.[9] What do we find in these texts (the letters of Pliny and the poems of Statius)? First of all we find a notion that we have already come across in the arts of

living of the Stoics and other philosophers, that is to say the notion of
concord* (*concordia*), extolled as the fundamental, determinant element
of marriage. For example, regarding a friend called Macrinus, Pliny
says that Macrinus and his wife have spent thirty nine years together
without quarrel or dispute.[10] And he writes to his own wife telling her
that all her conduct inspires in him "a hope full of confidence" that
their concord (*concordia*) "will be endless and grow daily."[11] Another
text, not by Pliny this time, but Tacitus, in the *Life of Agricola* where,
with reference to Agricola and his wife Domitia, he says the following, which is very interesting because in it we find the exact transposition of the notions we have encountered in the Stoics: Agricola and his
wife have lived all their life in a *mira concordia*, a remarkable, admirable
concord, and this *concordia* was assured *per mutuam caritatem*—one cannot
say "by charity," one finds the word *kharis* anyway: by a sort of mutual
affective bond.[12] This notion of *caritas* in Tacitus seems roughly intermediary between the notion of *kharis* I was talking to you about and
the notion of *eunoia*, the benevolence each has for the other. So they
have this mutual *caritas*, each preferring the other to him or herself. We
have already come across this idea in the Stoics of a competition in the
love of each for the other, and which means that each tries to think
of the other more than the other thinks of them. So competition in
the devotion, *caritas*, obligingness of each for the other. So that is what
characterizes *concordia* for Tacitus.

In these texts we also find the idea that this bond [of] *concordia*, this
bond between man and wife contains dimensions that are, of course,
those of the necessary complementarity of the couple, and then also
love relationships. Thus, in a poem concerning the death of Priscilla,
wife of Abascantius, Statius describes the marriage, the union, the
concord between Priscilla and Abascantius.[13] We can simplify the text
saying that it picks out three types of complementarity or bond. First,
a complementarity that I shall call rustic and that he emphasizes how
admirable it is since the couple, Priscilla and Abascantius, belonged to
important families and were very rich. Despite this, Priscilla prepared
his meals for him and looked after him like the wife of the simple
Apulian laborer.[14] That is to say the peasant model—the provincial

---

* The manuscript adds: "profound."

model of that old economic complementarity where the man works outside, the wife prepares the meals, in short all that we know from Hesiod[15]—is present even in a family as rich as the one Statius speaks of. It is present and it codifies the relations between man and wife. And the wife, as rich as she is, materially prepares, assures this rustic complementarity of the wife's role in relation to the man's.

We could also speak of a political complementarity. In the portrait given by Statius the wife is not directly concerned with politics, the which [moreover] would not have been wholly abnormal for a Roman matron. But it is precisely the way in which Statius represents the wife's relation to the political career of her husband that is interesting. The wife does not participate directly in her husband's political career, she does not help him, she does not support her husband's political life through her name, her own wealth, and her family and social status. It is not a case of the old model of the Roman matron, the daughter of an important family marrying someone and multiplying the powers of the two families. Statius gives the following definition: Most anxious, he says, Priscilla, the wife of Abascantius, eases the cares of her husband, who had great political responsibilities. She makes his zeal more active and his work lighter.[16] That is to say, through her affection, through all her affective bonds with her husband, her function is to form the accompaniment of mildness and tenderness that will enable him to lead the political life he has to lead, with that compensation at home, that re-balancing in the home, that is the friendship, tenderness, and affection his wife gives him. So there is a sort of direct connection between affective bonds on the one hand and political life on the other. The wife connects her affection, her love, to her husband's political life. She does not take part in it but is present in it through her affection.

The third complementarity, or the third type of bond, after the rustic and the political bond, [is] the amorous bond. And Statius stresses the fact that the wife of Abascantius has a *castissimus ardor* for him,[17] an extremely chaste ardor, a chastity that designates both the virtue, of course, that she would have opposed to any relationship outside her marriage,[18] but also the fact that the love for her husband was itself internally governed by chastity. She both loved him and loved him chastely. So that's the Statius text.

It seems to me that in Pliny too we find a description of the life of a couple in an aristocratic milieu, an interesting description for the type of collaboration it shows between conjugal relations and social activities. In letter 19 of Book IV (a letter addressed to a friend), Pliny draws the portrait of his own wife that contains an element concerning the connection between love and politics that is very similar to what is found a bit later in Statius. Pliny praises his wife because she loves him. And she loves him, he says, not because he is young, not because of his beauty, but because of his glory, *gloria*.[19] What does it mean to love Pliny for his *gloria*? He explains it in the course of the letter. He says: She reads my speeches, gives me advice, and is delighted when my speeches are applauded. Here again it is very interesting to see that Pliny's wife does not take part in his political career as the representative of an important family. It is not her wealth that will help his wealth, but the fact that, pursuing a political career that of course, at Rome, takes the form of glory, she will help him as a private rather than a social individual, by that sort of affectionate and permanent presence that is connected up to the orator's and politician's work itself. She reads his speeches, approves them, applauds them, and is delighted by applause.[20] So intra-conjugal love is connected to political life, but without the figure of the wife interfering in political life.

And then in another letter, here addressed directly to his wife, letter 5 of Book VII, he tells her that he misses her. He writes to his wife because she is absent, of course. He misses her, and to designate this feeling of missing her Pliny employs the clearly crucial word *desiderium*.[21] And to what is *desiderium* due? Two things. On the one hand: to the love I have for you and which means that I miss you.[22] And then equally: to the fact that we are never separated,[23] that our lives are constantly linked, that we live completely and always together, so that your absence gives rise to a desire, a *desiderium* whose basis is the fact that I miss your physical presence, that I miss you sexually. And then I miss you also in the absolute continuous daily life of our shared existence. And how does this *desiderium* that Pliny is going to describe manifest itself? Well, he says, at night, instead of sleeping I stay awake. I stay awake and think. What do I think about? I think of you, and I picture your body, your figure with an *imago* (image*). [...] [During] the day I pass my time wanting to find you. And at the usual or normal times when I come to see you in your room, I am carried there as if you were there, forgetting that you are not there, and I am forced to confirm your absence. Then I return to my apartment sad and aggrieved, as if you had shut me out and refused me entry to your room. All this, he says, this insomnia, these thoughts turned to you, this imagination that makes me picture your body during the night, this coming and going around your absence during the day, all this is torment.[24]

In this passage we find word for word what is found in the love elegies of the same period or an earlier period: the description of amorous behavior. Night, absence, the image, solitary reverie, absence [again], coming and going, the closed door, and so on, is the classic behavior of the lover [described] in erotic or love texts, but which, here, is re-invested in marital behavior and characterizes the relationship of a husband whose wife is absent. This conjunction of the classical, typical, traditional description of lover's behavior and marital absence seems to me to be characteristic of this notion—desire (*desiderium*)—which we cannot say appeared at this moment, but to which considerable importance should be given in this text because it will of course serve as the framework for a whole range of things. Take this notion of *desiderium* and see what elements Pliny brings into it: absence, lack, imaginary representation, the play of day and night, comings and goings, the closed door, exclusion, and so on. There are a number of fundamental elements here of what will constitute desire in all European literature, thought, I was going to say experience. From Pliny to Proust, from Pliny's wife to Albertine, there are a number of [recurring] elements and which bear precisely this name desire, with the link between *desiderium* and *supplicium*.[25] Desire is a torture. And how, in and through lack, is desire a torture? And how will Pliny console himself in this tormenting experience of desire? Well, he says, I can console myself only by going to the forum, by pleading, and by continuing my judicial and political activities. Imagine, he says ending his letter, what my life is when I have to seek rest in work (*requies in labore*) and comfort in cares.[26] This idea that he will seek consolation in work and the cares of public life is interesting

---

* Gap in the recording. One can hear only: well, I replace it with the image.

because one sees then that this letter, so with all the analysis of *desiderium*, of torture, and so on,* is exactly the opposite life, the reverse of the now established image of happiness, of the happy life. In the happy life one has a wife with whom one lives permanently, who is consolatory, who with her private affect supports, as it were, a public life in which she is not present, but which is bearable for the man only because he depends on her private consolations. That is the good life, the fine, happy life. The wife's absence reverses this life completely, and within the *supplicium* one has the worries, the misfortune of not having what one wants with one. And one is forced to seek consolation outside this field, which is obviously a false consolation. So Pliny's letter is interesting as a negative image of the good life and as a definition, by contrast, of that fundamental notion *desiderium*. Absence, desire. The great Western problematic of desire arises within the ideal image of the conjugal family and of the relation between man and wife as defined at that moment. At that point desire was not what makes one flee outside the family, it was rather inscribed in it and marked its solidity and strength.

The third characteristic that we can find in this bond of desire, of affection, this bond of both tenderness and sexuality within the family, is that it is seen to be indissoluble. In the texts I am talking about, those of Pliny and Statius, the length of marriage, the permanence of this *concordia*, are profoundly valorized elements. For example, Pliny emphasizes that the marriage of Macrinius and his wife has lasted for thirty nine years.[27] And one of Statius' poems evokes the long, unbroken chain of concord [between] Priscilla [and] her husband.[28] And I would like to take up here another poem by Statius in which he speaks of his own wife, and which is very odd.

So Statius is speaking about his own wife. He addresses himself to her and sings of the solidity of their union. He tells her: Venus who granted us her favor at the start of our life "will preserve his favor for us on the decline" (so he will have relations of affection with his wife until the end, but also relations of intense love), "I will not break a bond that I tighten every day."[29] The more one ages, consequently, the more one is attached, not despite Venus who has withdrawn from life, but rather thanks to Venus who is increasingly present throughout this

---

* Passage difficult to hear. One hears only: and then consolation in ...

existence. Evoking the land on which he was born, Naples, he says: This land created me for you for long years, it has made me your *socius*, your companion.[30] What is strange in this text is this:[*] after or rather in the middle of this exposition on the faithfulness of the two spouses to each other, among all the qualities of his wife that Statius evokes, among all her good actions, we read: "you still visit the ashes and manes of the first object of your affection, and embracing the remains of that friend of harmony, although wholly mine already, you renew your heartrending moans from the bottom of your heart."[31] I am quoting here a somewhat grandiloquent nineteenth century [French] translation,[32] but you can see straightaway what is involved. This admirable woman to whom Statius feels bound from birth to the end of his days, this *socia* with whom his whole destiny is bound up, had actually had another husband to whom, after his death, she continues to render homage and display the affection she had for him. That is to say the second marriage has absolutely not weakened the faithfulness of the woman to her first marriage, which seems to indicate that the matrimonial bond is considered so fundamental, so essential, and so in[dis]soluble[†] that even a second marriage does not exclude, suppress, or erase it, and that a wife who is a proper wife, a wife who has good feelings, must be able to love her present husband, as if totally bound to him, at the same time as she continues to love her previous husband, considering herself always bound to him.[‡] That is to say we arrive at the idea that the bonds are not just conceived as coextensive with life, but also as able to last beyond death and to form a sort of indefinite spiritual union.[§] In fact, the text of Statius is a bit enigmatic for us. It was certainly not so at the time, but one feels already that the problem of the second marriage cannot fail to be raised. It will be impossible for it not to be raised, and in

---

[*] M. F. adds: this is why I have quoted it.
[†] M. F. says: insoluble.
[‡] The manuscript notes: "discredit of divorce, which certainly remains a frequent practice, but is viewed disapprovingly."
[§] The manuscript adds:
"One finds an expression of this indefinite spiritual bond in the art of tombs: representation or effigy of the couple (a practice that will disappear in Christianity) with inscriptions like this: 'Maervius, inscription for his wife who lived twenty two years, three months, and thirteen days with him without quarrel. He asks the gods to render him to his spouse who lived with him with such concord until the fatal day. I beg you, ensure that I am not separated from her for longer.'"

actual fact it will be. There is, for example, the very strong reluctance of Marcus Aurelius on his remarriage after his first widowhood.[33]

I will skip all this, and let's say, to end this rapid review of some literary texts, [that] through [them] we see very clearly the idea of a marriage as an indefinite spiritual union that goes beyond life itself and will extend into the indefiniteness of time. We see the idea that these bonds of affection between husband and wife receive a religious guarantee. To Abascantius, whose wife has just died, Statius writes: "Jupiter has witnessed your pain, he reads the depths of your soul, and judges your feelings for her by your tenderness for a shade and by your faithfulness to the cult of the tombs."[34] It is here that we find the expression *castissimus ardor*: that extremely chaste ardor worthy of approval by the monarch [of the world]. Jupiter himself, the god, the divine power, not only consecrates marriage as an institution, but well and truly this feeling. And this feeling of attachment, of enduring affection of the couple for each other, is a way of giving homage to the god. So religiosity of this feeling internal to marriage.

These are some themes of the literature of this period, and so we come to this conclusion, or rather to the false conclusion, which would be the following. That there is no doubt a continuum that shows the very general existence of one and the same marital model, since we find in sum the same themes, or at least some of the same forms. On the one hand, in evidence of an institutional kind, [we see] the importance, the valorization, spread, and intensification of marriage in the general population, or at least in the social strata that have left traces at the institutional level. So spread and intensification of the matrimonial model [in these strata].[*] Intensification also of the matrimonial model in this literature (Pliny and Statius) and in philosophy, in the rules of life.[†]

So all this gives a perfectly coherent and at first sight entirely satisfying image. It is as if one has concordance: this is what happens in reality, this is how literature reflects it, and this is how philosophy expresses

---

[*] M. F. says: in them.
[†] M. F. adds: that I talked to you about last week (see above, lecture of 4 March, p. 189 sq.].

it. For once philosophers have expressed reality, we should be pleased. But you understand that it is precisely this that poses the problem. What does it mean to say: "The philosophers express it, and they do so because it happened in reality?" You really think that we are precisely missing the problem, a problem that is interesting, or anyway that interests me. And it seems to me that the so-called continuity I have just outlined, far from resolving the problem, far from supplying the solution, rather than allowing us to say: "Well that's it, if the philosophers say so, it is because it happened in reality ...," it seems to me that this is rather what we should stop to consider and question; and it seems to me that the idea that the philosophers have described what happened cannot be accepted.

It cannot be accepted for two reasons. For a reason that can be formulated in particular terms for this problem and that can be formulated more generally. [In] more particular terms: if in actual fact this new sexual and conjugal morality was a reality, if in actual fact things really did happen like that and the new conjugal morality that the Stoic and other philosophers proclaimed, prescribed, pointed to, and transformed into constantly repeated lessons became a reality, why was it necessary to say it, and to say it in a prescriptive form? Why transform into a rule of conduct, why present as advice for living well something that would have effectively already been established at the level of real behavior? Why would philosophers have been led to reproduce in the form of injunctions what was already given in reality? And you can see that the sort of sociological-historical explanation, which [appears] serious and realistic and consists in [explaining] the philosophers' model by the fact that that's how it was in reality, cannot hold up and that it is precisely because that's how it was in reality that the fact that the philosophers presented it as prescription becomes enigmatic. Or again the fact that the philosophers presented it as prescription allows one to suspect that it must not have been like that. In either case, there is a problem. And it is here that I will say more generally [the following]. I say it because it is a question of general methodology—more than of methodology, it is a question I think of general interest, in any case which is important for me.

It seems to me that the reality to which a discourse refers, whatever it may be, cannot be the raison d'être of that discourse itself. You will tell me that this is a statement of the obvious. It is a statement of the obvious when it is a matter of a prescriptive discourse. And of course, if there is a prescriptive discourse, if one states a law, if one says: "That is what must be done," one should not look at what is done for the reason one says "one must do it." Or at any rate, if one says it must be done, it is actually because reality is not like what one says. But I shall say that the same question, the same difficulty is posed and has to be posed regarding any discourse, in particular regarding discourses that claim to give orders and bring about what does not yet exist, but equally regarding discourses that claim to tell the truth, to say what is. The existence of a true discourse, of a veridical discourse, of a discourse with the function of veridiction, the existence of [such] a discourse is never entailed by the reality of the things of which it speaks. There is no fundamental ontological affiliation between the reality of a discourse, its existence, its very existence as discourse that claims to tell the truth, and the reality of which it speaks. In relation to the domain in which it exercised, the game of truth is always a singular historical event, an ultimately improbable event in relation to that of which it speaks. And it is precisely that singular event consisting in the emergence of a game of truth that we have to try to reconstruct. This is in any case the point to which any project of a history of the truth is attached. To do the history of the truth, of games of truth, of practices, economies, and politics of veridiction, [presupposes that one cannot in any way] be content with saying: If such a truth has been said, it is because that truth was real. One must say rather: Reality being what it is, what were the improbable conditions, the singular conditions that meant that a game of truth could appear in relation to that reality, certainly a game of truth with its reasons, its necessities, but reasons and necessities that are not simply the fact that the things in question existed? Telling the truth about something does not originate solely [in], is not explained or justified solely by the fact that this true thing is real. A correspondence between the truth and reality may well be established when, within a precise game of truth, one undertakes to know on what conditions one may say that a proposition is true. At that point it is no doubt legitimate to say: If such a proposition is true, it is because reality is such. It is because the sky is blue that it is true to say: the sky is blue. But, on the other hand, when one poses the question of how it comes about that there was a discourse of truth, the fact that the sky is blue will never be able to account for the fact that I say that the sky is blue.

And it is precisely here that the problem I want to pose regarding sexuality in general, and also regarding the *aphrodisia*, is situated: how were *aphrodisia*, how was that mixture of acts and pleasures that is so enigmatic for us and that the Greeks saw as part of their experience, taken up in a certain game of truth that objectivized them, imposed a certain analytical grid on them, and also modified one's experience of them? This is the problem I would like to pose regarding the *aphrodisia*, that I would like to pose generally with regard to sexual behaviors: how did it come about that in the West sexual behaviors were taken up in a certain game of truth that brought about not only the existence of a whole series of discourses, but [also], and constantly, modifications in one's experience of these behaviors? How did the appearance of these games of truth come about? How was it that right within that experience one has of oneself as subject in a sexual relationship, the obligation of truth, the possibility and necessity of telling the truth appeared? You see that it is the same problem that I wanted to pose with regard to madness, crime, and so on: how are games of truth connected up to real practices? And this is why analysis that consists in referring the game of truth to reality by saying: "The game of truth is explained because the real is such," however realistic it may seem to some, seems to me absolutely untenable and inadequate. Reality will never account for that particular, singular, and improbable reality of the game of truth in reality. And it is the deep rootedness of this game of truth in reality that must be recovered. Hierocles, beginning his treatise on marriage, said: *Ho logos peri gamou anagkaiotatos*, "The discourse on marriage is of all things what is most necessary."[35] The problem then is this: why was it necessary to speak about marriage so much and at such length, if in actual fact marriage was in reality what the philosophers said it ought to be? Why this need of *logos*, why is *logos* about marriage *anagkaiotatos*, the most necessary of all things?

I have gone on a bit too long. I am going to stop now and I will try to answer this question next week. Thank you.

1. On this point, see *Le Souci de soi*, ch., "Le rôle matrimonial," pp. 90-97; *The Care of the Self*, "The Marital Role," pp. 72-80.
2. Paul Veyne, "L'amour à Rome ...," in *La Société romaine*, p. 96: "Let us recall in a word that Roman marriage is not a public act; what am I saying, it is not even a juridical act: it is a de facto state marked by a ceremony."
3. As well as C. Vatin's book (see the next note), in the chapter of *Le Souci de soi* ("Le rôle matrimonial"); *The Care of the Self* ("The Marital Role") devoted to this subject, Foucault refers to the works of: J.-P. Broudehoux, *Mariage et Famille chez Clément d'Alexandrie* (Paris: Beauchesne & Fils, "Théologie historique" 11, 1970); J. A. Crook, *Law and Life of Rome*, 90 B.C.-A.D. 212 (London: Thames and Hudson, 1967); J. Boswell, *Christianity, Social Tolerance and Homosexuality* (Chicago: University of Chicago Press, 1980); S. B. Pomeroy, *Goddesses, Whores, Wives and Slaves: Women in Classical Antiquity* (New York: Schocken Books, 1975).
4. C. Vatin, *Recherches sur le mariage et la condition de la femme mariée à l'époque hellénistique*.
5. On this point see the article by E. Caillemer and G. Humbert, "Adulterium," in C. V. Daremberg and E. Saglio, eds., *Dictionnaire des Antiquités grecques et romaines* (Paris: Picard, 1877) vol. I, pp. 54-55. For a recent synthesis see G. Rizzelli, *Lex Iulia de adulteris. Studi sulla disciplina di adulterium, lenocinium, stuprum* (Lecce: Ed. del Grifo, 1997).
6. On this point see the arguments of C. Vatin in *Recherches sur le mariage et la condition de la femme mariée*, chapter IV, "L'évolution du droit matrimonial," section, "Actes juridiques," pp. 145-179.
7. See ibid., ch. IV, section: "Devoirs réciproques des époux," pp. 200-206.
8. P. Veyne, "La famille et l'amour ...," in *La Société romaine*, p. 108: "Under the Empire, there is no longer a question of it being understood that dissension may exist between husband and wife, since henceforth the very functioning of marriage is supposed to rest on harmony and the law of the heart. Thus a new idea arises: the 'couple' of master and mistress of the house."
9. See the analysis of these texts again in *Le Souci de soi*, pp. 97-99; *The Care of the Self*, pp. 78-80.
10. Pliny the Younger, *Lettres*, Book Eight, 5, ed. and trans., Anne-Marie Guillemin (Paris: CUF, Les Belles Lettres, 1967) p. 56; English translation B. Radice, *The Letters of the Younger Pliny* (London: Penguin Books, 1969), 5, p. 213.
11. Ibid., Book Four, 19, Fr., vol. II, pp. 38-39: "All this conduct inspires in me a hope full of confidence that our mutual affection will never end and grow daily (*perpetuam nobis majoremque in dies duturam esse concordiam*)"; Eng., p. 126: "All this gives me the highest reason to hope that our mutual happiness will last for ever and go on increasing by day."
12. Tacitus, *Vie d'Agricola*, VI, ed. and trans., Eugène de Saint-Denis (Paris: Les Belles Lettres, CUF, 2003 [1942]) p. 4: "Returned from Brittany to Rome in order to accede to the magistracy, he married Domitia Decidiana, of brilliant birth; this marriage put him in the light and in a solid position for the highest aims; they lived in an admirable concord, never ceasing to cherish each other (*vixeruntque mira concordia, per mutuam caritatem*), and each preferring the other to him or herself"; English translation M. Hutton revised by R.M. Ogilvie, "The Life of Julius Agricola," in *Tacitus. Vol. I: Agricola. Germania. Dialogue on Oratory* (Cambridge, MA: Harvard University Press, Loeb Classical Library, 1914) pp. 34-35: "From this field he passed on to the city to take up office; there also he married Domitia Decidiana, a woman of high lineage. The marriage proved at once a distinction and a strength to him in his upward path; their life was singularly harmonious, thanks to mutual affection and putting each other first."
13. Statius, *Silves*, V, I, ed., Henri Frère and trans., H.J. Izaac (Paris: Les Belles Lettres, CUF, 1944) pp. 173-183; English translation D. R. Shackleton Bailey [revised by C. A. Parrott, June 2015] Statius, *Silvae* (Cambridge, MA: Harvard University Press, Loeb Classical Library, 206, 2003) pp. 310-331.
14. Ibid., V, I, 121-126, Fr., p. 178: "She served him herself simple meals and plain cuts, and she invited him to follow the example of the master; the Apulian wife of a frugal peasant, or the one that the Sabine sun has tanned, who notices, in the glance already thrown by the stars, that the time is close when her husband will have finished his day, prepares in haste the table and the bed and listens for the sound of cart coming back"; Eng., pp. 320-321: "She herself serves him his frugal meals and temperate cups, and admonishes him by his master's

example; even as some thrifty farmer's Apulian wife or sun-tanned Sabine, when she sees the stars are peeping out and it's nearly time for her man to come home from the day's work, smartly sets up the table and the couches and listens for the sound of the returning plough."

15. Hesiod, *Works and Days*, trans., Dorothea Wender in *Hesiod and Theognis* (Harmondsworth: Penguin Books, 1973).

16. Statius, *Silves*, V, I, 119-120, Fr., p. 178: "She took care with anxiety (*anxia*) of her husband's cares, she both exhorted him and moderated him in his work (*hortaturque simul flectitque labores*); *Silvae*, pp. 320-321: "She cherishes anxiously her husband's cares, at once encouraging his labours and deflecting them."

17. Ibid., V, I, 41, Fr., p. 175: "Here indeed is the purest of attachments (*hic est castissimus ardor*)"; Eng., pp. 314-315: "This is passion at its most chaste."

18. Ibid., V, I, 55, Fr., p. 176: "You will draw to yourself a greater lustre by wishing to know only one bed"; Eng., pp. 314-315: "but greater the dignity that came from yourself—to know only one bed." See also ibid., verses 55-63 on Priscilla's fidelity.

19. Pliny the Younger, *Lettres*, Book Four, 19, p. 39: "For what she loves in me is not youth or beauty, which will fade and wilt, but glory (*sed gloriam diligit*)"; *Letters*, p. 126: "for she does not love me for my present age nor my person, which will gradually grow old and decay, but for my aspirations to fame."

20. Ibid., Four, 19, Fr., p. 11: "She has my works, she reads and re-reads them, she even learns them by heart. What anguish when she sees me on the eve of pleading, what joy when it is done! She manages to be kept informed of the approval and applause that greets me, of my success in the case. And when I give a public lecture, from behind a curtain nearby she listens with an avid ear to the compliments I am given"; Eng., p. 126: "she keeps copies of my works to read again and again and even learn by heart. She is so anxious when she knows that I am going to plead in court, and so happy when all is over! (She arranges to be kept informed of the sort of reception and applause I receive, and what verdict I win in the case.) If I am giving a reading she sits behind a curtain near by and greedily drinks in every word of appreciation."

21. Ibid., Book Seven, 5, Fr., p. 11: "You wouldn't believe how much I miss you (*incredibile est quanto desiderio tui tenear*)"; Eng., p. 187: "You cannot believe how much I miss you."

22. Ibid., Fr., "The reason is my love first of all (*in causa amor primum*)"; Eng., "I love you so much."

23. Ibid., Fr., "We are not used to being apart from each other (*non consuevimus abesse*)"; Eng., "we are not used to separations."

24. Ibid., Fr., pp. 11-12: "This is why a large part of my nights pass with me awaking picturing your image (*magnam noctium partem in imagine tua vigil exigo*), why in the daytime, at the times when I am used to coming to see you, my feet carry me themselves, as it is right to say, to your apartment, why, finally, sad, aggrieved and as if the door had been shut, I return from your empty room"; Eng., pp. 187-188: "So I stay awake most of the night thinking of you, and by day I find my feet carrying me (a true word, carrying) to your room at the times I usually visited you; then finding it empty I depart, as sick and sorrowful as a lover locked out."

25. Pliny does not employ the term "*supplicium*" here (no doubt suggested to Foucault by the French translation: "*torture*") but "*tormentum*" (see the following note).

26. Pliny the Younger, *Lettres*, Book Seven, 5, Fr., p. 12: "The only time when I am exempt from this torture is when I go to the forum and am absorbed in the trials of my friends. Imagine then my life when I must seek my rest in work, my consolation in troubles and cares (*aestima tu quae vita mea sit, cui requies in labore, in miseria curisque solacium*)"; *Letters*, p. 188: "The only time I am free of this misery is when I am in court and wearing myself out with my friends' lawsuits. You can judge then what a life I am leading, when I find my rest in work and distraction in troubles and anxiety."

27. Ibid., Book Eight, 5, Fr., p. 56: "She has been with him for thirty-nine years without a quarrel or a sulk (*sine jurgio, sine offensa*)"; Eng., p. 213: "they had lived together for thirty-nine years without a quarrel or misunderstanding."

28. Statius, *Silvae*, V, I, 44: "*iunxit inabrupta concordia longa catena*."

29. Statius, *Silves*, III, 5, 23-29 in *Stace, Martial, Manilus, Lucilius junior, Rutilius, Gratius Faliscus, Némésianus et Calpurnius: œuvres complètes, avec la traduction en français, publiées sous la direction de M. Nisard* (J. J. Dubouchet & Cᵢₑ, 1842): "It was Venus who united us in the flower of

our years; Venus will preserve her favor over the decline of life. Your laws Claudia (for was it not you who, from the first blow of love fixed my fickle youth subjecting it to the yoke of hymen?), your laws have found me docile and content, I will not break a bond that I tighten more and more every day"; *Silvae*, pp. 226-227: "For sure 'tis you whom Venus joined with me by a kind destiny in the springtime of my years and keep with me till old age, 'twas you that pierced me with my first wound, untouched as I was by wedlock and still a young wanderer, and yours were the reins I received in willing obedience and to this day press the bit that once in my mouth I shall never change."

30. Ibid., 106-107, Fr., p. 52: "this land gave birth to me for you, it has forever enchained my fate to yours (*creavit me tibi, me socium longos adstrinxit in annos*)"; Eng., pp. 232-233: "she created me for you, bound me to be your partner for many a long year."

31. Ibid., 51-55, Fr., p. 51; Eng., pp. 228-229 : "you still seek the ashes and shade of him that was, so you embraced the obsequies of your songful spouse, raining violent blows on your breast yet again, when you were already mine."

32. The translation is from 1842.

33. Marcus Aurelius was married in 145 to the daughter of Antoninus, Faustina, who is portrayed by some historians as dissolute and treacherous (see Dio Cassius, *Roman History*, LXXI, and *The Historia Augusta*). Nonetheless some personal testimonies evoke the authenticity and intensity of their love (*Meditations*, I, 17; *The Correspondence of Marcus Cornelius Fronto with Aurelius Antoninus*). After the death of Faustina in 165, Marcus Aurelius seems to have felt a profound pain, and the Senate offered impressive mourning ceremonies. He refused to remarry, out of love perhaps, at least he did not wish to complicate access to rule of his first son, Commodus. See Ernest Renan's fine presentation, "Examen de quelques faits relatifs à l'impératrice Faustine, femme de Marc-Aurèle—Lu dans la séance publique annuelle des cinq académies, le 14 août 1867," in E. Renan, *Mélanges d'histoire et de voyage* (Paris: Calmann-Levy, 1906).

34. Statius, *Silves*, V, I, 39-41, Fr. (Nisard ed.); *Silvae*, pp. 314-315: "The god who governs the reins of all the world and nearer than Jupiter disposes of men's doings, he marks it and sees you grieving; and therefrom he takes private proof of his chosen servant, in that you love the shade and pay tribute to her obsequies."

35. Hierocles in Stobaeus, *Florilegium*, 22, ed., A. Meinecke (Leipzig: Teubner, 1860-1863) vol. III, p.7: "*Anagkaiotatos estin ho peri tou gamou logos.*"

This page intentionally left blank

ten

18 MARCH 1981

[
The problem of redundant discourse (discours en trop).
∽ *The Christian re-appropriation of the Hellenistic and Roman
matrimonial code.* ∽ *Problematization of the relation between
discourse and reality.* ∽ *First explanation: representative
reduplication.* ∽ *Four characteristics of the game of veridiction
in relation to reality: supplementary, pointless, polymorphous,
efficient.* ∽ *Second explanation: ideological disavowal.* ∽ *Third
explanation: universalizing rationalization.*
]

TODAY I WOULD LIKE to concentrate a little on a historical and
methodological problem that I had trouble with last week and that
was in short the problem of redundant discourse (*discours en trop*).[*]
This is what I mean. It seems to me that from these rules of conduct,
from these arts of conducting oneself recommended by the Stoics, and
also from what the documents may show regarding the evolution of
matrimonial practice in at least some regions of the Mediterranean
world in the Hellenistic and Roman period, from this set of texts and

---

[*] M. F. adds:
You know, it's a bit tough giving a lecture every week. It doesn't look like much, but it's a
bit tough and there are times when one would like to stop, reflect a bit more, and make an
appointment [for] one or six months. This is to apologize if my reflections today are somewhat disjointed and disordered. I am aware of not having exactly and properly resolved the
problems that I have decided to pose.

practices it seems to me that it is fairly easily to extract what we will call, without stretching things, a code of conduct, a code of sexual conduct that is, moreover, easy to delineate. First, this code of sexual conduct is organized around marriage, which becomes the exclusive place in which sexual conduct is authorized. Monogamous marriage of course, which in principle therefore excludes any sexual relationship other than that of husband and wife. Even inside marriage itself sexual relations are not so free, they contain something perilous, dangerous, impure, and so on. Inside marriage itself these sexual relations must be codified a second time as it were by two requirements: on the one hand, the purpose of sexual intercourse must be procreation; on the other hand, sexual relations must constitute the particularly intense expression of affective bonds that form the fundamental fabric of husband-wife relations within conjugality. That, in short, is what we have been able to extract from the texts and practices.

First remark: obviously you can easily recognize this code of good sexual conduct. We precisely began the analysis of this whole problem with the famous model of the elephant I spoke to you about. In the naturalists of the first two centuries CE, you recall, for example in Pliny the Elder, a bit later in Aelian, there is a both naturalistic and moralistic description of the elephant's good manners. The elephant is offered to human conduct as the blazon of good matrimoniality and chastity; the elephant that is precisely monogamous, that has sex with his female only with a view to procreation, and whose attitude towards its female is one of attachment, affection, and so on. This model of the elephant is thus manifestly the fabricated transposition into natural history of this code of sexual conduct indicated by practice on the one hand and the texts on the other. We have seen the passage of this model of the elephant from the accounts of Antiquity to those of Christianity. We found it in *Physiologus*, a text of late Antiquity that until the end of the Middle Ages was the great stock, the great store of moralizing fables drawn from the animal world. We also find this fable of the elephant in sixteenth-century naturalists like Aldrovandi. And finally it is developed in the *Introduction to the Devout Life* of Saint Francis of Sales. So we can say that we have a model of sexual conduct that, without fundamental modification, is passed on within, let's say, pagan reflection and practice, from the first two centuries CE.

And then it continues on up to the seventeenth century and finally the modern period.

What is interesting moreover, is that this transit, which I have skimmed over in this way through the rather marginal fable of the elephant, can be plotted in its development, its passage much more finely in, for example, the first Christian texts concerning marriage. It should be remembered that, to start with at least, the Christians did not attribute great importance to the problem of sexual behavior. It needed the dualist challenge, the Gnostic challenge for the problem of sexual behavior to arise within Christianity. But if you take, for example, the texts of the first apologists, if you take the *Didakhē*,[1] if you take the texts of apologists like Justin,[2] for example, you see that they take up very exactly the forms of pagan morality, and they refrain from forming a divergent sect within society, noting that they have exactly the same type of matrimonial and sexual behavior as any respectable philosopher. They are married, they have only one wife, of course, they have sexual relations with their wife only in order to have children, and as far as possible they do not remarry when they are widowed. This then is what we find in the Christian literature of the second century, and also in the first great theorist, if not of sexual behavior, at least of married life, Clement of Alexandria; we will no doubt return to this next week.[3] We can already say that its attitude is characterized by three major principles. First, of course, the rejection, the polemic against the dualist, Gnostic attitude let's say of the Marcionites[4] and Valentinians,[5] who rejected all procreation. In the different sects they did not necessarily reject sexual intercourse, but they rejected procreation inasmuch as it was the instrument of the continuation of the very material world from which it was necessary to free oneself.* Against these dualists, these people inspired by Gnosticism, Clement of Alexandria develops a whole polemic that reveals then two [other] things: the affirmation of the spiritual values peculiar to absolute continence, to a chastity that characterizes at the very least the final† stage of the spiritual journey; but at the same time the acceptance of marriage, regarding which, in *The Instructor*

---

* The manuscript adds: "from the flesh."
† The manuscript indicates: "the third."

(*Paedagogus*), Book II, Chapter 10, he gives a series of indications on what marriage should be, and on what the relations between husband and wife, and sexual relations between husband and wife should be.[6] And here,* Clement of Alexandria's formulation for married life is more or less, word for word, what could be found in the Stoics, in Plutarch, in the moralist philosophers of the first and second centuries. The only modifications are some supplementary prohibitions whose Semitic, Hebraic origin is clear, but which absolutely do not change the economy of this codification—prohibitions concerning sexual relations with pregnant women, for example, or during menstruation.[7] These two things are not found in ancient morality, in ancient codification, they are found in Judaism. Clement of Alexandria picks them up here, but the whole is of the same type. So one has here a massive transfer that can be followed step by step, from the texts of Antipater, Musonius Rufus, Seneca, Plutarch, Epictetus, a transfer that takes place through the intermediary of the Apologists, of Clement of Alexandria, into Christianity. So that we can say that at the level of the codification of sexual conducts, Christianity inherited a system that was already formed and established.

If one were to define a sexual ethics by the codifying framework that [determines] the permitted and the prohibited, in other words if a sexual ethics were nothing other than a system of prohibitions and tolerances, then one could say that the morality, the sexual ethics of Antiquity passed bag and baggage into Christianity. But it is precisely this problem of the bag and baggage that I would like to pose, that is to say the problem of the accompanying discourse that surrounds this codification. For again, if it is true that the codifying framework is the same, there are clearly considerable differences between the discourses within which this codification is presented, [between] texts like those precisely of Musonius Rufus and the Stoics and those of Saint Augustine three or four centuries later. Exactly the same system of prohibitions and permissions is transcribed within what could be called a philosophical or theological terminology, analysis, or discourse that is entirely different. To give you an example and get you to grasp the point of the problem, take the two major principles found in the Stoics: sexual intercourse must be linked on the one hand to a procreative end, and on the other to the existence of intense affective bonds between husband and wife. If you take the *De bono conjugali*, written by Saint Augustine right at the start of the fifth century, 401, you find the same two principles, that Christianity will take up moreover, develop, and pass on from century to century, with merely a different formulation, a different justification. First the imperative of procreation,[8] but here Saint Augustine does not formulate the imperative in terms of the interests of the city, nor even exactly in terms of the interdependence of the individual and humankind. Certainly humankind is involved, but in an entirely different way, inasmuch as it is a question of bringing humankind, through procreation, to its fulfillment and perfection, bringing humankind to a point of development such that fulfilled humanity will then be able to await the parousia and return of Christ. Second, sexual intercourse had its place in what the Stoics called matrimonial *krasis*, that is to say the substantial physical, bodily unity of individuals, which has to bind them body and soul. And in Saint Augustine sexual intercourse is justified in marriage on the basis of that notion in need of examination, the *debitum conjugale*, that is to say conjugal debt.[9] Out of charity, one owes sexual intercourse to the other so as to avoid his exposure to sin if left alone to his own concupiscence as it were. So there is transposition of the same codifying kernel (affective bond [and] procreation are still the two fundamental elements), but within an accompanying discourse that is not the same.

So there is a problem: in a history of sexual ethics can we regard the codification as a sufficiently important element on its own, the transfer, transpositions, and modifications of which can simply be followed without taking account of the discourse that accompanies it, or regarding the accompanying discourse as only its theoretical, philosophical, and conceptual logistics, its surroundings, clothing, or transcription as it were within a theoretical system that actually is not really important for knowing in what sexual experience strictly speaking consists? I would like to show you—this is the whole meaning of these somewhat plodding analyses—that of course these accompanying discourses are not just the theoretical clothing of a codification. In fact, we should be able to find through them—this is what I would like to try to grasp—the very form of an experience: the type of relationship that there may be between subjectivity [and] the codification of conducts, the relationship of truth that the subject establishes to himself through his relationship to the codification of his own conduct. Subjectivity, truth, and codification of conduct appear, not when one looks simply at the main theme of the codification and its transformations, but when, along with this framework of codifications, one takes account of the discourse that accompanies it. So I would now like to tighten up a bit more this relationship between codification, accompanying discourse, experience, and subjectivity.

You recall the givens of the problem as it presented itself, and what I had difficulty with last week. So we had a first given, which was that literature (philosophical literature, the literature of moralists, of spiritual guides) that recommended marriage and that, with regard to marriage, prescribed a whole set of precise conducts concerning the relations between man and wife, their affective bonds, and sexual relations. This was the first given: a fact of discourse. The second given of the problem, identifiable through a somewhat different documentation, was the existence of a practice, or rather of an evolution, a transformation of practices, namely, so far as one can identify them: the real spread of matrimonial practice in a large part at least of the Hellenistic and Roman world, and the development within this matrimonial practice of conjugal bonds quite similar to those [mentioned] in the prescriptive texts. The third given that needs to be taken into account is that this real practice of marriage, however in line with what can be read in the prescriptive texts, can in no way be seen as the result, the effect, or the absorption of this set of prescriptions. The real practice cannot be regarded as the effect of these prescriptions for a whole range of reasons, primarily these: even if the philosophers claim to address everyone and humanity in general, the advice on life that they give could be heard and followed only by a very small social group, by a cultural elite. Now the documents available to us concerning matrimonial practices show that these new marriage practices were not only taken up by fairly large groups but, it seems, by classes that were far from forming the social, political, and cultural elite of society. As far as it can be established, the rather popular, peasant type of models of matrimonial behavior seem to have been gradually imposed by an upsurge from below. So the idea that a model defined philosophically within a theoretical discourse was gradually imposed on society from above no doubt has to be rejected. We should recall that the great theorization of matrimonial life found in the philosophical texts are mainly from the first century BCE, but especially the first and second centuries CE. Now it seems that the phenomena I am talking about—the spread of matrimonial practice, the strengthening of bonds—are attested before this period, from the second century BCE. Consequently, we cannot hold that the philosophical texts were at the origin, that they formed the framework of the new behavior.

At this point a problem of method arises: what is one to make of the existence of this redundant discourse? What is one to make of this speculation that took place around marriage, of these discourses that contain at least two profoundly linked aspects: on the one hand, a set of principles of codification and, on the other, a whole speculation, anyway a whole analysis in terms of truth concerning the nature of marriage, the true bond between the spouses, what true love should be, and so on? Faced with the existence of a practice prior to the discourse and a discourse that seems to reduplicate it, I think we can define, we can pick out three possible attitudes, three methods, three ways of situating and explaining the existence of this redundant discourse, this strange discourse that appears to reproduce a real practice in the form of codification and to give in terms of truth what seemed to be already established at the level of behavior.

The first explanation, the first type of possible analysis would be explanation by representative reduplication. This is the simplest attitude. One could say that the moralists, the philosophers have really only registered, only transcribed in the form of prescriptions a real process. Claude Vatin, one of the historians who has precisely studied the problem of marriage in the Hellenistic world, mainly in Egypt and Alexandria, adopts this approach in an analysis that I do not criticize, that seems interesting to me [but that], while valid as method for the history of practices, is inapplicable to the problem I would like to pose, which is [that of] the history of discourses of truth. This is how Claude Vatin deals with this problem of the existence of a social practice and of a subsequent discourse that seems to reflect and simply reproduce it. He says that basically we find two major models of philosophical discourse on marriage. The Platonic model, which is also the model moreover that we find in certain Cynics, is that of women and children held in common. [According to him,] the philosophers outlined, proclaimed, explained, and justified this model, [but] it remained entirely a dead letter. With the Stoic model, on the contrary, the model I have talked about in these sessions, with matrimonial sexuality and so on, we find a correspondence in reality. Far from remaining a utopia or dead letter, reality shows and proclaims how close this model is to what actually took place. And why, asks Claude Vatin? Quite simply this proves that philosophers are not able to determine reality. Philosophical theories, he says, have no hold on it,[10] and if something in the real world corresponds to the Stoic model it is quite simply because the model only follows that reality. The new ethics of marriage, he says, "is not the result of philosophical speculation; on this point literature only reflects the change of moral customs. The idea that the couple is an elementary and autonomous unit was formed on contact with realities, in the progressive dissolution of familial and political institutions, bringing with it that fragile hope that the will of two beings pushed by a mutual love to live together create a harmony that goes beyond them."[11] In short, philosophical prescriptions would translate an already established practice and would express, would manifest an already formed sensibility, mentality. I do not think this kind of analysis should be treated offhandedly. No doubt, to start with, because such an analysis certainly contains a large part of the truth. Insofar as this philosophical literature put itself forward, with the Stoics, as a direction of life, as a way of prescribing the appropriate way for people to behave, it certainly could not exist, it could not keep going [remaining] completely outside the field of individuals' real practices. It could not have been accepted at all if it were not in some way rooted in an already established practice. But it is not on this that I would like to focus reflection. It seems to me that an historian's approach* may well be justified by such a conception of philosophical reflection as the reproduction within theoretical discourse of an already established practice. It may well be justified for the following reason.

---

* M. F. adds: an historical approach.

When the problem posed by the historian is how people actually behave and how in actual fact they practiced marriage, what type of relations men and women might have had, and so on, then in order to investigate what the actual behavior of people was it is entirely legitimate to take as starting point, as the first elements anyway of the main theme, the more or less theoretical, speculative, programmatic representations of life, and then, starting with these texts and through a series of cross-checks based on a different type of documentation, to see if anything actually corresponded to them. And one can as it were get back from the theoretical, speculative formulation, from the later fact of discourse, to a prior practice of which the fact of discourse would to a certain extent bear the trace. So a perfectly legitimate approach if one takes the analysis in this direction.

But you can see that this approach—which once again cannot be criticized when it is addressed by an historian whose object is to determine the real behavior of individuals—is no longer legitimate if one tries to take the analysis in another direction: when one no longer asks what the reality is that philosophical or theoretical or speculative discourse may convey or of which it may bear the trace, but what is the reality of which the discourse consists. In other words, if one takes the discourse as a documentary instrument for rediscovering the reality of which it speaks or to which it refers, then the historian's approach I have just been talking about is entirely acceptable. But if one questions the discourse in its existence, not in its function as document, but in its existence as monument[12] (in the fact that it exists, in the fact that it has in actual fact been uttered), if one wonders about the reality of the discourse then one cannot be content with affirming that it is the things that have been said that account for the fact that they have actually been said. We must turn our attention to, we must stumble on this reality of discourse removing the postulate [that] the function of discourse is to represent reality.

⚜

Reality does not contain in itself the raison d'être of discourse. At least, the reality at issue in the discourse cannot by itself account for the existence of the discourse that speaks of it. To suppose this is to resort to what could be called the logicist dodge, a logicist dodge consisting in using the criterion of verification as an explanation of existence.* That a proposition is true may be established by establish-

---

* On these important methodological problems the manuscript contains some variants and supplements to the discourse uttered by Foucault in his lecture. For this reason we quote them in their entirety:

"You must have found that I have been making no headway for a long time on this question of method only to say in the end some fairly known things. The reason for this is that I would like once again to delimit what constitutes the (somewhat buried) point of departure, the (still remote) point of arrival, and the central focus around which the research I am talking about revolves and of which I can give you no doubt only the somewhat historicizing result.

Let's say schematically: no doubt ontological astonishment is that there is being and [not] nothing; epistemic surprise is that there is truth, I mean that there is a game of truth and error and not just a game of desire and aversion, of love and hate, of the useful and the harmful, of the effective and the ineffective.

Why this game of truth and error in addition to the real?

But isn't this the same thing as the critical question?

Undoubtedly not.

For the critical question bears on the possibility that there is truth, and [on] the conditions given once and for all for there to be truth.

Epistemic surprise does not bear on the possibility of knowledge itself, but on the existence of the true/false game, and on the indefinite history of this game that continues, recommences, is transformed, shifts.

And I would like merely to mark here some characteristics.

(1) It is improbable in the sense that one should never consider that it is necessarily, immediately entailed by the reality to which it is applied; one should always regard it as coming in addition, not from another world or from a mind foreign to reality, but through a series of encounters and conjunctions that one has to try to determine each time. Why the game of true and false with regard to madness? Why a game of true and false with regard to the movement of bodies, and why a game of truth with regard to secret desires of our soul? One must hold to this surprise and never consider that the being of something is ever sufficient to account for the game of T[rue]/F[alse] that concerns it.

(2) It is unprofitable. I mean that its index of utility is extremely weak. Simply recall: all the reality that passes outside the game of true and false, of how little true, definitive and finally established truth is produced by this game of the true and false; and recall the fantastic efforts made for this game and, in this game, the individual achievements, the normative systems that have been established to govern them: dogmas, sciences, opinion, institutions responsible for watching over them (producing, reproducing, imposing): education, scientific institutions, Church; the economic, social, and political costs. When one thinks of what has been done for the truth (from the Wars of Religion to nursery schools) and of how little truth that can be said and of the even fewer things that being true gives us a hold on reality, one may well say that the game of T[rue]/F[alse] has been a formidable expenditure in human history, but [has been] only pure expenditure.

(3) To say that it is a pure expenditure or that it is unprofitable does not mean that it is without effect; more precisely it is unprofitable in the sense that it does not open up, or opens so little access to reality, but it is not without effect inasmuch as it [*illegible*] as game of the true and false in reality; more clearly and taking some examples: we know that economic knowledge gives only an extremely limited hold on the reality it claims to show and we do not even know very well what reality is thereby shown; and yet the existence of a game of T[rue]/F[alse] in this domain of things and practices has modified a considerable number [of things]; the appearance of a "strategic knowledge" has had incalculable effects on military practice and on ways of killing people. All human practice is connected with games of truth,

ing that things are such as the proposition establishes them. But it is not because things are such that the discourse in which the proposition is found will exist as reality. This is the first point, and, in a questioning that fastens on the reality of discourse, the way, the reason why one cannot accept an analysis in terms of documentary locating like that I have just [stated], a documentary locating that once again is legitimate for the historian, [but] not for that form of history that wonders about the reality of the discourse. [We should wonder] about the fact that in addition to things, there are discourses; [we should pose] this problem: why, in addition to reality, is there truth? If we can say that ontological astonishment consists in wondering: "Why is there being rather than nothing?" well, I will say that there should be an epistemic astonishment, an epistemic surprise that should always be kept as keen as possible, and which is: why then, in addition to reality, is there truth? What is this supplement that reality in itself can never entirely account for, and which is that truth comes into play on the surface of reality, in reality, right in the depths of reality—not by a

---

it is this insertion of games of truth in human practice that is the bearer of essential effects and not the opening of these games of truth onto reality.
Latent relation of connection [versus] sagittal relation of representation.
(4) The relation of connection with domains of practices is not everywhere analogous; the regime of veridiction may have for example a role of evasion, marking, displacement; in relation to the reality to which it is linked, it may also have the role of rationalization that I have just been talking about, but there is no general and uniform function; it is an individual case every time. And conversely, the same regime of veridiction may be connected to different practices in which it does not have the same role; veridictions are polyvalent.
Objection: you do not cite the sciences.
(5) But this is because these games of truth are very polymorphous; in Science ("la" science), we know this well, the games of truth of genetics cannot be superimposed on those of algebra or particle physics. But above all there are many other games of truth: it is a singular historical circumstance that has placed Science ("la" science) in pole position in relation to the other games of truth; so there are non-superposable games of truth.
(6) The analysis of these games of truth of these regimes of veridiction may therefore be thought in a very diverse way.
One may ask for an account of their relation to reality: do they indeed manifest reality in its truth? But one has to realize that this will in fact consist in submitting them to criteria and other games of veridiction; it is a matter of asking what effects of obligations, constraints, encouragements, limitations have been created by the connection of determinate practices with a game of T[rue]/F[alse], a regime of veridiction itself marked by particular characteristics. So we may speak of a political history of truth. But one really must understand: it is in no way a matter of saying that scientific knowledge depends on certain institutions of political power that give rise to its existence, inflect its results in terms of its interests. What is involved is showing the effects and results of the interaction between the forms of a practice and [illegible]; to what obligations the subject of this practice is bound when the division of the T[rue]/F[alse] plays a role in it; to what T[rue]/F[alse] obligation is the subject of true discourse bound when it is a matter of a definite practice."

logic or necessity internal to the reality itself within which this truth operates, but by something else that is the supplement of truth to the reality of the world? The reality of the world is not its own truth to itself. Or at any rate, let's say that a true thing's reality is never the factual reason why the truth of this thing is said in a discourse. When I speak of this epistemic astonishment that consists in asking oneself: why, in addition to reality, is there truth? I do not mean truth understood as the truth of a proposition, but as a certain game of true and false, a game of veridiction that comes to be added to reality and that transmutes, transforms it.

I think we need to recall at least two or three things about this game of veridiction that I am trying to question. First, this game of veridiction is then a supplement in relation to reality. Removing the logicist postulate that would consist in wanting to deduce the game of true and false from reality. Second, this game of true and false is [not only] a supplement, but I would say that it is unprofitable in that one cannot deduce this game from a simple economy that would make it effective in relation to the domain on which it operates. When one considers the considerable deployment of the game of true and false and how little actual, effective, and useful truth humanity has been able to get from this game, when one compares what costs, what economic, political, social, human costs it needed, what sacrifices and wars even in the strict sense of the term these games of veridiction, of the true and false called for, and when one sees what the economic or political benefit has been of the truth found through this game of true and false, the difference is such that one can say that on the scale of human history the game of veridiction has cost much more than it has yielded. Supplementary game: removing the logicist postulate. Unprofitable game: removing the utilitarian and economic postulate.

A polymorphous game also, inasmuch as there is not only one game of the true and false, just one game of veridiction. After all, science is only one of the possible games of true and false. You know indeed that without doubt the game of truth and falsity peculiar to science cannot be defined in its unity, and that it is not possible to speak of science but that one should speak of different so-called scientific games of true and false in terms of borders that are always both difficult to establish and changing. So: supplementary game, unprofitable game,

game that cannot be regarded as unitary and for the analysis of which, consequently, the postulate of scientificity should be removed.

Finally, the fourth characteristic of the game of veridiction: I should say that, while being supplementary, unprofitable, not unitary, not fundamentally, essentially scientific, this game is not however without effect. This game of truth and error, of true and false, these regimes of veridiction have effects in reality, effects that are not due to the fact that the truth is produced by these games of veridiction. What is important is not [the] sagittal relation of the game of veridiction to the true thing it says. What is important is the connection between the game of veridiction, the regime of veridiction, and the reality in which it is inserted or to which it refers. There is connection between games of truth and error, between regimes of veridiction and reality. It is in this connection that the real effect of games of truth and error is inscribed (*se marque*). And the analysis of regimes of veridiction may be called political analysis of the truth inasmuch as it involves show-ing the reciprocal effects of the connection between human practices and the regimes of veridiction related to them. This is precisely what I would like to try to analyze in the case of sexual behavior: what are the real effects that have actually been inscribed (*marqués*), produced, induced by the games of veridiction applied to sexual behavior? It seems to me—and this is the point of the analysis—that the real effects induced by the game of veridiction on sexual behavior clearly pass by way of the experience of the subject himself, finding his own truth within him and in his sexuality. The effect of regimes of veridiction on sexual behaviors is inscribed (*se marque*) in this subjectivity-truth relationship. But this was the same problem with regard to madness, illness, crime, and so on. This is the general perspective that got me interested [in], and stuck on this problem of the supplement of a dis-course accompanying sexual behavior, in paganism as in Christianity.

Let us return to the main line of our analysis, after this generaliz-ing parenthesis. So I have tried to show how the very existence of this discourse accompanying the code needs to be taken into account as a problem, [for] it cannot just be treated as a kind of sign, an element indicating the reality to which it refers. There is another kind of analy-sis that would consist in saying: we are not doing what those rather naive historians do who say: "That's it, we can stop our reflection when

we have shown that philosophers' discourse reflects real practice."
These slightly more subtle people say that they do not follow those
historians and do indeed hold that discourse has an entirely different
function that that of reflecting reality. Not only has it another function
than that of reflecting reality, they argue, but in fact its function is not
to reflect it. In other words, this second type of analysis that I refer to
consists in trying to locate the reality of the discourse in that which
excludes the discourse from the reality it is supposed to formulate or
express. [This analysis] consists in situating the reality of the discourse
in what it does not express of reality, or in what it denies of it. That
is to say, instead of dissolving the existence of the discourse in what
could be called the representing transition of reality, as in the previous
method, this method consists, and you will recognize it easily, in setting
up discourse rather as the very form of the non-representation of reality.
In this kind of perspective one can see very well how the discourse of
philosophers and moralists on marriage might be analyzed. From this
perspective, which thus makes discourse the element by which reality
is not expressed, we can see very well how the analysis might be and in
actual fact has been undertaken. We could say the following: historical
reality shows that in the domain of matrimonial practice there was a
complex process of a break up of family institutions, of a weakening of
the tight social and hierarchical structures of the city, a weakening of the
political power shared by at least some of the citizens, the formation of
a new, monarchical, autocratic type of political power in the Hellenistic
world, in the imperial, the Roman world, and so on. The withdrawal
of individuals into conjugal life as the only remaining stable social form
that could be maintained without the support of that relative autonomy of the cities and the strengthening of conjugal life [that follows
from this are] nothing other than the effect of that destruction of the
ancient social fabric. And in what does the moralists' discourse consist
in relation to this real process? It consists in representing this real process in such a way that the real itself is evaded. And if in actual fact the
moralists of the first and second centuries seem to represent an already
established practice purely and simply in terms of code, it is because in
reality, in this repetition, there was something essential. This essential
something was that reality was precisely not expressed, and that, under
the apparent repetition in the form of code of an already established

reality, the main, strong, lively point, the strategic, central element of reality was evaded, and all the phenomena of dislocation of the city's economic-political structures were concealed thanks to the displacement of the analysis. In this discourse marriage does not appear at all as the real effect of a real break up of social structures, but is re-transcribed as an obligation, so not as a fact but as an obligation linked to certain constraints that appeared at the level of ideality. That is to say, what made marriage necessary and obligatory is the ideal bond of each individual to the whole of humankind, to that ideal reality that humankind is for each individual. In this way, in this transfer towards ideality, in that reproduction of a real practice in obligation, the living heart and cutting edge of reality was thereby evaded[*] [...].

I think this analysis, which takes the form of ideological disavowal, nevertheless raises some difficulties and, whatever its interest, cannot be maintained. [This is] for two reasons, the first specific and the other more general. Specifically, it does not seem that this kind of analysis really accounts for what happens, for the relationship between philosophical discourse and the reality it talks about. In fact, when we look more closely, what is it that is hidden in the philosophers' discourse, what is hidden in the reality of the actual practice? Have the philosophers hidden the generalization of marriage? Not at all. Have they hidden the forms that marriage should take, the requirements it imposes, the obligations to which it binds individuals? Absolutely not, the philosophers only stress them even more emphatically than anybody. Do the philosophers conceal or evade the renunciations entailed by this monogamous marriage that has become the exclusive place of sexuality? Not at all, they insist rather on all that such a marriage necessitates in the way of self-mastery and control of one's desires, of renunciation of pleasures, and so on. The discourse of the philosophers expresses all this reality, and the analysis in terms of ideology cannot deny the presence of this reality. When one looks at how ideological analysis proceeds one sees that in fact it consists in saying that what is hidden in a discourse like that of the philosophers is, what? Not the reality itself of the practice. What would be hidden according to those

---

[*] Gap in the recording: All that can be heard is: anchors in the reality of an already established practice.

who make this kind of analysis is that the generalization of conjugality is linked to a certain cause, which is the break up of the family. Or the analysis in terms of ideology tries to show how the philosophers' discourse conceals the weakening of social bonds, as this weakening would be the cause of the intensification of the private and dual relationships between husband and wife. So that, in fact, we see that in pursuing such an analysis the ideological disavowal of the discourse does not bear on the reality the discourse speaks about so much as on the cause that the ideological analysis attributes retrospectively and hypothetically to reality. In the analysis that denounces the unspoken of a discourse, one recognizes that a discourse is ideological in the fact that it does not speak of the same causes as the one analyzing the discourse. And by this one comes back, in an inverted form, to the idea that the existence of the discourse is always a function of the relationship of the discourse to the truth. Analysis in terms of ideology always reveals the discourse studied as fallen, alienated, deceptive in relation to what would be the as it were original, authentic essence, function, and nature of the discourse faithful to its being, which is the discourse that tells the truth. Basically, ideological discourse always appears as alienated discourse in relation to the right discourse, the discourse that speaks the truth. Ideological analysis consists in dialecticizing, through the mechanism of deception, the movement of the false and the need to hide the old and stubborn logicist principle, namely that the truth of what is said must ultimately account for the reality of the discourse. And [the fact of saying] that one can account for the reality of the discourse by the non-truth of what is said does not change the sovereignty of this logicist principle that is found again underlying analysis in terms of ideology.

There is a third possibility of analyzing the relations between reality and the supplementary discourse that is as it were added to it. This is no longer the theory of reflection, of the reflection, and it is no longer analysis in terms of ideology; it is analysis in terms of [rationalization]. [This would consist in saying:] of course, discourse does something else entirely than purely and simply represent reality, something else entirely than evade reality. Discourse works effectively on reality, and it does so by transforming it. By transforming it how? By transforming it through the very process of the *logos*, that is to say by the process of rationalization. We could say, and in a sense it would be entirely plausible and no doubt the direction one would be tempted to take most spontaneously: basically, what did the Stoic philosophers do in relation to this practice that pre-existed them and that in a sense they only repeated? They did not want to represent it and they did not want to hide it. They well and truly wanted to transform it. They tried to transform this, in a sense insular, discontinuous practice that was spreading progressively in the social body, no doubt from different social classes, maybe also from geographically distinct regions, they tried to transform it into a sort of universal rule of conduct. Or, they tried to connect together in a coherent system diverse behaviors, some of which concerned the juridical status of women, the slightly more egalitarian relationship that was established between the woman and her husband, the widespread obligation to marry, and so on. The Stoic philosophers [attempted] to present all these elements, which were not entirely connected to or necessarily entailed each other, as forming a logical, unshakeable, indissociable unity. If you marry, you must be faithful. If you have to be faithful sexually, this sexual fidelity must depend also on an affective fidelity. In short, from diverse behaviors they formed a coherent, internally systematic logic. Finally, we may say that the Stoic philosophers, in presenting this conception of marriage, tried to present an as it were radical, absolute model of phenomena that no doubt were rather trend-setting forms of behavior, behavior tending towards a bit more marital fidelity, a bit more marital affection, a bit more marital solidity. [They] presented this as a radical model: one must marry, one may marry only once in one's life, and when one is married one may have sexual relations only with one's wife, and so on. In short, they generalized local phenomena, they systematized dispersed phenomena, they radicalized underlying movements.

All this constitutes what could be called the rationalization of behavior. So we might say: the role of this redundant discourse has not been to reflect reality or to hide it, but to rationalize it. Here again we should pause for a moment and tell ourselves that without doubt this is not exactly satisfying. To start with, of course, because such a conception—let's say the Weberian method of analyzing discourse—implies all the same a quite arbitrary and problematic sense of reason. After all, why would it be more rational to impose absolute rather than relative conjugal fidelity? Rather than wanting to impose such a difficult, rigorous, absolute fidelity, everything suggests that social order is much more easily guaranteed by supple rules of fidelity, relative fidelity, a sieve form of fidelity. Why would the model of monogamy and conjugal fidelity be rational? Why would it be rational to make marriage a rational obligation for everyone and not simply a circumstantial rule that would permit people, that would advise them to marry if it is truly useful, indispensable, or if they want to (but especially that they do not feel forced to)? Finally, saying that these general principles are rational principles implies a very strange conception of marriage or reason, or maybe of both. And anyway, one also has to wonder why it would be rational to rationalize reality and practice. This is the general question that has to be asked when one speaks of the rationalizing functions of discourse (of discourse of truth, of knowledge, and so on) in relation to real practices. Do you really think that when a practice is rationalized it is in some way more rational (*raisonnable*) than when it is not? Is wanting to rationalize reality not the most absurd undertaking? We know full well that Laws, prescriptions, opinions, and advice on life with regard to marriage, affective relations, and sexual fidelity never have any effect etc., and that if things really have come about it is not because orders and advice were given. Procedures of rationalization have a very weak index of effectiveness. We know full well that, in any case, whether it is marriage or something else, reality does not function according to a rational order. Reality, at least if one means [by this] human practices, is always inadequate, always poorly adapted; it is always in the interstices between laws and principles on the one hand, and real behavior, effective conduct on the other, it is always in this interplay between the rule and what is not in accordance with it that things happen and things hold. If, by chance, reality were in actual fact rational, or if the rationalization of conducts, practices, and behavior were really to have an effect that made conducts adequate to the schema of rationalization offered to them, you know full well that they would immediately cease to be real.* What permits reality to exist, what permits reality to hold together, what permits human practices to continue to exist in their "own economy,"† is precisely the fact that they are not rational, or that there is always a necessarily insuperable distance between the schema of rationality offered to them and the reality itself of their existence. To invoke rationalization as if it involved a self-evident necessity clearly cannot constitute an answer. It is not because reality is real that there is *logos* in it, and it is not because it is necessary to rationalize it that one needs discourse.

The game of true and false that thus accompanied, enwrapped the great codification of sexual behaviors in the first two centuries CE, this regime of veridiction that accompanied the regime of jurisdiction of the code, this discourse of truth cannot be explained and analyzed in terms of reflection, ideology, or rationalization. So it is necessary to take up the problem again, to try to take up the material anew in order to try to see why and how this regime of codification, of jurisdiction—ultimately simple, ultimately fairly "effective,"‡ at any rate sufficiently solid to be passed on for centuries, for millenia from the Alexandrian, Hellenistic, or Roman period almost to our period—was accompanied by a regime of veridiction, by a game of true and false. This is the question that we have to try to understand, telling ourselves that, and this is precisely the point of the analysis, if the regime of jurisdiction, the system of codification did indeed remain the same, it is the regime of veridiction formulated by the Stoics that was transformed, not by Christianity on its appearance, but by processes internal to the Christianity of the fourth century. So it is the function and effects peculiar to the regime of veridiction found in paganism, comparing it with the effects of veridiction peculiar to Christianity, that I will try to explain to you in the last lectures. That's it, thank you.

---

* The manuscript offers the following formulation: "It is fortunate that the rational has such a slight hold on reality: it is this perhaps that permits the latter to exist."
† M. F. notes: in quotation marks.
‡ M. F. notes: in quotation marks.

1. *Le Doctrine des douze apôtres* (*Didachè*), vol. 1, ed. and trans., Willy Rordorf and André Tuilier (Paris: Éd. du Cerf, "Sources chrétiennes," 248, 1978); English translation Maxwell Staniforth, "The Didache," in *Early Christian Writings. The Apostolic Fathers* (Harmondsworth: Penguin Books, 1968). This text was studied by Foucault with reference to the problem of baptism and penance in *Des gouvernement des vivants*, pp. 101-106 and pp. 171-173; *On the Government of the Living*, pp. 103-108 and pp. 174-176.

2. Justin, *Apologie pour les chrétiens*, ed. and trans., Charles Meunier (Paris: Éd. du Cerf, "Sources chétiennes," 507, 2006); English translation by Alexander Roberts and James Donaldson as, Justin Martyr, *The First Apology of Justin Martyr*, in A. Cleveland Coxe, ed., *The Ante-Nicene Fathers. Translations of the Writings of the Fathers down to A.D. 325. Vol. I: The Apostolic Fathers with Justin Martyr and Iranaeus* (Grand Rapids, Michigan: W. B. Eerdmans, 1985). On the problem of marriage see ch. XV, Fr., pp. 167-169; Eng., pp. 167-168. Justin's book is studied by Foucault with reference to the problem of baptism in *Du gouvernement des vivants*, lecture of 6 February, pp. 101-106; *On the Government of the Living*, pp. 103-108.

3. Again, Foucault announces a more developed analysis of the doctrine of marriage in Christianity that he will not have time to put forward.

4. Marcionism is a heretical current of early Christianity deriving from the thought of Marcion (end of the first and beginning of the second century CE). The Marcionite doctrine (which has many features in common with Gnosticism, notably its dualist ascesis) is founded on the postulate of two divine principles: the Old Testament God (evil God of anger and the Law, Demiurge at the origin of the existence of matter) and Jesus Christ, God of the New Testament (God of love and liberation). By virtue of this dualist ontology, according to which the flesh depends on the evil principle, Marcionism rejects the dogma of the incarnation of Christ and the resurrection of the body, condemns marriage and reproduction, imposes a life of chastity and continence (abstention from meat, wine, entertainments, and so on). See Tertullian, *Against Marcion*, trans. Peter Holmes, in *The Ante-Nicene Fathers*, vol. 3, ed., A. Roberts and J. Donaldson (Grand Rapids, Michigan: Eerdmans, 1980).

5. Valentinus was a Gnostic philosopher and theologian of Egyptian origin of the second century CE. His doctrine, condemned by the official church as heretical, is complex, mixing Neo-Platonist and mystical conceptions, ancient Egyptian soteriology and Greek cosmology, Gnostic theses and Pauline theology. The whole universe derives, through the emanation of couples of aeons, from the union of a single and unique immaterial divine principle (the Father, or Abyss) with the feminine principle (Thought, or Silence) that coexists in it. Matter is born at the lowest point of the chain of emanations because of the "fall" of Wisdom that, for having wanted to contemplate the first principle directly, beyond its powers, falls away from it and gives origin to evil and the passions. The Demiurge creator is the Old Testament God, while Jesus Christ is the one who reveals the gnosis to men, enabling them to undertake the path of wisdom and liberation, and to reunite with the original Monad. The Valentinians deny the incarnation of Christ and the resurrection of the body. On account of their spiritualism, they can devote themselves to an ascetic life rejecting the life of the body, and so procreation as means of propagation of matter and evil. See Tertullian, *Against the Valentinians*, in *The Ante-Nicene Fathers*, vol. 3.

6. Clement of Alexandria, *Le Pédagogue*, livre II, ch. x: "Distinctions à faire à propos de la procréation," ed. and trans., Claude Mondésert (notes by Henri-Irénée Marrou) (Paris: Éd du Cerf, "Sources chrétiennes" 108, 1965) pp. 176-179; English translation William Wilson, *The Instructor (Paedagogus)* in *Ante-Nicene Fathers*, vol. 2, ed., A. Roberts and J. Donaldson (Grand Rapids, Michigan: Eerdmans, 1994), pp. 259-263.

7. Ibid., II, X, 92, Fr., pp. 176-179; Eng., pp. 259-263 [Most of this chapter is left untranslated in the English edition; G.B.].

8. Augustine, *Le Bien du mariage*, VIII, 8, in *Œuvres de saint Augustin, Opuscules II: Problèmes moraux*, ed. and trans., Gustave Combès (Paris: Desclée de Brouwer, 1937), p. 43: "That procreation of mortals that is the end of marriage (*ista generatio propter quam fiunt nuptiae*)"; English translation Rev. C. L. Cornish *On the Good of Marriage (De bono conjugali)*, in *A Select Library of the Nicene and Post-Nicene Fathers of the Christian Church*, vol. III: St. Augustin, ed., Philip Scaff, (Grand Rapids, Michigan: W. B. Eerdmans Publishing Company, 1988) p. 403:

"Thus ... this mortal begetting, on account of which marriage takes place." See also, ibid., XXIV, 32, Fr., p. 79: "procreation is the sole aim of marriage" [Unable to find English translation on basis on French reference given; G.B.]

9. Ibid., VI and VII, 6, Fr., p. 37: "performance of conjugal duty (*debitum conjugale*) is exempt from all sin"; Eng., p. 402: "to pay the due of marriage is no crime."

10. C. Vatin, *Recherches sur le mariage et la condition de la femme mariée à l'époque hellénistique*, p. 275.

11. Ibid.

12. Foucault here takes up a conceptual distinction that he first problematized in *L'Archéologie du savoir* (Paris: Gallimard, "Bibiliothèque des sciences humaines," 1969) pp. 13-15; English translation A. M. Sheridan Smith, *The Archaeology of Knowledge* (London: Tavistock Publications, 1972) pp. 6-7.

This page intentionally left blank

eleven

25 MARCH 1981

*The spread of the matrimonial model in the Hellenistic and
Roman period. ∽ The nature of the discourse on marriage:*
tekhnai peri bion. ∽ *Definition of* tekhnē *and* bios. ∽ *The
three lives.* ∽ *Christian (or modern) subjectivity and Greek* bios.
∽ *From paganism to Christianity: breaks and continuities.*
∽ *Incompatibilities between the old system of valorization and the
new code of conduct.* ∽ *Adjustment through subjectivation: caesura
of sex and self-control.*

THE PROBLEM I CONCENTRATED on last week was more or
less this: it can be seen that a certain model of sexual behavior was
actually common in the Hellenistic and Roman periods, a model
of sexual behavior organized essentially around marriage—around
the principle of monogamy, of course, but also of conjugal fidelity.
And in this model of conjugal fidelity, the model of sexual behavior
involved the preeminence of procreation and affective bonds between
the persons. This model, which is surprisingly moralizing when one
compares it with the idea one traditionally has of ancient morality, is
also singularly advanced when one thinks that it was taken up more
or less as such in Christianity. When I say "more or less as such," I put
stress, of course, on "more or less." There are notable differences, such
as, for example, the much greater value attached to total abstinence or
virginity in Christianity. One should also stress that in Christianity

the very way in which sex will be integrated into matrimonial relations will be in a way less restrictive, less austere than what can be found in the Stoic model at least. Finally this singularly advanced model is also endowed with remarkable stability, since all in all it is found from the Hellenistic period in which it began to spread up to [the period of what] is traditionally called bourgeois morality—something of course that should encourage some to pose [some] problems of history and method when they refer the matrimonial model of sexuality either to Christianity itself, or to capitalism, or to bourgeois morality, and so on. But let's leave that discussion.

The problem I would have liked to concentrate on is this: this model—which while not characterizing the behavior of every individual in Hellenistic and Roman societies, left traces, marks of its dynamism and development across these societies—was accompanied, redoubled, reproduced within a whole lengthy discourse insisting on the intrinsic value of marriage, the need to marry, on how to conduct oneself [and] define the place of *aphrodisia* in marriage, on the true nature of matrimonial *erōs*, and so on. In short, a discourse that seemed to repeat reality in terms of prescriptions. It is with regard to the relationship between this reality and the discourse that [appeared] to reproduce and reconstitute it in terms of prescriptions that I have tried to pose a problem of history and a problem of method: should one see this relationship between discourse and the reality it seems to reproduce as a relationship of representation? Is it a matter of a sort of ideological evasion of reality? Is it a matter of an effective rationalization or programming of reality? It seemed to me* that none of these methods and hypotheses could be regarded as very satisfactory, whether one tries to apply them to the problem in question or envisage them in themselves and with the postulates they entail. Since we have only two lectures left I would like to indicate at least the general principle of a solution, to indicate in what direction it seems to me one might advance in order to elucidate this problem a little.

The first step to take, or the first point to which one should devote oneself, apply oneself in the approach (before wondering about what these discourses said or wanted to do, their role), is of course

---

* M. F. adds: this is what I have tried to show you [see above, lecture of 11 February p. 123 sq.].

[to attempt] to take a bit seriously what these discourses were, or at any rate what they claimed to be. You know, I pointed it out at the start, all these discourses that [extol] matrimonial life, that [want] to indicate both that one ought to marry and how to conduct oneself in marriage, etcetera, absolutely do not put themselves forward, or not exactly, as rules, as a code. Nor, moreover, do they present themselves as a purely theoretical discourse on the essence of marriage, the essence or nature of *aphrodisia* or good conduct. They are neither codes, nor exactly prescriptive systems, nor theoretical systems. They present themselves as *tekhnai* (techniques) *peri ton bion* (whose object is life). We need to dwell on this notion of "techniques for living," that is to say something different from "rules of conduct,"* from what a code of behavior would be. They are techniques, that is to say ordered procedures, considered ways of doing things that are intended to carry out a certain number of transformations on a determinate object. These transformations are organized by reference to certain ends that are to be reached through these transformations. To carry out transformations on a determinate object with a view to certain ends is the Greek definition, and let's say the definition in general of *tekhnē*, of technique. *Tekhnē* is not a code of the permitted and prohibited, it is a certain systematic set of actions and a certain mode of action.

It is not very difficult to define *tekhnē*. The notion of *bios* (life), however, is much more difficult to pinpoint for us. When the Greeks speak of *bios*, when they speak of that *bios* (that life) that must be the object of a *tekhnē*, it is understood that they do not mean "life" in the biological sense of the term. This goes without saying.[1] But what do they mean? I think the difficulty we have in grasping exactly what the Greeks mean when they speak of *bios*, of that life with regard to which it is necessary to deploy *tekhnai*, [is due to] two things. On the one hand, the sharp Christian division (fundamental scansion) between life [in]† this world and life [in]‡ the hereafter no doubt brings about for us the loss of the unity and immanent sense of the Greek *bios*. [On the other hand,] for us a life [is defined] by a social division,

---

* In quotation marks in the manuscript.
† M. F. says: of.
‡ M. F. says: of.

according to profession and status. The Greek *bios* is neither profession nor occupation, and no more is it something marked out by reference to salvation, to the relationship and opposition between this world and the other world. Both the notion of salvation and that of status risk somewhat obliterating for us the sense of the Greek *bios*.

To grasp something of this word, very rapidly because it would of course need an immense study, I would like to recall a text from Heraklides Ponticus that was quite fundamental and of which all we have are some re-transcriptions by Cicero, Diogenes Laertius, etcetera.[2] Heraklides Ponticus tries to define the three forms of life, while attributing this idea moreover to Pythagoras, the value of this attribution and Pythagorean reference is not important. Heraklides Ponticus—I am quoting from the version given by Diogenes Laertius in the chapter precisely devoted to Pythagoras, Book VIII—says: Life is like a *panegyris*, some come to compete in the struggle, others to trade in the market, and others for the spectacle. Well, in the same way, in life (*bios*) some are born slaves to glory, others are greedy for wealth, but the philosophers pursue the truth.[3]

This [passage] is interesting because [it constitutes] the quasi statutory, quasi basic text defining the three major forms of life traditionally recognized in Greek thought. The political life [is led by] those who are slaves to glory—which corresponds to those at the games, the *panegyris*, who come to compete. There are those who come into life avid for wealth—which corresponds to those who come to the *panegyris* to trade. And then there are the philosophers who pursue the truth—these correspond of course to those who, in the *panegyris*, come to enjoy the show. Struggle, market, spectacle. Politics, chrematistics, philosophy: these are the three modes of life. I shall not dwell on the different meanings of these notions and the relation between them. My problem is the meaning of the word "life" in each of these forms, whether it be political, chrematistic, or philosophical. What seems to me characteristic in this analysis or definition by Heraklides Ponticus and in the panegyric metaphor that he gives to pinpoint, to fix the meaning of these three lives is that the life is not defined by the nature of the occupations so much as by what one desires, what one wants to do, what one seeks. What defines the political life is not so much that one is, for example, lawyer, strategist, or judge. What defines the political man in the text of Heraklides Ponticus is not his career. Similarly, what makes a man lead the *bios chrēmatistikos*, the life of wealth, is not that he is rich, it is not even exactly that he spends his time making deals. Even if in actual fact he spends his time making deals, this is only a consequence of his pursuit of wealth. Similarly being a philosopher is, of course, not having the profession of philosopher. It is not even possessing the truth. It is seeking it, or more precisely setting it as one's aim. In short, what characterizes the *bios*, in the three forms we know or even in general, independently of the forms in which it may be defined, is not then the status, the activity, what one does, and not even the things one handles. It is the form of relationship that one decides to have to things, the way in which one places oneself in relation to them, the way in which one finalizes them in relation to oneself. It is again the way in which one inserts one's own freedom, one's own ends, one's own project in these things themselves, the way in which one as it were puts them in perspective and uses them. Consider this metaphor of the *panegyris*, which is interesting. The *panegyris* is then this festival where many people come together and many things take place. It is the same festival for all. And what will define the *bios* is the end one sets for oneself when one comes to the festival, it is the way in which one will put in perspective, perceive those different choices that are common to everyone and that characterize for them [alone] the *panegyris*.

It seems that the Greeks did not know what subjectivity is, or that they did not have the notion of subjectivity.[4] There is certainly nothing in Greek that exactly matches our notion of subjectivity. But what comes closest to what we understand by subjectivity is this notion of *bios*. *Bios* is Greek subjectivity. And here again what prevents us from really grasping this sense of *bios* is the fact that we have what may called a Christian framework for codifying and thinking subjectivity. Despite everything, and even apart from explicit Christian references, we have a great model of subjectivity. This great model of subjectivity is constituted, first, by a relationship to a beyond; second, by an operation of conversion; third, by the existence of an authenticity, a deep truth to be discovered and that constitutes the foundation, the base, the ground of our subjectivity. It is this that Christianity has no doubt constructed over the centuries. It is this no doubt that we find again even now in our conception of our subjectivity. Let us say that the more religious forms lay greater stress on the pole of the hereafter, while we, in the epistemological bonds that we have with ourselves, rather place our relationship with our own subjectivity on the side of the discovery of an authenticity. Existence of a hereafter, a sort of absolute end, valid for all and beyond each of us, beyond our history, and which must however polarize our existence; the need to drag ourselves from what we are in order to turn towards this region of the essential; finally, through this movement that polarizes us towards the beyond and makes us convert with regard to the first immediate form of what we are, the possibility of discovering ourselves, of discovering in ourselves what authentically we are. We have here, [with these three determinations,] the general matrix of Western and Christian subjectivity.

The Greek *bios*—or this subjectivity, this thing that up to a point we may compare with our own subjectivity—[first] is not thought in terms of a beyond or an absolute and common end (*terme*), it is thought in terms of aims that each sets themselves. Second, unlike Western and Christian subjectivity, Greek *bios* is not defined by reference to the possibility or the injunction of a conversion, but by a continuous work of self on self. And finally, Greek *bios* is not defined by a relation to a hidden authenticity that has to be discovered in the very advance towards the absolute end and the movement of conversion; it is defined as the approach, or rather the indefinite or finite search in the very form of existence for an end that one both does and does not reach. So let's say that the discourses we are dealing with, and which present themselves as techniques of life (*tekhnai peri ton bion*), are at bottom procedures of constitution of a subjectivity or of subjectivation, and this is how they should be understood. They are not ideologies that attempt to conceal a code any more than they are the rationalization of a code. It is the definition of the conditions on which one will be able to insert as it were the individual's *bios*, the individual's subjectivity, within a code.

Second remark of a general kind. The best way to recapture something of what is expressed in these arts, in these *tekhnai*, these procedures of subjectivation, is I think to compare them to what takes place in Christianity, inasmuch as Christianity also had its *tekhnai peri ton bion* (its techniques of life). And here a few words of historiographical reminder to situate properly the problem and the way I have tried to construct this analysis of *aphrodisia* and the techniques of life regarding them in the Greek period. When one basically accentuates the let's say jurisdictional aspects, the elements of code that frame and give the outline of the permitted and forbidden regarding the *aphrodisia*, I think a certain number of things appear, a certain historical scansion is possible.

First, it appears that the discourses—specifically Stoic discourse, and more generally philosophical discourses in the last centuries BCE and the first centuries CE within ancient civilization—obviously, clearly and evidently, anticipate what will constitute the Christian doctrine of the flesh. At any rate, let's say that one has here a continuum that, through a certain number of minor modifications, allows one to offer a scansion going from the Hellenistic period up to a date x in Christianity. In any case, Christianity will not contribute any fundamental modification at the level of this philosophical discourse on the rules, the codes. This continuity between the moral doctrine of the philosophers and the moral doctrine of Christianity has been noted by historians for a least a century.*

Second, for twenty or thirty years, historical doctrine, which no longer focuses on the philosophical doctrines but on social, institutional, and suchlike phenomena, has shown† that, within society itself there was also a certain continuum, that is to say the appearance, the development of a certain model of conduct in the Hellenistic period that will be maintained and continue in Christianity. Here again, continuum. The appearance of Christianity absolutely does not bring about a break. Very recently, both with Bailey's book,[5] which appeared about a dozen years ago, and especially with Boswell's book, which came out in the United States last year,[6] we can pick out another type of continuum, or another way of establishing this continuity between Christianity and paganism. This does not consist in the model of moralization attributed to Christianity being discoverable well before the appearance of Christianity and in ancient civilization. It would consist in adopting the opposite approach and saying that many of the things seen as typical of ancient morality and opposed to Christian morality

---

* The manuscript adds: "In which the Stoics said what the Christians were going to do. Doctrinal anticipation."

† M. F. adds: this is what I reminded you of two or three weeks ago.

are found in Christian morality; [thus] tolerance of homosexuality. Boswell has shown how you find [this tolerance] as it were without fundamental modification up to the high Middle Ages and practically up to the eleventh and twelfth centuries.

So a whole set of signs tend to show that the long history, from the Hellenistic period to a period [in the course of which] Christianity asserted itself, imposed itself, and ends by becoming if not a universal, at least an official, deeply implanted religion, should be considered as a whole, an *ensemble*, a continuum. Fundamentally, Christianity really modified neither the models of behavior nor even the theoretical discourse that one may hold. Obviously the problem is where we should put the break. Should this *ensemble* start from the second and third centuries BCE, or should one wait until the Roman Empire? Should we look for the later break, towards the seventh-eighth century, at the time of the Carolingian Empire? Or should we wait, as Boswell suggests, until the eleventh-twelfth century? Well, all this is another problem. The fact remains that a certain consensus among historians would suggest the existence of this continuum.

I think in fact that if we adopt the point of view of the code, if we look at things from the angle of jurisdictions, of systems of prohibitions and tolerances, we may well accept a schema like this. But if we take precisely that history of the *tekhnai peri ton bion*, of technologies of life, technologies of the self, technologies of subjectivity, then we have to modify this schema, or to intersect the schema the historians establish regarding codes with another schema that marks other scansions, [notably] a scansion that specifically does not coincide with the appearance of Christianity but with a certain mutation within Christianity, which is situated in the third or fourth century, that is to say when Christianity has been led to replace the techniques of self, the technologies of subjectivity developed by classical or late antiquity, with new technologies of subjectivation, new *tekhnai peri ton bion*. And here an important phenomenon marked the appearance of what is not a new code of sexual behavior but forms the matrix of a new experience, what Christians call the flesh. It is to the comparison of these two major sets of techniques of life, of technologies of subjectivation that I would like to devote the rest of the time remaining to me.

✤

Today, I would like to identify something in these discourses I have
analyzed that may bring to light a certain specific technology of subjectivity or subjectivation. I would like to return to the starting point
for a moment. I started with the analysis of a text from Artemidorus, a
text of dream analysis, of method for interpreting dreams and grasping
their significations. Artemidorus' analysis made it possible to pick out
a sort of ethical perception regarding *aphrodisia*; it made it possible to
pick out two major principles [in terms of which] a certain value can
be attributed to the *aphrodisia* and to hierarchize them in relation to
each other, to define in short those which are good and those which
are bad, those worth more than the others and those worth less. The
two criteria at work in this analysis of Artemidorus were the following.
First, what I have called a principle of socio-sexual isomorphism that
posits the following criterion: the value of a sexual relation between
two persons depends, in part at least, on the type of social relation
between these two persons. For example, a sexual act that one practices
with a prostitute is perfectly acceptable inasmuch as the social relation between a free man and a prostitute is a relation that is in itself
accepted; the sexual relationship finds its exact place within the social
relationship. In the same way, sexual relations with a male or female
slave are acceptable [on the same criterion]. They do not circumvent
[the social relation], they do not ignore it, they do not reverse it. They
find their place in it, and it is the isomorphism of social and sexual
relations that makes sexual relations acceptable. And then, the second
great principle that we noted was the principle of activity. That is to
say that here too the value of a sexual act is, to an important extent,
determined by the activity of the subject involved in it. The more the
subject is active, the more the act is worth. The active pole defines the
positive value of the sexual act; passivity defines its negative value.

  These two criteria come together in the privilege of the male act. The
male act—basically in the form: penetration, ejaculation—constitutes on
its own the nature, the very essence of all the *aphrodisia*. To the point
that the feminine act, feminine pleasure were seen, analyzed, and considered in Greek thought (not only in moral thought, but in the physicians themselves) exactly on the model of the masculine act. There was

only one essence, one nature of the sexual act, which was that defined by the masculine sexual act. It was the social privilege of the male exercising his rights and the value of an activity that sanctions and rewards pleasure that allowed sexual acts to be valorized.

Let us now return to this new code of matrimonial *aphrodisia* that I have spoken about. You see that there is divergence and even incompatibility between this schema of valorization and this code. In fact, there is no longer any socio-sexual isomorphism in the new code. Why? Quite simply because, unlike previously, marriage is no longer integrated as one component amongst others in the general field of social relations. [On the one hand,] marriage constitutes an absolutely specific relationship heterogeneous to all the others, [and its] particular intensity means that it cannot be superimposed on any other social relationship. There is an insularization of marriage that means it is no longer one relationship amongst others: it is the absolutely privileged relationship around which others can be distributed but having a lesser intensity and a different nature. On the other hand, the new code of conduct shows us that the *aphrodisia* must be localized in marriage and nowhere else. Consequently you can see that the valorization of sexual relations can no longer take place according to the general grid of social relations. There can no longer be any socio-sexual isomorphism but rather a very marked social and sexual dimorphism, a dimorphism with marriage on one side—as specific relationship, the only one in which *aphrodisia* should occur—and then the other social relations that should be without *aphrodisia*. Let's say that there is a sort of "de-aphrodization"* of the social field in favor of the matrimonial relationship and of it alone. So you can see that there is incompatibility between the new code and the old traditional principle of socio-sexual isomorphism.

The second modification, the second incompatibility appears with regard to the other principle of valorization of sexual relations. This is that in the new code of conduct, in the new schema of sexual conduct, the principle of activity understood as male activity, as physical and social domination over the male or female partner, is no longer applicable since, in the matrimonial relationship, which should now

---

* In quotation marks in the manuscript.

be the only relationship entitled to sexual relations and *aphrodisia*, the husband's rights and obligations tend towards a certain equality with those of the wife. In any case, these rights and obligations must be rather strictly limited so that there can be a sort of reciprocity between husband and wife. Let's say that according to this new code the greatest pleasures of marriage, its benefits, are tied less to the exclusive and wholly dissymmetric activity of the husband than to the attractions and advantages of life as a couple and of a certain reciprocity of the two partners in this life together, even if many basic inequalities remain. Consequently, we see a code of behavior appearing and spreading that can no longer be inhabited by the same values. Or: one has a system of values to whose existence Artemidorus still attests at the end of the second century, and this code of values cannot be regarded as compatible with the code of behavior that is, moreover, attested.

So, to that extent, I think we can better understand the role of these philosophical discourses, or more exactly of these arts of living, these *tekhnai peri ton bion* that the philosophers, moralists, spiritual directors, or masters of existence tried to develop. These arts of conduct, these techniques of life will find their place between the spreading code of behavior and the surviving system of values. [...]* What [use] will these arts of conduct have? To adjust the code, tone it down, add some concessions such that it can find its place within the major term of reference of values I have been talking about? It could well be. For example, in the casuistry of the sixteenth and especially the seventeenth century, we clearly see how the casuists developed a whole art of adapting a code to an emerging system of values that was [inadequate to] the structure of the existing code. But this isn't what the philosophers, and particularly the Stoic philosophers are doing in the *tekhnai peri ton bion*, since far from mitigating the code, far from adapting it as far as possible to a system of values, they radicalize it. From a tendency to monogamy, conjugal fidelity, a degree of equality between spouses, etcetera, they make an absolutely rigorous system. They do not adapt a code to a system of values but radicalize it. It might also be thought that the role of philosophical discourse, of arts of living, will above all be to modify the system of values somewhat, to soften it, to color it in such a way

---

* Gap in the recording.

that it can incorporate the emerging code. These phenomena often occur. For example, in the Christianity of the High Middle Ages, the strictly Germanic code of incest, absolutely unknown to the Christians, Romans, or Greeks, was successfully integrated by a whole series of elaborations, of modifications to the system of values. But nor is this what the arts of living propose to do in the period and form of which we have spoken. For these arts of living it is absolutely not a matter of modifying the system of values. On the contrary, no-one values the principle of isomorphism or the principle of activity more than the Stoics and these authors of arts of living. Look, for example, at how the Stoics insist on the married individual as citizen of the world, at the fierceness with which they denounce any *pathos* in the individual's life, and you see clearly that they rigorously maintain both the principle of activity and the principle of isomorphism. In fact, it seems to me that the role of these arts of conduct is to prescribe procedures of the transformation of the subject himself, procedures of transformation that should permit him to stay within the code, to practice and accept it while keeping at the same time the two great axes of valorization (isomorphism and activity) I have talked about. It is the modification of the relation of self to self, it is the modification of the way in which one constitutes oneself as subject that will enable one to conduct oneself according to a code and according to values that are in fact incompatible with each other.

Two questions. First: what are the procedures of transformation of the subject that are actually put forward in these arts of living? Second: why was it necessary to adapt these two *ensembles*, to accept this code and maintain this system of values? So, first set of questions: what are the procedures of transformation of the subject, of the elaboration of self by self that are put forward in these codes? It seems to me that we can propose three major lines of elaboration and transformation of the subject. First, definition of a caesura in the individual's relationship to his sexual identity, or definition of two modalities of the individual's relationship to his own sex. Second, the constitution of the *aphrodisia*, of the pleasures of sex as the privileged object of the relationship of self to self, and this in the form of desire. Third, the constitution of an affective domain closely related to sexual pleasure.

Under these three abstruse formulae, this is what I mean. First: constitution of a double relation of the individual to his own sex. When we look at these analyses that focus on married life, we are struck by one thing, which is of course the constantly reiterated affirmation that one can be married, bound intimately to one's spouse, in a symbiotic relationship, a relationship of *krasis*, and at the same time a citizen, a citizen in one's city but a citizen of the world in general, a citizen in the midst of all humankind and a member of humankind. This general affirmation is doubled by a whole series of precise advice and views concerning this existence on two registers. Register of life with the wife, register of symbiosis, of the inner, private world. And then, on another side, public regime, external regime, collective regime. The dual relationship and the plural relationship are the precise, constant objects of these arts of marriage. We can say that with the Stoics in general, or at any rate with this kind of philosophy or advice on life, we see emerging very clearly and for the first time the distinction between public and private life, or [ex]ternal* world and private world. What strikes me is not so much or not only the existence of this double register, but the effort made, through these arts of living, to show the individual how he can be on both registers at once, how he can and must find himself exactly at the point of intersection between the dual relationship that characterizes marriage and the plural relationships that characterize his social life [...]†

In Xenophon, in the pseudo-Aristotle's *Economics*, and so on, there was a schema of complementarity that was already thoroughly established and that said that the wife, the household, the inner and private life, must [be] in some way complements to public life. There was also the old schema, the old model of the Roman aristocracy in which the relationship of man to wife is not just the other, complementary side to what happens in public life, but this (private, dual) relationship to the wife is the condition of existence of public life, although radically different from it. And conversely, public life does not appear as something different and complementary to private life, but something that has its form and effect in private life itself. With, as point of articulation and element that allows conversion from one to the other, precisely the man, the husband, the citizen, who is on both registers and

---

* M. F. says: internal.
† Gap in the recording.

who finds the condition of his public life in his private life, and finds in his public life the elements that will enable him to assert and assure his authority in the private world. You remember, I laid some stress on that letter of Pliny's that described his relationships with his wife.[7] His wife—in the private relationship, the relationship of symbiosis, in that intimate, affective relationship of love for her husband—loves him as a public man with a career, who makes speeches, who has an activity on which she gives her judgment, for which she feels admiration, and to which she also gives her help, her advice, and so on. And so it is as private man, relying upon this private life, this private relationship that he will be able to lead his public life. And his public life is actually present within his life as a couple, his dual relationship with his wife. So, rather than a dissociation, what strikes me is the connection between these two registers, with the man situated at the point of intersection of the two systems of relationships.

But within the two bundles of relationships, the man, who is then the husband, the male, the free citizen, and so on, will have to have two different types of relationship to the fact that he is a man, to his virility. In the dual relationship, his virility will be the possibility—better: the duty—to have sex with his wife. He will thus be a male individual inasmuch as he will have a sexual activity within this relationship. And in the social field, he will again be a man, a male, but this time his identification as man, as male will no longer be linked to the exercise of a sexual activity, but simply to the fact that in society he actually has a status as man, a status of masculinity that excludes or is independent of the exercise of his [sexual] activity itself. In other words, we will see an essential split appear in the individual's relationship to his own sex between the activity of male and the status of masculinity. If the man is actually able to assure the continuity between the social and the familial, if between these two types of relationship (dual and plural) there is the continuum I was talking about, this is because there is an individual who is the hinge. This individual can be the hinge and assure the family-society continuum, the socio-sexual continuum only on condition that he has as it were two sexes: a statutory sex and a relational sex, a sex-status and a sex-activity. It is this split that is the condition of the socio-sexual continuum, it is this split that is the condition in terms of which family relationships and social relationships will, despite their heterogeneity, nevertheless be able to form one and the same relational continuum, one and the same relational field around the man who thus now has a statutory sex and a relational sex. You will tell me that ultimately this is all rather simple.* Yes of course, but precisely it needed a whole elaboration of the individual's relationships to his own sex for it to be possible. What is it exactly that characterizes the socio-sexual continuum in the old system of conduct and before this elaboration I am talking about? It is precisely that the individual, when he was a man, when he had a social status as a man in society, had by this very fact the possibility and right to exercise his status as man in the form of a sexual activity. Since there were around him slaves, prostitutes, youths, and so on, his situation as a man, as masculine individual in the social field immediately and directly entailed the possibility of exercising his sexual activity. In the period I am talking about (the first and second centuries), what the arts of living tried to establish, within the subject himself, is the recognition of the double signification of sexual identity. To be a man is to have the status of man in the social field, but it does not entail the existence of an activity and the right to a sexual activity. It is simply in marriage that the status as man will take, firstly and originally, the sense of sexual activity from which, in this same marriage, the status of man follows. This split, this reversal of the relations between status of man and virile, sexual activity is one of the main objectives of the arts of living.

The second major operation that these arts of living tried to constitute consists in making the *aphrodisia* (pleasures, sexual activities) a privileged object of the relationship of self to self, and this in the form of desire. To get you to understand quickly what is involved, I would like to return to a text from Musonius Rufus regarding the symmetry of adultery by husband or wife.[8] You recall this fundamental text that— not for the first time, because there was already a formulation of this

---

* The manuscript gives the following comments:
"It seems very simple ... But it is very important, and difficult to attain. Look at what happened for the women: the uncoupling between sex-activity [and] sex-status did not come about so quickly. And the consequences were important for homosexuality: so long as sex-activity was not separate from sex-status, every time the latter was at stake (for example in the army, in pedagogy, in political life), sex-activity was linked to it. It did not interfere with it, it doubled it, accompanied it, was inseparable from it. And then when sex-activity and sex-status were distinguished, all aspects of sex-status have to be finished off (or made invisible)."

kind a bit before, but no matter—establishes in the strongest way the man's obligation to be faithful to his wife, since the wife is faithful to the man. You recall his argument: How could one permit a man to have sexual relations with his slave when the wife is absolutely not permitted to have sexual relations with a slave, which would be the most shameful thing in the world? And since it is shameful for the wife, it is just as much so for the man. Therefore the man must not commit adultery. Or rather Musonius Rufus shifts the very notion of adultery, which previously was the act of taking someone else's wife. Henceforth, any sexual relationship outside marriage falls under adultery and becomes impossible. How does Musonius Rufus establish this symmetry? He could establish it—Stoic doctrine would allow this—in as it were strictly juridical terms: since there is no natural inequality between human beings, everything required of some (namely wives in general) should be required of others in the same way. Musonius Rufus and the Stoics generally have the theoretical instruments for establishing a formal and juridical symmetry. Now when we look at the texts of Musonius Rufus we see that this is not at all what he says.[9] Or rather, this is not the argument by which the principle of symmetry of obligations is established. Let's assume in effect that a husband, unlike his wife, permits himself sexual relations with a male or female slave. In such a case he would show that he is not master of himself. He would show that the desire for sexual relations is so intense and violent in him that he cannot control it, that he is not master of himself. Now in the conjugal relationship, in that dual relation between husband and wife, if it is true that husband and wife are in a position of juridical equality, the fact remains nonetheless that the husband's real role is to be the wife's guide, it is for him to show the right way, to show how to live, to give the living example of the way of living. [Musonius Rufus] uses the word "pedagogise (*pédagogiser*)": the husband must *paidagōgein* his wife.[10] Now how could he establish this relationship of mastery over his wife if he was unable to establish it over himself? Strangely, Musonius Rufus does not establish the principle that there must be a reciprocity and equality of obligations between man and wife on the basis of juridical equality but on the basis of moral inequality.[11] Since, up to a point, the man must be guide and master of his wife's life, he must be master of himself; which at best restores the ultimately rather fictitious character of the principle of juridical equality and its rather minor importance in the analysis. It is interesting also to see how the inegalitarian relationship of man and wife is absolutely at the heart of the analysis.

But it is especially interesting to see that it is the relationship of mastery of self over self that becomes fundamental, and more exactly foundational of the relationships of mastery that one individual has to exercise over another. It is on condition of being master of himself that he will able to be master of the other. There is something very new here compared with the old, absolutely classical theme of mastery of self that is found in Plato, Socrates, and so on. Actually, in the ancient system or practice of mastery of self as defined in Xenophon regarding Socrates, or in Plato, self-mastery is as it were the kind of limit, or measure that defines which desires one will be able to realize, which desires one cannot realize. And it is when one is master in the sense of free and active citizen that one is master vis-à-vis a boy to whom one teaches the truth, how to live, and so on. In is in this relationship of mastery, and as it were at the end of this relationship of mastery, that a limit must be fixed that will be defined by self-mastery. Self-mastery rounds off as it were the profoundly dissymmetrical character of the relationship of mastery one has with others. The mastery one has over others must be limited by self-mastery. On the other hand, in the system I am talking about—which is roughly the Stoic system, but you find it in other authors who are not specifically Stoic—the relationship of mastery* is not what sets a limit to mastery over the other, what prevents its abuse or excess. The relationship to self becomes the prior condition for having the right to mastery over others.† And consequently controlling oneself, having mastery over one's desire in particular—I will come back to this—is the fundamental condition. I say "having mastery over one's desire" because it is precisely this that makes the difference. In the old system of mastery, or in the old morality, the old technique of self, the problem was not so much mastering one's desire. Or rather, what was required was to exercise a mastery such that one's desire did not drag one into things that were not acceptable. The example on which the Stoics, and after them all Patristic literature will reflect, is

---

* The manuscript notes: "*sōphrosunē*".
† The manuscript adds: "Hence the importance of modesty."

the famous example of Socrates and Alcibiades. Socrates does not con-
summate any sexual relationship with Alcibiades, he abstains from a
physical relationship with Alcibiades, not because he does not desire
him, but because, desiring him, he thinks that he must renounce this
sexual pleasure if he wants to have the pedagogical relationship with
Alcibiades that he hopes for.[12] But the desire remains. For Socrates it
is not a matter of rooting out his desire from the depths of himself. He
has to show to himself and to others how, given the nature of this desire,
he can exercise self-control in relation to the acts to which this desire,
this *epithumia*, might lead him. The word Xenophon uses precisely
with regard to Socrates is *egkrateia*,[13] that is to say the courage involved
in standing firm when confronted by the enemy. In the Platonic sense
of the term, *egkrateia* is precisely, desire being what it is, being able to
resist it.[14] But it is an entirely different matter in the new technology of
the self that the books of existence, the treatises of existence, of conduct
put forward. The problem is making self-control the first and fundamental condition. It is not: "How can I prevent my desire from leading
me beyond what I want?" but: "What can I do so as to not desire, how
can I root out *epithumia* from myself?" This appears very clearly in the
texts of Epictetus in which he takes up the idea of the sage who succeeds
in resisting. But the sage Epictetus presents is not, like Socrates, someone who while desiring, resists his desire. It is Epictetus who, seeing a
beautiful woman or an attractive youth, does not even experience any
desire.[15] And it is here, in this self-mastery exerted at the very root of
the *aphrodisia*, that is to say at the level of *epithumia* itself, it is in the
eradication of *epithumia* that self-mastery manifests itself. This is how
the Stoics, or these theorists, these directors of existence are able to
maintain the principle of moral valorization of activity in the new code.
It will be exactly maintained in the sense that activity will always be
the main criterion enabling definition of the positive value of something
or of an act, but this activity is no longer defined in a relational way as
the activity of one on another (maintaining one's action on and domination of the other); it will be exercising one's domination of oneself,
acting on oneself, on the part of oneself that is desire. So it seems to me
that neither the affirmation or reaffirmation of the existence of a code,
nor the elaboration of a system of values that could as it were package
and justify the code itself, entered into the main objectives of these arts

of living, these *tekhnai peri ton bion* put forward by the philosophers of the first and second centuries. What they wanted was to establish, within an interplay, a distance, an incompatibility between an already established, traditionally maintained system of values and a developing code, the fundamental elements enabling the subject to be transformed in such a way that he can live in this code of conjugality, this code of matrimonial *aphrodisia*, while still maintaining the value of socio-sexual continuity and procedures of valorization by the principle of activity. The main aim was this transformation of self so as to live within a code, so as to implement the code within a system of values.

There are still two small questions to resolve. I will mention them rapidly next week: that of the constitution of a closely related regime of pleasures; and then that of why in fact it was necessary both to accept the code and maintain a system of values. Two questions that I will deal with briefly, and then I will show you how Christianity took up and transformed these techniques of existence, these techniques of subjectivation I have talked about.*

---

* The manuscript gives the following conclusion to the session, which announces the developments of the lecture of 1 April:

"The arts of conducting oneself define another way of constituting oneself as subject of pleasure. Pleasure can no longer be regarded and experienced as the other side or effect of activity. Dangerous effect to which it is necessary and sufficient to subject to measure. It is a matter of nullifying pleasure (at least as end or positive element) in conjugal *aphrodisia*, and of replacing them with a different type of sensation of benevolence, kindness, willingness, gratitude, etc.; of opening up a whole set of movements of the soul that are no longer held in the simple schema: desire/pleasure, but integrated into this new type of experience: matrimonial *erōs*. In short three operations: freeing the individual from the socio-familial status that he was called upon to recognize as his sole fundamental mark; leading him to recognize himself as single element and on the border of heterogeneous relations; displacing the relationship between mastery of self and mastery of others, and fixing not the appropriate limit of the latter but the moral condition of the former; in short, defining a certain relation of self to self.

Questions: why was it necessary to maintain these values of isomorphism and activity, even though the code seemed incompatible with them? why was it necessary to accept the code if one wished to maintain these values?

The model was imposed from below (urbanization; role of provincial elites). As for the values, they were those of the traditional aristocracies. And the role of philosophers, who certainly addressed themselves to the aristocracy, was to show by what technologies of self one could practice the code while maintaining the same values. He could maintain his pre-eminence as free man, male, etc., in the social field while being married, on condition of making the caesura of relational sex/sex status. He could maintain the value of the principle of activity on condition of action on self as validation of action on others. He could find in these male activities the reward of pleasure on condition of opening up a new field of sensibility in himself."

268 SUBJECTIVITY AND TRUTH

1. See above the clarification already in the lecture of 14 January regarding the difference between *bios* and *zōē*.
2. This comparison is found in three texts, Cicero, *Tusculan Disputations* V, 9; Diogenes Laertius, *Lives of Eminent Philosophers*, VIII, 6 (this is the text Foucault cites in Festugière's translation), and Iamblichus, *Life of Pythagoras*, 58.
3. Foucault uses here the Festugière translation in A. J. Festguière, "Les Trois Vies," in *Études de philosophie grecque* (Paris: Vrin, Bibliothèque d'histoire de la philosophie, 1971) p. 118: "Life, he [Pythagoras] said, is like a *panegyris*. Just as some come to the latter to compete in the struggle, others to trade in the market, and the others, the best, to enjoy the show, so in life, some are born slaves of glory, others are greedy for wealth, but the philosophers pursue the truth"; English translation by R. D. Hicks and Diogenes Laertius, *Lives of Eminent Philosophers* (Cambridge, MA.: Harvard University Press, Loeb Classical Library, 185, 1925) vol. II, Book VIII, 8, pp. 326-328: "he compared life to the Great Games, where some went to compete for the prize and others went with wares to sell, but the best as spectators; for similarly, in life, some grow up with servile natures, greedy for fame and gain, but the philosopher seeks for truth."
4. Foucault may be thought to be alluding here to Heidegger who made the metaphysics of subjectivity, the determination of being-in-the-world as subject, begin with the Cartesian *ego* (*res cogitans*), Nietzsche and his will to power representing the last stage of this metaphysics.
5. D. S. Bailey, *Homosexuality and the Western Christian Tradition* (London: Longmans, Green, 1955).
6. J. Boswell, *Christianity, Social Tolerance, and Homosexuality* (see above p. 223, note 3).
7. See above, lecture of 11 March, p. 212 sq.
8. See above, lecture of 25 February, p. 160 sq.
9. In his lecture of 25 February (see above pp. 166-167) Foucault accepted this interpretation of equality in juridical terms that in the end he therefore does not accept.
10. The verb "*paidagōgein*" is used by Musonius Rufus, but its object is the desires felt by the man himself: *Prédications*, XII, 8, p. 95; *Reliquiae*, XII, p. 66: "However, I do not think one will judge men worse than women, nor less able to discipline their desires"; *Musonius Rufus*, pp. 87-89: "And yet surely one will not expect men to be less moral than women, nor less capable of disciplining their desires."
11. Ibid., Fr., "It is fitting that men be superior to women if it is true that they think it right to rule them. If however they appear more uncontrolled, they will also be more immoral. Now that a master sleeps with his slave is proof of lack of control and nothing else, what demonstration is needed?"; Eng., p. 89: "In fact, it behooves men to be much better if they expect to be superior to women, for surely if they appear to be less self-controlled they will also be baser characters. What need is there to say that it is an act of licentiousness and nothing less for a master to have relations with a slave? Everyone knows that." See *Le Souci de soi*, pp. 201-202; *The Care of the Self*, pp. 172-173.
12. The scene is recounted by Plato in *The Symposium*, 217a-219e. Regarding what Foucault calls the "test" of Socrates and its meaning, see *L'Usage des plaisirs*, p. 265; *The Use of Pleasure*, p. 241.
13. Xénophon, *Mémorables*, I, 5, 4, ed. and trans., Louis-André Dorion (Paris: Les Belles Lettres, CUF, 2000) p. 41: "Is it not the duty of every man who regards self-control as the foundation of virtue (*tēn egkrateian aretēs enai krēpida*) to assert it first of all in his own soul?"; English translation E. C. Marchant, Xenophon, *Memorabilia* in *Xenophon IV: Memorabilia, Oeconomicus, Symposium and Apology* (Cambridge, MA: Harvard University Press, Loeb Classical Library 168, 1923) pp. 66-67: "Should not every man hold self-control to be the foundation of all virtue, and first lay this foundation firmly in his soul?"; ibid., II, 1, 1, Fr., p. 1: "I have the impression also that when he held the talks that will follow, he urged his companions to exercise a control over their desire (*ekgrateian pros epithumian*) for food, drink, sexual relations, and sleep"; Eng., pp. 80-81: "In other conversations I thought that he exhorted his companions to practice self-control in the matter of eating and drinking, and sexual indulgence, and sleeping, and endurance of cold and heat and toil."
14. On this notion see the chapter "Enkrateia" in *L'Usage des plaisirs*, pp. 74-90; *The Use of Pleasure*, pp. 63-77.
15. Epictetus, *The Discourses*, vol. II, Book III, 3, 14-19, pp. 32-35.

twelve

1 APRIL 1981

Situation of the arts of living at the point of articulation of a system of valorization and a model of behavior. ∽ *The target-public of techniques of self: competitive aristocracies.* ∽ *Historical transformation of the procedures of the distribution of power: court and bureaucracy.* ∽ *Re-elaboration of the principle of activity and socio-sexual isomorphism in marriage.* ∽ *Splitting of sex and doubling of self on self.* ∽ *Cultural consequence: fantasy of the prince's debauchery.* ∽ *Problem of the government of self of the prince.* ∽ *Subjectivation and objectivation of* aphrodisia. ∽ *The birth of desire.*

SO THIS IS THE last lecture for this year. Last week I tried to interpret the arts of conducting oneself in matrimonial life, analyzed in the previous lectures, by taking their avowed intention literally. This was to be *tekhnai peri ton bion* (techniques concerning life, techniques of life). These techniques of life occupied a place between a model of conduct on the one hand, and principles of valorization on the other; principles of valorization that can be found in a text like that of Artemidorus that I analyzed right at the start of the course.[1] These principles, [namely:] the principle of socio-sexual isomorphism and the asymmetric principle of activity, made it possible to assess the *aphrodisia* and their respective value, to hierarchize them in relation to each other. So, on one side there are these principles of valorization, the text of Artemidorus proving that in the second century CE they are still perfectly recognized, valid, applied, and brought into play. And on another side there are these models of behavior that are not just theoretical models but, according to the documentation studied by historians, do seem to have been real practices.

These models of behavior emphasized the insularity of marriage, its specificity, its singularity in the field of all social relations. They entailed an exclusive localization of *aphrodisia* in the matrimonial relation alone and, finally, bonds that, while being absolutely unequal, appealed to a certain reciprocity of obligations between husband and wife. So there were these models of behavior and this old set of principles of valorization. It seemed to me that the role of the arts of living— this set of philosophical, moral, and medical prescriptions that we have studied—was not to convey, impose, disguise, or rationalize the models of behavior. No more was their function to modify the principles of valorization. It seemed to me that they could be defined as technologies permitting the modifications of oneself that are necessary and sufficient for, on the one hand, acceptance of these models of behavior and, on the other, maintaining the intensity of the old schema of valorization of conducts and *aphrodisia*. The technologies of self, consequently, [are situated] between a model of behavior and a system of valorization.

But then—this is where I stopped last week—obviously the question arises: why should it be necessary to adapt a model of behavior to a system of valorization in this way? What necessity makes their adaptation obligatory? What makes it indispensable to keep this system of valorization and this model of behavior together? If there is some kind of incompatibility between them, if it is so difficult to hold them together, why should they be kept together? If the model of behavior I have talked about really was widespread in certain social strata, if it had increasingly taken hold in real practice, why was it necessary for the system of values, that old system I have been talking about, to be maintained? Why was that old system of values not simply brushed aside by the real extension of this new model of behavior? And, conversely, if that old system of values was solid, consistent enough, if it still had enough actuality to maintain itself until the end of the second century, how did it come about that the model of behavior was in actual fact widespread? How did it come about that it actually developed if it was in contradiction or entailed incompatibility with the system of values? So, if we situate these arts of living, these technologies of self between a system of values and a model of behavior, we still have to explain, or try to understand, why the simultaneous maintenance of both had to be at the price of a new technology of self. I would like, by referring to the work of Paul Veyne,[2] to indicate what seems to me could be a line of historical analysis of this problem, before drawing some conclusions from this year's study.

Consequently, a general principle of explanation. Some facts should be kept in mind. I mentioned [the first] when speaking about the model of matrimonial life[3]—I am not talking about the theoretical model, but of what was, according to the documents studied by historians, a sort of tendency of the evolution of matrimonial practice. This model of matrimonial life, which involves the single, exclusive character of conjugal life, an intense and stable relationship between husband and wife, and makes the couple a sort of fundamental relationship for individuals, a form of life, is not implemented in the big, important families. The important families of the Greek as well as Roman aristocracy had an entirely different idea of marriage, as well as a different marital practice. For these families marriage was a complex operation that corresponded to political, economic, and social objectives. [For them] the fundamental raison d'être for marriage was a system of alliances. Marriage organized around the privilege of the dual relationship, around the couple relationship, corresponded rather to the behavior of poor, or at any rate less privileged classes. The movement of the spread of this dual model of the conjugal relationship should be seen as arising from below. It is not a sophisticated, culturally rare model, arising in some elite, formed at the top, which is then developed and gradually takes hold in more and more classes. It is rather a centripetal model going from the countryside to the towns, from the province to the metropolis, from the Empire's provinces to the capital, from the less favored classes to the elite. So it is a centripetal movement from the bottom up.

We should also keep in mind that this philosophical teaching—traces, expressions of which are found in texts like those of Musonius Rufus, Seneca, Plutarch—with the theoretical model of marriage that it conveyed, was obviously not addressed to the popular strata. This theoretical model was not in principle [intended for] a broad stratum of population, even though in actual fact it sometimes had the appearance of diatribes. In fact this model of instruction was addressed by philosophers, pedagogues, spiritual directors, or life guides to an elite who could of course read and write, an urban elite, a both political and economic elite; the elite that, precisely, was not immediately won over by the new model of matrimonial life. It is easy to see the reason why such a lesson was found to be necessary for this social class, and for this class in particular; it is in fact that this social class, this traditional aristocracy, was most attached—as it were structurally, politically, culturally—to the old system of valorization. Actually, these aristocracies were traditionally (and this is where I refer to Paul Veyne's works) what are called "competitive aristocracies."[4] That is to say that social status, share of political power, and pre-eminence in the social field were determined in a shifting form of constant rivalry between clans, families, clan chiefs, and heads of families. Power—political power, economic power, social power generally—was exercised through a permanent process of jousting between a very limited number of individuals representing or belonging to a few families. This jousting between a very limited number of individuals characterized the political, cultural, social, and economic life of the cities of Greece and the Roman republic, even if, here and there, institutions gave the people or democratic bodies some possibility of arbitration and choice—but always [amongst] this limited number of contesting families and individuals in constant competition and conflict over this power in all its forms. If it is true that the Greek or Roman aristocracy, the Greek or Roman elite should basically be defined as a competitive aristocracy, then it is understandable that as such it was attached in its sexual behavior and mode of life to a system of values, to procedures of valorization [corresponding to socio-sexual]* isomorphism and [the] asymmetric principle of activity.

   This aristocracy was generally attached to those processes of valorization that made use of social hierarchies, of individual, family, or clan pre-eminence [in the] social field in general, of status differences as principle of regulation, principle of modelling, and also principle of

---

* M. F. says only: social.

assessment of each individual behavior. And this same aristocracy, as a competitive aristocracy, was also attached to a principle of valorization that relied upon the rights of activity, the exercise of authority and power, the bringing into play by individuals of social and status pre-eminence. As you see, inasmuch as this aristocracy was a competitive aristocracy, its mode of life could not fail to manifest itself in an attachment to a system of valorization whose principles I have given you and which appeared in the traditional ethics of *aphrodisia*.

Now this competitive aristocracy clearly found itself threatened, its power and the exercise of this power reduced, first of all by the development of the Hellenistic monarchies in the post-Alexandrian world, and then by the Roman imperial system—[which] modified its situation considerably. They limited the space of competition as it were, the possibilities and real effects of this competition inasmuch as, anyway, power, the exercise of power was fixed in a relatively definitive way and in a way blocked by the existence of a monarchical, dictatorial, imperial or whatever, type of power. So: a considerable limitation of the space of competition, of its stakes, [of] its effects, and the definition of another mode of exercise of power, of another way of marking the status of individuals in terms of their place. The individual's place, the status he was given were defined in relation to the sovereign in institutions of the court, which was unknown to the Greek cities of course, and to republican Rome. The court is therefore the procedure by which the individual's status is fixed on the basis of the person of the sovereign, of his choices and decisions. And court rivalries and conflicts, the confrontations that take place in the space of the court are clearly very different in their forms, effects, and rules of the game than the competition and political jousting that could unfold in a city like the Greek city-state or an organization like republican Rome. So the differentiations, the valorization of individuals take place, on the basis of the sovereign, through the court. Or—this is the second formula that will obviously be especially developed in the Roman world—they take place in relation to, thanks to, or through administrative mechanisms (what in our terminology will be called: the bureaucracy).

The distribution of power, the hierarchization of individuals, the definition of their position, and so on, thus takes place either on the basis of the court phenomenon, or on the basis of the organization of the bureaucracy. The old organization—the space of competition, rivalry, and jousting that characterized the procedures of distributing power, of fixing the individual's place in society, of their individual valorization or the valorization of their acts—is thereby completely modified. As a result, marriage, the conjugal relationship, private life are induced to play a particular role in this old competitive aristocracy that is now formed, distributed, divided up in these new spaces of the court and bureaucracy. In effect, this old aristocracy is forced to accept the model of behavior proposed by the new social strata from below, from the provinces—the social stratum that occupies the administration and forms its staff, its cogs, and that, with the emperors from the provinces, also begins to occupy the court. [Furthermore,] this model of behavior becomes absolutely necessary for this aristocracy when the space in which marriage—social, familial marriage, marriage between clans, marriage-alliance—loses its signification, undoubtedly its economic value, but especially its political value.

At the same time as it is reduced to no more than coexistence, a dual relationship between a man and a woman, this old marriage alliance, a component in a complex social strategy, becomes the site where one will be able to promote, preserve, put into practice and exercise the old principles of valorization (activity and socio-sexual correlation) that can no longer govern the rest of behavior, but subject to certain conditions.[*] [On the one hand,] the principle of socio-sexual correlation has to be reorganized entirely around the relationship between husband and wife, private life and public life, the couple [being situated] at the pivot between a dual, intense, personal, affectively highly charged relationship and a social life. And, on the other hand, the principle of activity has to appear as a principle, if not of asceticism, at least of self-control. It is in this way that these new arts of living, conveyed—or rather: formulated, developed—by philosophers on the basis of a real and widespread behavior, get their raison d'être from the fact that they are addressed to a very specific stratum of the population, to that aristocracy that still survived in its self-consciousness, in the recognition it had of itself only on condition that its old system of values be maintained and it actually be able to judge, to gauge its own behavior on the

---

[*] M. F. adds: that I tried to explain last week [see the lecture of 25 March, p. 249 sq.].

basis of it, although henceforth it can no longer be applied to the social field, to the field of politico-social differentiations to which it had been applied hitherto. This system of valorization must actually be implemented in marriage and only in marriage, on condition of course—this is what was necessary—that there was a whole reorganization of the relationship of self to self. Philosophical discourse was proposing, was conveying techniques, actually presented as such, precisely in order to be able to live, to accept the modes of behavior proposed and imposed from outside, techniques that literally rendered them livable.

These techniques of self that literally enabled subjectivation of a code, permitted valorization of a self-enclosed conjugality, constituting it as the highest possible purpose to be found in the social field, and valorization of the relation of equality or at any rate reciprocity between man and wife as the highest form of activity that the individual could exercise on himself. But for this two things were needed*: first, that in the individual's relation to his own sex, sex-status—through which and in the kinds of which the individual figures as simple masculine element in social relations—is clearly distinguished from sex-activity, the sex-relation, the actual sex relationship to the other—which henceforth can only manifest itself in the private and dual relationship of marriage. This caesura between sex-status, neutralized [on the level] of activity, and sex-activity, invested entirely in a dual relation, this caesura that, as it were, cuts in two masculinity, the *androsunē* the Greeks spoke of (since anyway it is always a morality of men that is at work), has now become indispensable. In the art of living, in the technologies of self, it is what makes possible the adjustment of modes of behavior and system of valorization. One had to be masculine in the social field and male in the dual conjugal relationship.† The second condition was that sex-activity not be simply the exercise by right as it were of a preeminence, a status, a position of mastery over a partner—whether wife, slave, boy, or prostitute ... Sex-activity no longer had to manifest itself as the exercise of an activity of general mastery in relation to any individual. It had to manifest itself as mastery, for sure, but as self-mastery. Sex had to

---

* M. F. adds: I come back to what I was telling you last week [see above lecture of 25 March, p. 262 sq.].
† The manuscript notes: "There are now two sexes, more exactly two modes of subjectivation of sex, two ways of recognizing that one has a sex."

become the object of a mastery that the individual exercised on himself, he had be the object of an activity of control, of limitation.

Briefly,[*] I will say that the integration of the code of conduct [linked to the] old system of values, the integration of two heterogeneous elements, concerning a determinate social class that was roughly the sociocultural elite of the period, could take place within a same mode of life only through a technology of self that was defined by philosophers who were precisely guides of life, spiritual directors, a technology that entails: first, a splitting of sex, or rather of the subject's relationship to his sex; second, a doubling of self on self as object of one's own activity. Splitting of sex, doubling of self on self: this is what appeared as the stake, the objective of these techniques that the philosophers offered, and so something entirely different from the ideological dressing of a code or the rationalization of a practice.

Now I think we need to move on to the consequences and, as far as possible, conclusions. Among the consequences, I will begin with perhaps the most anecdotal. It seems to me that this interpretation, this analysis anyway that I have tried to give you of the function of the arts of living, of the technology of the doubling of self and splitting of sex, may be confirmed in the following way. We could put opposite, alongside, in counterpoint to these philosophical technologies of good sexual behavior, of these technologies that take up and resituate the problem of the *aphrodisia*, a theme that is no longer philosophical, but historical, historiographical, and that was very important in the Roman Empire, especially in the second half of its history. This is the grand old theme of the prince's debauchery, a theme that was widespread in an historical and political literature and of which Suetonius,[5] Tacitus of course,[6] and then especially the *Historia Augusta*,[7] provide some flamboyant examples; a literature, we should not forget, that stemmed [from], was inspired by, was more or less (and often more than less) directly driven by a political opposition to imperial power, an opposition of

---

[*] M. F. adds: of course, I would not like you to take this schematization literally, it is just to summarize.

the senators; the senatorial class still representing, at the end of the
Roman Empire, that aristocracy that seems to me to have been the
target and site of formation, point of application at any rate, of these
technologies of conduct I have been talking about. This same senatorial
class, this same socio-cultural elite on which was imposed—or which
imposed on itself in order to accept the new code—that technology
of self, threw back at, projected onto the emperor (who, through his
existence and that of the imperial system was found to be the reason
why [the senatorial class] was induced to apply to itself this technology of self) the fantasy, the dream, the fear of precisely the prince's
completely dissolute sexual activity, his debauchery. This is certainly
not the first time of course that the [sexual] activity of the person who
exercises power became a moral and also a political theme. After all,
it has only been for a short time—and I am not even sure it has been
established—that the sexual activity of the person exercising power has
no importance. In any case, in the Greeks, and well before the period
to which I am referring, well before the development of the technologies of self, we [note] the importance, the concern with this problem
of the relation between the exercise of power and the sexual activity
of the one who exercises it. We would find many traces of it. It is the
very traditional figure that you see appearing from the literature of the
fourth century BCE, and you find it in Lucian,[8] Plutarch,[9] Maximus
of Tyre[10]; the figure of the tyrant who, from the height of and on the
basis of his power, abuses his power in order to exercise a violent, illegitimate, literally desecrating sexual activity. The theme of the tyrant's
rape of the free boy or girl is a constant and centuries-old theme in
Greek literature. But this is not the sole aspect. There are even some
very precise things on this in Greek legislation, such as, for example,
the effective and legal prohibition against the exercise of power by
anyone who led a debauched life in their youth and had been, in short,
a prostitute—although "prostitute" no doubt needs to be given a fairly
broad sense. There is a very interesting example of the application
of this law in the defense speech, the speech by Aeschines in *Against
Timarchus*.[11] Timarchus was one of his political adversaries, a friend of
Demosthenes who had performed an important political function as
envoy to Philip. When he comes to have his mission ratified by the
people's Assembly, he is attacked by Aeschines, who explains that of

course Timarchus' mission cannot be ratified and that he should be
deprived of any possibility of exercising political power in the future.
Why? Because he had had a shameful youth. When he was young he
had passed through the hands of a number of men with whom he had
lived and who had supported him, who had kept him as we would say.
So it is not exactly that he had been a professional prostitute but the
fact that he had entered into precisely those discredited relationships
with men that I talked about in another lecture. That is to say he was
found to have been in a feminine, passive position, and so on. The fact
of having been, in an absolutely explicit and as it were statutory way,
the passive and consenting object of another's desire, of having actually
occupied, with everyone's knowledge and in everyone's eyes, this position of femininity, or more generally passivity, disqualified him in the
name of the law itself from any political activity. He could no longer
exercise power over others when he was found to have been in that
position in his youth. Not only did he not have the right to exercise
power, but he did not have the right to speak, the right to political
speech. Which goes to prove that even in classical Greece, in classical Athens, the sexual activity of someone who exercises power is an
essential problem.

In the historical literature of the Roman Empire we find again, of
course, this question of the relation between the exercise of power and
the sexual act, but in a quite different form. For the senatorial class,
for those in whose eyes the emperor is an object of mistrust, especially
when his policy is contrary to [some principles] of this class, the bad
emperor, the one whose power can only be bad and fundamentally
tainted, is the one who does not apply to himself this technology of the
subject I have spoken about, and which now characterizes the virtuous
life. The emperor, the person who exercises power, the person who is
precisely above society in a position of absolute mastery without symmetry, the person who can exercise this power absolutely in the entire
social field, who claims moreover to deify himself and be something
other than a man, is precisely the person who refuses, evades, renounces
the technologies of the subject. So there are good emperors, whose
model is of course Marcus Aurelius, the emperor-philosopher, that is to
say the one who exercises control of self by self in the exercise of power
itself and whose position, power, and political authority are constantly

taken up, controlled within his own subjectivity by the mastery he exercises over himself.[12] And opposite these, the bad emperors basically manifest themselves by their sexual debauchery.

We can decipher the major figures of the prince's debauchery, of imperial debauchery—that commonplace of Roman political history— precisely on the basis of the two aspects of the technology of the subject I have talked about. The bad emperor is first of all one who does not practice the separation of the two relationships that the individual must have to his sex, who completely mixes up the statutory sex that makes him a masculine individual and male sexual activity. The bad emperor is one who asserts his sexual activity in his imperial functions, in the very exercise of political power over others, and constantly mixes, confuses political authority and sexual activity. The bad emperor is one who violates the wives of senators, who makes his favorites ministers, and so on. As a result, and at the same time, this non-separation leads him to trample on the principle of behavior of conjugal fidelity (he kills his wives, constantly divorces, and so on) as well as overturn the principle, the old value of social isomorphism. Not content in effect with abusively exercising his political and social preeminence in order to [feed] his sexual activity, he as it were scorns and tramples underfoot his own political and social preeminence. He exposes himself at the circus, frequents vile places, disguises himself as a prostitute in order to practice his debaucheries. And another aspect of the technology of the self that the bad emperor disregards, scorns, and tramples underfoot is that he does not exercise self-mastery on himself, he is someone in whom one does not find that doubling of self on self in the form of mastery. Of course, the bad emperor allows himself to be carried away by his passions, but he goes much further: he turns the very principle of activity around, becoming a passive subject, willingly, spectacularly, theatrically passive in the eyes of all. He is the plaything of his wife: this is Claudius.[13] And above all he is sexually passive: he marries his freed slave, like Nero.[14] And, of course, at the summit of all this, of the bad emperor, there is the figure of Elagabalus, in which a politically monstrous power and an unnatural masculine nature are closely linked.[15] Monstrous power since, for Rome, Elagabalus represents the model of the oriental monarchy, of the monarchy without law, reliant upon religious mystique rather than a political system of law—for the

West, the East has not only functioned as the pole of the most profound wisdom, it has also always represented the pole of the worst power. Elagabalus is the pole of the worst power. Unnatural masculine nature since, as a priest, he was castrated—this is all the false mastery of the eunuch who, having renounced his own masculinity, takes advantage of it to move on to the worst debaucheries, to sexual activities not orientated, controlled, or ordered by his masculinity. This is a debauchery of pure passivity, and finally it is the total reversal of the social field since Elagabalus prostitutes himself in the streets. The figure of the absolute prince, priest of an oriental religion, castrated and prostitute, whatever the historical truth of this—that is something else—reveals indirectly, but very clearly I think, the ethical and political value of these "arts of living properly"* I have talked about.

You will say that all this is of no great importance. I have reminded you all the same for two reasons. On the one hand, it seems to me that in this representation of the bad emperor, of the debauched prince, one finds again the reverse projection of those technologies of self that were imposed or being imposed in the same period. And inasmuch as this anti-imperial literature on the debauched prince very clearly, manifestly, and certainly had a senatorial, aristocratic origin,[16] one sees through this how especially intense this problem of the technology of self was for a certain class of society. The prince's debauchery is a subject that has preoccupied me for some time for this reason. [Subjacent to] this theme of the prince's debauchery one finds the following theoretically, politically, and historically important question: what is the prince as an individual subject? What is the prince insofar as he is self, insofar as he has a relationship to himself? In our political thought, for a number historical reasons that are easy to figure out, the question especially posed since the sixteenth and seventeenth century is not so much what the prince is as individual, as subject and object of himself, but the opposite question: what is the relationship between subjects (in the political sense) and sovereignty? How, to what extent, up to what point, on what basis can the individual, individuals, you and me, be the foundation or the source of a political sovereignty? Trying to find the root, the deep-rootedness of the principle of sovereignty

---

* In quotation marks in the manuscript.

in the individual subject has been the philosophical, juridical, and political problem of Western thought since the sixteenth-seventeenth centuries. But before and independently of this, an entire, both historical and moral literature—a literature that we read now at the simple level of anecdote or somewhat tiresome moralism but that was actually a "political" literature (in quotation marks inasmuch as I am not sure that the blunt application of the word is entirely valid in this domain), a serious literature—persistently raised the problem of the sovereign as subject, that is to say as having a relationship of self to self. This question appears very clearly in Roman historical literature precisely at a time when what is, after all, a new notion of the absolute sovereign has just appeared.* It appears at this point through these moral portraits of princes, which are, I think, an entirely serious genre and were to remain so. It would be wholly insufficient to see this anecdotal, individual, biographical description of princes as a rather naive monarcho-centrism driven merely either by the animosity of some towards the emperor, or by a moralism. A fundamental problem is posed in this way: that of the prince's power over others and of the technology of himself, of the prince as subject, of the prince inasmuch as he governs others and has to govern himself.

It seems to me that the problem of governmentality—government of self by self and government of individuals by each other—in its generality, in all its dimensions, in its two major aspects, appears quite clearly in this literature that, once again, is not a moralizing literature. You are obviously familiar with the considerable work of Marc Bloch on the character of the king as individual endowed with religious powers in the Middle Ages, especially in France and England, as individual who, sacralized by the exercise of his power, above humanity and other people, occupies an absolutely specific and mediating position between God and men—God's delegate, Christ-king, Christlike figure ... Marc Bloch has studied all this very well.[17] But I wonder whether one could not study the character of the king [from the opposite angle]. Not in that position above men where he exercises a Christlike power

---

* The manuscript notes:
  "A question that arises I think under the Roman Empire, for reasons I have just given, with the two figures: Elagabalus, Marcus Aurelius (for him it was a real problem that he really wanted to deal with)."

or mediation between the gods and men, not the king as endowed with those religious powers that manifest themselves, for example, in the [gift]* he has been given to cure scrofula, to perform miracles (the thaumaturge king). I wonder whether one could not study what I would call the king's sub-power, his non-mastery of himself, the prince insofar as he is passive in relation to himself and insofar as that passivity manifests itself by a whole range of things that may be the sick king (the wounded king, the unfortunate king—Arthur, the Parsifal theme, and so on—also the mad king, Charles VI of course) as well as the king prey to his passions, the debauched king. And obviously there will be the series of English kings, the series of French kings who thus appear in the form not so much of the sub-man but of the one who, in the exercise of power and responsible for governing others, is not able to govern himself: the passion-prince.

Much more important from the methodological point of view are the comments I want to make now. It seems to me that in this analysis of the technologies of self, that you may have found rather plodding and slow, we may even so grasp an historically important moment in this history that would be the history of subjectivity—of subjectivity understood as the set of processes of subjectivation to which individuals have been subjected or that they have implemented with regard to themselves. It seems to me in fact that we see here the point of intersection where two processes, linked to each other but distinct from each other, are articulated. First, in the technology of self I have talked about regarding sexual pleasures, the *aphrodisia*, we see the formation of an as it were singular, permanent, subjective bond between the individual and his own sex as principle of activity. More clearly, we started from the notion, enigmatic for us, somewhat difficult to recapture, of *aphrodisia*, which we translated as: sexual acts, pleasures of love, and so on. This notion of *aphrodisia* was basically characterized precisely by the fact that in Greek thought there did not exist a sort of uniform and continuous category permitting the definition of something like sexuality, or what Christians called the flesh. The Greeks knew neither sexuality nor the flesh. They knew a series of acts called *aphrodisia*, which fall within the same category and involve the same type of behavior,

---

* M. F. says: power

the same practices of the body, and so on. But in any case, these are *aphrodisia*, sexual acts, and not something like the flesh, like sexuality. It is impossible to find [the equivalent] of these categories in Greek thought. *Aphrodisia*, in the Greeks, *veneria* in the Latins, is an activity.[18] It is not a property, a feature of nature, it is not a dimension of subjectivity, it is a type, a series of acts characterized by their form, by the violence of the desire that traverses them, by the intensity of the pleasure one experiences, and by the fact that it is an activity that, due to this violence of desire and intensity of pleasure, is in danger of escaping itself and losing control of itself. Such were the *aphrodisia* in Greek culture and thought. But as soon as one [establishes] a technology of self in which the subject must define two modalities of relationship to his sex—a statutory modality coextensive with the field of social relations and a relational modality coextensive solely with conjugality—as soon as this caesura has to be carried out, maintained, ever renewed, in short, when the subject must not exercise his male activity where he has to manifest himself solely as having masculine status, a permanent relationship is established between the individual and this caesura, this split that separates in him the two aspects of his sex. Henceforth there is a constantly reactivated relationship between the individual and his own sex. Of course, in this Greek and Roman philosophy of the first and second centuries, this permanent relationship is not yet defined as what Christians will call the flesh and the moderns sexuality. This permanent relationship does not yet have a name, because it is not yet conceptualized. It remains nevertheless that we have here, in this caesura and its maintenance, the principle, the establishment of the possibility of a permanent and fundamental bond between the subject and [the principle of]* his sexual activity. Sexual activity becomes, through this, a permanent dimension of subjectivity, or of one's relationship of self to self. Let's say roughly and schematically that the Greek *aphrodisia* were a series of acts in which one essentially related to others—hence the importance precisely of the social field. Now, through this technology, the possibility and necessity emerges of a relationship in which the *aphrodisia* (sexual acts or the category to which they belong) will not appear as just a class of acts, but as a specific aspect, a dimension

---

* Inaudible passage; reconstruction according to the manuscript.

of the individual, a permanent [relationship] of himself to himself. What previously characterized the *aphrodisia* (a relationship of self to others) will now be internalized or projected onto the subject himself and become essentially relationship of self to self. So: subjectivation of sexual activity or transition from a subjectivation that had the form of acts to a subjectivation in the form of a permanent relationship of self to self.

The second process I would like to talk about, and which seems to emerge from the establishment and development of these technologies of self, is a process that I would call, again somewhat schematically, objectivation. These *aphrodisia*, as we see them in Greek thought and culture, were therefore, first characteristic, a series of natural acts. But I just reminded you that these acts were inhabited by, undermined as it were, at any rate threatened from within by a violence, an impulse that was paradoxically both natural and excessive. These acts, these *aphrodisia* thus called for a quantitative measure fixed by a certain regime of use. One has here two important notions: the notion of use and the notion of regime. As they are acts, one has recourse to them on occasion, and this is the *khrēsis aphrodisiōn*.[19] Because of the internal threat due to their characteristic natural excess, this *khrēsis* of the *aphrodisia*, this resort to the *aphrodisia*, this enactment had to be subject to a limitation that was the regime, a regime defined in terms of the needs of the body or anything else.

The idea that the *aphrodisia* are acts to which one must have recourse, and that this must be moderated by a certain regime, will clearly no longer have a place in the new technology of self. The idea of a regime of the use of the *aphrodisia*, of recourse to the *aphrodisia* cannot remain. On the other hand, there will be a different form of regulation in which the problem will be to carefully separate sex-status and sex-activity, in which it will be necessary to limit the latter to a dual, conjugal relationship, and in which there will have to be a permanent relationship between sex-activity and sex-[status].[*] To that extent the problem is no longer the limitation of recourse to these acts according to a quantitative kind of regime. A permanent consideration of self by self is needed, sexual activity, or rather the principle, the very root of

---

[*] M. F. says: sex-subject.

sexual activity has to be constituted within oneself as the object of a control, of a permanent observation that will assure the split I have talked about. So it is no longer simply a matter of moderating one's own activity, but of working on oneself, of becoming an object for oneself in such a way as to carefully assure this division and control. We have here the principle of what could be called objectivation. Just as with regard to subjectivation a short while ago, we must not go too quickly or anticipate. What is formed in these technologies of self is not yet a developed relationship of knowledge. It is entirely characteristic, in fact, that in these techniques of self-examination I talked about last year, which are developed especially from the first two centuries CE and evidence of which is found in Seneca, Marcus Aurelius, and so on, in these kinds of knowing of self by self, we find nothing like those subtle and suspicious decipherments that will be found, for example, in Christians with regard to the flesh.[20] In the examples of self-examination [presented by these authors], and this is typical, it is practically never a question of sex, but much rather of ambition, anger, behavioral reactions. The problem of sex, the problem of sexual desire becomes the main and central component of the examination of conscience only with the spiritual techniques of the fourth-fifth centuries in Christianity. In Seneca, in Marcus Aurelius, it is practically never a question of this. Anger is the main object of self-examination,[21] that is to say precisely the problem of the individual's status in relation to others and of the exercise of power over others. But if we cannot speak of an objectivation or of a relationship of knowledge and analysis developed with regard to the relationship of self to self in what concerns sex, it remains nevertheless that when one defines self-control as the necessary condition for effectively carrying out the division between sex-status and sex-activity, when the individual is required to have a permanent relationship of self with the very principle of his sexual activity, one will have something entirely different than a quantitative regime concerning the use of the *aphrodisia*. One will have the kernel, the primordial and elementary form of a relationship of self to self that is at the same time a relationship of objectivation.

In what form does this point of objectivation appear and is it defined before developing as object of knowledge? What is the element on which one has to exercise one's mastery in this relationship that is the kernel and matrix of a knowledge relationship? What is at issue? Not sexuality; the Greeks and Romans do not know what this is. Not the flesh, which corresponds to an entirely different type of experience. What is the element it is necessary to control, the element regarding which it is necessary to establish at least the principle of a relationship of knowledge because it is the point of departure of sexual activity? What is the subject, the principle of activity that must be mastered and the element that must be known as object? It is of course *epithumia*, the desire that means that as subject of sexual activity I am constantly tempted, led, brought to make my sexual activity overflow onto my status as individual endowed with a sex. It is this desire, this *epithumia* that I must check and master, that I must observe and take into account at its source in order to assure myself that I will be able to establish, maintain, and renew throughout my behavior the caesura necessary to the relation I have to my own sex. Desire is isolated as the element that will anchor the subjectivation of the *aphrodisia*: it is in the form of desire that I will establish the permanent relationship I have to my own sex. And it is in the form of desire, of *epithumia* that what requires to be controlled, mastered, and known in me will be objectivized. You will say that I am situating the appearance of *epithumia* as the fundamental element very late, when all Greek philosophy, at least since Socrates and Plato, made *epithumia* a central problem. But in fact *epithumia* in Plato and Greek philosophy is a general instance of my soul in relation to which I am passive.[22] That *epithumia* becomes the form par excellence of the manifestation in me of the very principle of sexual activity is something new and characteristic of these technologies of self that develop from the first and second centuries. What these technologies of self do in fact is extract that moment, that element of desire from the *aphrodisia*, and from them in a privileged way. The *aphrodisia* were acts in which the movements of body and soul were interdependent, bound to each other in a sort of paroxysmal unity. Desire was only an aspect of the manifestation of an organic mechanism: the accumulation of humors. This desire was linked to a pleasure that was itself the side, the soul's side, of an activity, a mechanism of spermatic expulsion. This model formed a tight unity. Body, soul, pleasure, desire, sensation, mechanism of the body all formed a sort of package that was the paroxysmal bloc characteristic of *aphrodisia*. What the technology of self

does, what these *tekhnai* I have talked about do—and here again I sche-
matize a great deal—is as it were extract, isolate from this whole (this
paroxysmal bloc of which the *aphrodisia* consist) the *epithumia* (desire),
considerably reducing the importance of the act, which now is no
longer the expression, the manifestation, the very essence of the *aphrodisia*. The important element is no longer the act itself, but, before the
act, the desire. Reduction of the act. Reduction also of pleasure, which
now is no more than an ultimately relatively inessential aim and end.

The bloc of the *aphrodisia* is thus dismantled. The solid unity that
existed, that was defined and described (desire, act, and pleasure forming a whole) will be broken up, and there will be a re-centering as it
were towards the before, upstream, a re-centering of the whole problem of the *aphrodisia* around desire, a re-centering that already appears
very clearly in Epictetus and Marcus Aurelius when they shift the only
terms in which they really speak of sexual conduct towards the aspect
of *epithumia*: do I desire? For Epictetus or Marcus Aurelius, I shall
really be master of myself, I shall really be pure, *egkrateia* will be realized for me, not when, [like] Socrates, I am able to forgo the sexual act
with Alcibiades even though I desire him, but when I desire neither
the most beautiful women nor the most handsome boys, even when I
see them.[23] Everything is thus re-balanced on the side of *epithumia*, on
the side of desire that becomes the essential, fundamental, formative
element, the form in which I really have a permanent relationship with
my sexual activity.[*]

We soon find all this again in the spiritual techniques of early
Christianity, according to which desire is the fundamental element,
and it is on desire, if grasped at its root or on its appearance, that one

---

[*] The manuscript adds:
"It is this question of the subject of desire that will run through the West from Tertullian
to Freud ... But it remains to show how, as well as the subjectivation of the *aphrodisia*, the
objectivation of the subject of desire is elaborated in Christianity. Thus the subject of desire as
object of knowledge appears in the West. We have passed from the old problematic: how not
to be carried away by the movement of desire that brings me and attaches me to pleasure? to
this other problematic: how to disclose myself to myself and for those close to me as subject
of desire?"

must bring to bear all effort. Hence the notion of concupiscence that is the basic element, the target, the real object of all Christian technology concerning sex. It is no longer the *aphrodisia*, it is no longer that paroxysmal act, it is the birth, the budding, the first point, the first pin-prick of desire, it is concupiscence, it is temptation. And this is, of course, what will be found again later: breaking up of the bloc of the *aphrodisia* and emergence of desire. At what price? At the price of the sidelining or relative neutralization of the act and of pleasure, of the body and of pleasure.

Consequently you see that it would be entirely imprudent to want to do a history of sexuality that takes as its guiding theme [the question]: how and under what conditions has desire been repressed? It has to be shown rather how desire, far from having been repressed, is something that was gradually extracted and emerged from an economy of pleasures and bodies, how it was actually extracted, how and in what way it was around and regarding desire that all the operations and all the positive or negative values concerning sex crystallized. It is desire, alone, that ended by confiscating all that was previously joined together in that unity of desires, pleasures, and bodies. And this is how the fundamental question of desire and the subject of desire gradually emerged, inasmuch as desire was in actual fact the form in which the problem of the *aphrodisia*, that is to say of sexual acts was objectivized and subjectivized (objectivized and subjectivized at the same time, the one because of the other).

Desire is in actual fact what I would call the historical transcendental on the basis of which we can and should think the history of sexuality. So: emergence of desire as the principle of subjectivation/ objectivation of sexual acts. I think I can stop there and simply remind you how, on these conditions, it would be somewhat inadequate and entirely insufficient, with regard to the extent and complexity of the problems, to want to do a history of sexuality in terms of repression of desire. Rather, on the basis of a history of technologies of self, which seem to me a relatively fruitful point of intelligibility, on the basis of a history of governmentalities—governmentalities of self and others— we should show how the moment of desire was isolated and exalted, and how, on the basis of this, a certain type of relationship of self to self was formed that has itself undergone certain transformations, since we have seen it developed, organized, and distributed in an apparatus (*dispositif*) that was first that of the flesh before becoming, much later, that of sexuality. That's it, thank you.*

---

* The manuscript ends in the following way: "The arts of conducting oneself define another way of constituting oneself as subject of pleasure: pleasure must no longer be regarded and experienced as the other side or effect of an activity. Dangerous effect on which it is necessary and sufficient to impose a measure. It involves nullifying pleasure (at least as end or positive element) in the conjugal *aphrodisia*, and of replacing it with another type of sensations that are benevolence, obligingness, willingness, gratitude, and so on. Opening a whole set of impulses of the soul that are no longer held in the simple schema: desire, pleasure, but are integrated in the type of experience new to the time: matrimonial *erōs*."

1. See above, the lectures of 21 and 28 January.
2. See P. Veyne, "La famille et l'amour sous le Haut-Empire romain" (see above, p. 45, footnote 16).
3. See above, lecture of 11 March, p. 207 sq.
4. In "La famille et l'amour," p. 91, P. Veyne talks of the transition from a "competitive aristoc-racy" to a "service aristocracy."
5. See the lives of Tiberius, Caligula, Claudius, and Nero in Suetonius, *Lives of the Caesars*, trans., Catherine Edwards (Oxford: Oxford University Press, Oxford World's Classics, 2000).
6. See especially his descriptions of the Julio-Claudian emperors in Tacitus, *Annales*, trans., Pierre Wuilleumier (Paris: Les Belles Lettres, 1974-1994); Tacitus describes the debauchery of Tiberius in the following way, VI, 1, p. 86: "The fervor [of his crimes and debaucheries] seized hold of him to the point that, following the example of kings, he defiled free born young people with his depravities. And it was not only the physical beauty and grace that stimulated his passion, but in some the simplicity of childhood and in others the images of ancestors"; English translation John E. Jackson, *Tacitus Volume IV: Annals*, Books 4-6, 11-12 (Cambridge, MA: Harvard University Press, Loeb Classical Library, 312, 1937) p. 155: "... the sins and lusts whose uncontrollable fires had so inflamed him that, in the kingly style, he polluted with his lecheries the children of free-born parents. Nor were beauty and physical charm his only incitements to lasciviousness, but sometimes a boyish modesty and sometimes a noble lineage."
7. See, "Life of Elagabalus" in *Historia Augusta*, Vol. II, trans., David Magie (Cambridge, MA: Harvard University Press, Loeb Classical Library 140, 1924).
8. See Lucian, "The Downward Journey or The Tyrant" in *Lucian*, vol. II, trans., A. M. Harmon (Cambridge, MA: Harvard University Press, Loeb Classical Library, 54, 1915).
9. See in particular "Life of Romulus" in Plutarch, *Lives*, Vol. I, trans., Bernadotte Perrin (Cambridge, MA: Harvard University Press, Loeb Classical Library, 46, 1914) and "Life of Sulla" in *Lives*, vol. IV, trans., Bernadotte Perrin (Cambridge, MA: Harvard University Press, Loeb Classical Library, 80, 1916). With regard to Sulla, Plutarch asserts (30, p. 426): "Naturally, therefore, his conduct fixed a stigma upon offices of great power, which were thought to work a change in men's previous characters, and render them capricious, vain, and cruel."
10. See Maximus of Tyre, *Dissertation*, trans., Jean-Isaac Combes-Dounous (Paris: Bossange, Masson & Besson, 1802) in particular "Qu'est-ce que l'amour de Socrate?" (vol. II, "Dissertation XXIV", pp. 41-58: "Periander, tyrant of Ambracia, took his pleasure with a young Ambracian. This was entirely illegitimate. It was a shameful passion rather than love. Blinded by his power, Periander engaged in his lovemaking with his Ganymede carelessly while drunk. Drunkenness sometimes neutralizing Periander's transports. This turned the young man into the tyrant's assassin: legitimate punishment for an illegitimate passion. [...] A young Athenian was loved both by a simple citizen and by the tyrant of Athens [probably Hipparchus, son of Pisistratus]. One of these passions was authorized by equal-ity of condition. The other was founded on violence, due to the tyrant's power. The young man was moreover truly handsome, and very worthy of being loved. He disdained the tyrant and gave his love to the private man. Angered, the tyrant sought only to molest both of them"; English translation M. B. Trapp, "Socratic Love" Oration 18, in Maximus of Tyre, *The Philosophical Orations* (Oxford: The Clarendon Press, 1997) p. 160: "Periander, tyrant of Ambracia, had a young man of that state as his boyfriend, but since their relationship was immorally constituted, it was a matter of lust not love. Confident in his power, Periander made a drunken joke against the youth, a drunken joke that had the effect of putting a stop to Periander's bullying, and turning the youth himself from a love-object into a tyranni-cide. That is how unjust love is punished. ... An Athenian youth had two lovers, an ordinary citizen and a tyrant; the former was a just lover because the two of them were of the same social status; the latter was unjust because of his superior authority. But the youth himself was truly handsome and worthy of love; accordingly, he spurned the tyrant and gave his affec-tion to the ordinary citizen. The tyrant in his rage devised various ways of insulting both of them."

11. See Aeschines, "The Speech Against Timarchus" in *Speeches*, trans., C. D. Adams (Cambridge, MA: Harvard University Press, Loeb Classical Library, 106, 1919). Regarding this speech see *L'Usage des plaisirs*, pp. 239-241; *The Use of Pleasure*, pp. 217-219.

12. On Marcus Aurelius, on the principle of the emperor's self-control, see *L'Herméneutique du sujet*, pp. 192-193; *The Hermeneutics of the Subject*, pp. 200-201.

13. Suetonius, "The Deified Claudius," 29, in *Lives of the Caesars*, p. 186: "At the mercy of these freedmen and his wives, as I said, he acted not as an emperor but as a servant, dispensing magistracies, military commands, immunities, and punishments, according to the interests and indeed the passions and loves of one or other of them, while he for the most part remained unaware and ignorant."

14. Suetonius, "Nero," 29, in *ibid.*, p. 210: "He prostituted his own body to such a degree that, when virtually every part of his person had been employed in filthy lusts, he devised a new and unprecedented practice as a kind of game, in which, disguised in the pelt of a wild animal, he would rush out of a den and attack the private parts of men and women who had been tied to stakes, and, when he had wearied of playing the beast, he would be 'run through' by his freedman Doryphorus. With this man he played the role of bride, as Sporus had done with him, and he even imitated the shouts and cries of virgins being raped."

15. See, "Life of Elagabalus," V-VII, in *Historia Augusta*, vol. II, pp. 115-119: He lived "in a depraved manner ... indulging in unnatural vice with men ... For who could tolerate an emperor who indulged in unnatural lusts of every kind, when not even a beast of this sort would be tolerated? ... He took money for honours and distinctions and positions of power, selling them in person or through his slaves and those who served his lusts ... He also adopted the worship of the Great Mother and celebrated the rite of the taurobolium ... He would toss his head to and fro among the castrated devotees of the goddess, and he infibulated himself." Among ancient sources, see also: Herodian, *History of the Empire*, Vol. II, trans., C.R. Whittaker (Cambridge, MA: Harvard University Press, Loeb Classical Library, 455, 1970) Book V, 5-8; Dio Cassius, *Roman History*, Vol. IX, Book 80, trans., Earnest Cary (Cambridge, MA: Harvard University Press, Loeb Classical Library, 177, 1927) p. 457: "[Elagabalus] had planned, indeed, to cut off his genitals altogether, but that desire was prompted solely by his effeminacy." Foucault may have known the astonishing (although historically little documented) work by Antonin Artaud, *Héliogabale ou l'Anarchiste couronné*, in *Œuvres complètes*, vol. VII (Paris: Gallimard, 1967).

16. In Roman historiography, the debauched prince is the prince who does not respect the appearances of a "republican" government, who does not agree to show respect and deference towards the Senate by getting it to participate (albeit formally) in the government of the State. Since the reforms of Augustus, imperial power is constituted as absolute power, progressively emptying the political action of the senatorial class of meaning. And yet, the political "fiction" of a government belonging to the *senatus populusque Romanus* remains. The nostalgia for republican liberties runs through in fact the Latin literature and historiography of the first centuries CE—whose authors are almost always senators—and ends up setting off, if not an effective revolt (the coups d'état become the prerogative of soldiers, praetorians in particular), at least the posthumous *damnatio memoriae* of the "bad princes." On this see P. Veyne, *L'Empire gréco-romain* (Paris: Seuil, 2005).

17. See M. Bloch, *Les Rois thaumaturges. Étude sur le caractère surnatural attribué à la puissance royale particulièrement en France et en Angleterre* (Paris: Gallimard, "Bibliothèque des histoires," 1961).

18. For a definition of *aphrodisia* as dynamics, see the chapter "Aphrodisia" in *L'Usage des plaisirs*, pp. 47-62; *The Use of Pleasure*, pp. 38-62.

19. *Khrēsis aphrodisiōn*: the use of pleasures, the title Foucault will give to the second volume of the *History of Sexuality*.

20. See *Du gouvernement des vivants*, lecture of 12 March 1980, pp. 232-240; *On the Government of the Living*, pp. 237-246, as well as *L'Herméneutique du sujet*, lectures of 27 January 1982, pp. 157-158 and 24 March, pp. 461-464; *The Hermeneutics of the Subject*, pp. 162-164 and pp. 481-484.

21. See above footnote 10.

22. See Plato, *The Republic*, IV, 439d sq., VIII, 558d-559c; *Phaedrus*, 237d-238c and 246a sq.

23. Epictetus, *The Discourses*, Vol, II, Book III, iii, 14-19, pp. 32-35.

This page intentionally left blank

COURSE SUMMARY[*]

¹UNDER THE GENERAL TITLE of "Subjectivity and Truth," it is a matter of beginning an inquiry regarding instituted modes of self-knowledge and their history: how was the subject established, at different times and in different institutional contexts, as a possible, desirable, or even indispensable object of knowledge? How were the experiences that one may have of oneself and the knowledge that one forms from it organized through certain schemas? How were these schemas defined, valorized, recommended, imposed? It is clear that neither appeal to an original experience nor the study of philosophical theories of the soul, the passions, and the body can serve as the main axis in such an investigation. The most useful line to follow for this inquiry seems to be what could be called "techniques of self," that is to say the procedures, such as no doubt exist in all civilizations, that are recommended or prescribed to individuals for fixing, maintaining, or transforming their identity in terms of certain aims and thanks to relations of self-mastery or self-knowledge. In short, it is a matter of placing the imperative of "know oneself," which seems to us to be so characteristic of our civilization, in the broader questioning that serves as its more or less explicit context: What is one to do with oneself? What work is to be carried out on oneself? How is one to "govern oneself" by exercising actions [in which] one is oneself the objective, the

* Published in the Annuaire du Collège de France, 80ᵉ année, Histoire des systèmes de pensée, année 1980-1981, 1981, pp. 385-389, and in DE, 4, pp. 213-218; "Quarto," vol. 2, pp. 1032-1037. An earlier version of this summary by Robert Hurley appears with the title "Subjectivity and Truth" in M. Foucault, EW, 1, pp. 299-304.

domain in which they are applied, the instrument they make use of, and the subject that acts?

Plato's *Alcibiades*[2] can be taken as starting point: the question of the "care of oneself," *epimeleia heautou*, appears in this text as the general framework within which the imperative of self-knowledge gets its meaning. The series of studies that can be envisaged starting from this could thus form a history of the "care of oneself," understood as experience, and thus also as technique for developing and transforming that experience. Such a project is at the intersection of two themes treated previously: a history of subjectivity and an analysis of forms of "governmentality." We embarked upon the history of subjectivity by studying the divisions carried out in society in the name of madness, illness, and delinquency, and their effects on the constitution of a rational and normal subject; we also set about it by trying to identify the modes of objectivation of the subject in forms of knowledge like those concerning language, labor, and life. As for the study of "governmentality," it answered to a double objective: to make the necessary criticism of current conceptions of "power" (thought of more or less confusedly as a unitary system organized around a center that is at the same time its source, and that is driven to continual expansion by its internal dynamic); to analyze it rather as a domain of strategic relations between individuals or groups—relations in which what is at stake is the conduct of the other or others, and which, according to the case, institutional frameworks in which they develop, social groups, and epochs, resort to various procedures and techniques; the studies already published on confinement and the disciplines, the courses devoted to raison d'État and the "art of government," and the volume in preparation, with the collaboration of Arlette Farge, on the *lettres de cachet* in the eighteenth century,[3] constitute elements in this analysis of "governmentality."

So the history of the "care" and "techniques" of self would be a way of undertaking the history of subjectivity: no longer, however, through the divisions between mad and non-mad, sick and non-sick, delinquent and non-delinquent, no longer through the constitution of fields of scientific objectivity in which the living, speaking, laboring subject is given a place; but through the organization and transformations in our culture of "relations with oneself," with their technical armature and knowledge effects. And in this way one could take up the question of "governmentality" from a different viewpoint: the government of self by self in its connection with relations with others (as one finds in pedagogy, advice on conduct, spiritual direction, the prescription of models of life, and so on).

⚜

The study undertaken this year delimited this general framework in two ways. By an historical limitation: we studied what was developed in Hellenistic and Roman culture as "technique of life," "technique of existence" by the philosophers, moralists, and physicians in the period from the first century BCE to the second century CE. And also by a limitation of domain: these techniques of life were envisaged as applying only to the type of act that the Greeks called *aphrodisia*, for which our notion of "sexuality" is clearly a very inadequate translation. The problem was therefore this: how, on the eve of the development of Christianity, did the philosophical and medical techniques of life define and regulate the practice of sexual acts, the *khrēsis aphrodisiōn*? We can see how far we are from a history of sexuality organized around the good old repressive hypothesis and its usual questions (how and why is desire repressed?). It is a matter of acts and pleasures, not of desire. It is a matter of the formation of self through techniques of life, not of repression by prohibition and the law. It is not a matter of showing how sex was held back, but of how the long history that links sex and the subject in our societies began.

It would be completely arbitrary to link the first emergence of the "care of oneself" with regard to sexual acts to this or that moment. But the division proposed (around techniques of self in the centuries immediately preceding Christianity) has its justification. It is certain in fact that in the Hellenistic and Roman period the "technology of self"—reflection on modes of life, on choices of existence, on the way to regulate one's conduct, to set ends and means for oneself—underwent a considerable development, to the extent of absorbing a good part of philosophical activity. This development cannot be dissociated from the growth of urban society, from new distributions of political power, or from the importance taken on by the new service aristocracy in the

Roman Empire. This government of self, with its specific techniques, finds its place "between" pedagogical institutions and religions of salvation. This should not be understood as a chronological succession, although it is true that the question of the training of future citizens seems to have aroused more interest and reflection in classical Greece, while the question of the afterlife and the hereafter raised more anxiety in later periods. Nor should one think that pedagogy, government of self, and salvation formed completely different domains and put to work different notions and methods; in fact there were numerous exchanges and a certain continuity between them. It remains nonetheless that the technology of self that was intended for the adult can be analyzed in the specificity and extent it acquired in this period on condition that one distinguishes it from the shadow that the prestige of pedagogical institutions and religions of salvation succeeded in casting over it.

Now this art of government of self as it developed in the Hellenistic and Roman period is important for the ethics of sexual acts and its history. It is here in fact—and not in Christianity—that the principles of the famous conjugal schema, which has had such a long history, were formulated: exclusion of any sexual activity outside of the relationship between husband and wife, procreative purpose of these acts, at the expense of pleasure as an aim, and affective function of sexual intercourse in the conjugal bond. But there is more: in this technology of self we also see the development of a form of disquiet with regard to sexual acts and their effects, the paternity of which is too readily attributed to Christianity (when not to capitalism or "bourgeois morality"!). Certainly, the question of sexual acts is far from having the importance then that it will have later in the Christian problematic of the flesh and concupiscence; the question of anger, for example, or of reversal of fortune, certainly occupies a much more important place for the Hellenistic and Roman moralists than sexual relations; but, although sexual relations are far from occupying first place in the order of concerns, it is important to note the way in which these techniques of the self connect the regime of sexual acts to the whole of existence.

⚜

In this year's course we have focused on four examples of these techniques of self in their relation to the regime of *aphrodisia*.

1. The interpretation of dreams. Artemidorus' *Oneirocritica*,[4] in chapters 78-80 of Book I, constitutes the fundamental document in this domain. The question it raises does not directly concern the practice of sexual acts, but rather the use to be made of the dreams [in which] they are represented.

This text is concerned with determining the prognostic value they are to be given in everyday life: what favorable or unfavorable events can one expect on the basis of the type of sexual relation presented by the dream? This kind of text obviously does not prescribe a morality, but through the play of positive or negative significations it ascribes to the dream images it reveals a whole interplay of correlations (between sexual acts and social life) and a whole system of differential evaluations (hierarchizing sexual acts in relation to each other).

2. Medical regimes. These directly propose to establish a "measure" for sexual acts. It is noteworthy that this measure hardly ever concerns the form of the sexual act (natural or not, normal or not), but its frequency and timing. Only quantitative and circumstantial variables are taken into consideration. The study of Galen's great theoretical edifice clearly shows the connection established by medical and philosophical thought between sexual acts and the death of individuals. (It is because each living being is destined to die but the species must live eternally that nature invented the mechanism of sexual reproduction); it also clearly shows the connection established between the sexual act and the considerable, violent, paroxysmal, and dangerous expenditure of the vital principle it entails. The study of regimes, strictly speaking (in Rufus of Ephesus, Athenaeus, Galen, Soranus), shows, through the endless precautions they recommend, the complexity and subtlety of the relations established between sexual acts and the individual's life: extreme sensitivity of the sexual act to all the external or internal circumstances that might make it harmful; huge range of effects of each sexual act on every part and component of the body.

3. Married life. There were a great many treatises on marriage in the period considered. What remains of Musonius Rufus, Antipater of Tarsus, or Hierocles, as well as the works of Plutarch, not only shows the valorization of marriage (which, according to historians, seems

to correspond to a social phenomenon), but a new conception of the matrimonial relationship; to the traditional principles of the complementarity of the two sexes necessary for the order of the "household" is added the ideal of a dual relationship enveloping every aspect of the life of the two partners and establishing definitively personal affective bonds. Sexual acts must find their exclusive place inside this relationship (hence Musonius Rufus's condemnation of adultery,[5] no longer as an offense against a husband's privileges, but as a breach of the conjugal bond that binds the husband as well as the wife). Sexual acts must thus be ordered by reference to procreation, since this is the end given to marriage by nature. And finally they must comply with an internal regulation required by modesty, reciprocal affection, and respect for the other (the most numerous and valuable indications on this last point are found in Plutarch).

4. The choice of loves. The classical comparison of the two loves— that of women and that of boys—left two important texts from the period under consideration: Plutarch's *Dialogue on Love*[6] and Lucian's *Amores*.[7] Analysis of these two texts attests to the persistence of a problem with which the classical period was very familiar: the difficulty of giving a status and justification to sexual relations in the pederastic relationship. Lucian's dialogue ends ironically with a precise reminder of those acts that the erotics of boys sought to elide in the name of friendship, virtue, and pedagogy. Plutarch's much more developed text reveals mutual consent to pleasure as an essential element in the *aphrodisia*; it shows that such reciprocity in pleasure can exist only between a man and a woman; better still in conjugality, where it serves to renew regularly the marriage pact.

1. The Summary is preceded by the following: "This year's course will be the object of a forthcoming publication. For now, then, it will be sufficient to give a brief summary."
2. Plato, *Alcibiades*, trans., M. Croiset (Paris: Les Belles Lettres, "Collection des universités de France, 1925); English translation by W. R. Lamb in *Plato*, vol. XII (Cambridge, MA: Harvard University Press, "Loeb Classical Library", 1986).
3. M. Foucault and A. Farge, *Le Désordre des familles. Lettres de cachet des archives de la Bastille au XVIIIᵉ siècle* (Paris: Gallimard-Julliard, coll. "Archives" 91, 1982).
4. Artemidorus, *La Clef des songes. Oneirocriticon*, trans. A. J. Festugière (Paris: Vrin, 1975), Book I, ch. 78-80, pp. 84-93; English translation Daniel E. Harris-McCoy, *Artemidorus' Oneirocritica: Text, Translation, and Commentary* (Oxford: Oxford University Press, 2012) pp. 136-151.
5. Musonius Rufus, *Reliquiae*, XII: "Sur les *Aphrodisia*," ed., O. Hense (Leipzig: B.G. Teubner, "Bibliotheca scriptorum Graecorum et Romanorum" 145, 1905) pp. 65-67; English translation by Cora E. Lutz in Cora E. Lutz, "Musonius Rufus, 'The Roman Socrates,'" *Yale Classical Studies*, 10 (Cambridge: Cambridge University Press, 1947).
6. Plutarch, *Dialogue sur l'amour*, 769b, in Œuvres morales, vol. X, ed. and trans., R. Flacelière (Paris: Les Belles Lettres, CUF, 1980); English translation W. C. Helmbold, *The Dialogue on Love*, in Plutarch, *Moralia*, vol. IX (Cambridge, MA: Harvard University Press, Loeb Classical Library, 1961) pp. 427-428.
7. Lucian (attrib.), "Affairs of the Heart (Amores)" trans., M. D. MacLeod, *Lucian Volume VIII* (Cambridge, MA: Harvard University Press, Loeb Classical Library, 432, 1967) pp. 230-233.

This page intentionally left blank

COURSE CONTEXT

*Frédéric Gros**

THE COURSE DELIVERED BY Michel Foucault in 1981 is the first
of a series of lectures focusing on Greek and Roman antiquity that will
continue until *The Courage of Truth.*[1] It will not be forgotten however
that the whole of the first course at the Collège de France (1971) was
devoted to the judicial procedures of ancient Greece.[2] Subsequently the
ancient reference is no longer invoked except as simple counterpoint to
the study of Christian governmentalities (governmentality of the city
state as opposed to pastoral governmentality in 1978;[3] ancient direction of existence as opposed to Christian spiritual direction in 1980[4]).
Now in 1981, Foucault finds a theoretical anchorage point in classical
Greek, Hellenistic, and Roman writings that allows him a renewed
conceptualization of subjectivity and truth.

1. A HISTORY OF TECHNIQUES OF SELF
AND PRINCIPLES OF VALORIZATION

This important change of periodization is actually accompanied by a
decisive decentering towards the problematic of subjectivity. However,
this does not take the form, as has been said too often, of a "return to
the subject," but of a genealogy of Western subjectivity. The first lecture
of *On the Government of the Living* announced, as general problematic

---

* Frédéric Gros is Professor of Philosophy at the University of Paris Est Créteil. He has edited
Foucault's last courses at the Collège de France (*The Hermeneutics of the Subject; The Government of Self
and Others; The Courage of Truth*). His most recent work is *Le Principe Sécurité* (Paris: Gallimard, 2012).

framework, the study of the connection between "exercise of power" and "manifestation of the truth."[5] That said, through the study of Christian penitential techniques, the analyses quickly focused on "truth acts,"[6] that is to say the subject's obligations to state a truth about himself. These Christian practices of confession certainly required his initiative (he was their "active agent"[7]), but they remained framed by a general imperative of obedience and self-sacrifice.[8]

In 1981, Foucault refocuses his research on what on what he now calls "techniques of self," defining a specific plane that is not deduced directly from the "government of men." The expression is not found in the 1980 lectures where it was a matter of only techniques of "direction," of "test of the soul" or "the philosophical life."[9] The term appears for the first time in the autumn following the lectures at the Collège de France (lectures at Berkeley and Dartmouth in October and November 1980[10]). The problematization of these "techniques of self," often distinguished from techniques of production, communication, and domination,[11] enable Foucault to problematize a subject that is not merely traversed and informed by external governmentalities, but constructs a definite relationship to self by means of regular exercises. The subject takes on an ethical consistency that is peculiar to it and also irreducible to what moderns understand as a psychological inner depth. This depth is historical through and through, structuring for the individual a particular experience of himself that determines his relationship to the body, others, and the world. The Greek term privileged by Foucault to designate this volume of a relationship to self that is configured by an ethical labor is *bios*, classically distinguished from *zōē*. The Greek *bios* designates a plane of immanence capable of taking a determinate form and orientated by practical objectives.[12] Foucault recognizes that it is the study of the ancient problematization of sexuality that facilitated his discovery of the domain of techniques of self.[13] Actually, sexuality in the Ancients is not, as it is for the moderns, the secret ground of a psychical identity that can be objectivized by introspection or positive knowledge. It is thought of as a natural dynamic that needs to be framed by rules of use and an ordered dietetics.

Beyond this perspective opened on a new consideration of the subject as nucleus of ethical transformations, the study of the sexuality of the Ancients also made it possible, as had already been the case with the first volume of the *History of Sexuality*, to remove the obstacle of prohibition as the obligatory reading grid, to overturn "the repressive hypothesis."[14] In 1976, the aim was to emphasize the productivity and positivity of a power that above all should not be reduced to the negative functions of illusion and submission. Something different is at stake in 1981. It is a question, rather, of revealing, for the subject, an ethical system structured by a game of preferences. The classical history of sexuality analyzes, for a given historical period, on the one hand, cultural prohibitions (codification of the permitted and the forbidden) and, on the other, practical tolerances (series of adjustments carried out by actual behavior), thus juxtaposing the description of what it is said one must do and what one actually does. But for Foucault, prohibition is not a pertinent reading grid inasmuch as it corresponds to an historically dated structuration of sexuality (it is in the Middle Ages that the "general juridification" of human behavior is carried out[15]) and so cannot be meta-historically valid, except by falling victim to what Foucault calls an "illusion of the code."[16]

So, the study of the sexual experience of the Ancients undertaken in 1981 does not dwell on searching, through sexual representations or behaviors, for a distribution of prohibitions or the delimitation of zones of tolerance. It endeavors rather to describe "principles of valorization," lines of structuration of an "ethical perception,"[17] the codified elements of which are only ever somewhat floating endpoints of a graduated system of preferences. The problem is not to find what was permitted or forbidden, but to understand on the basis of that general system which sexual behavior could be valorized to the detriment of another. In *Subjectivity and Truth* this study relies mainly on a text from the second century (Artemidorus's *Oneirocritica*[18]), which makes it possible to draw out two major principles that are, for Foucault, authentic vectors of historicity since it is their questioning by imperial Stoicism that will change the direction of the history of subjectivity. We should note again that this sexual ethics is developed from the exclusive point of view of the adult, married, free man. The first principle is a "principle of activity" in the name of which the passive position in the sexual act is broadly discredited. Sexual pleasure recognized as "good" is that which is experienced in and through the activity of the adult man. This sexual morality, focused on this active moment, is said by Foucault, in the lecture of 28 January, to be "non-relational." Nevertheless, this does not mean that in this ethical framework no limit is placed on the enjoyment once it is deduced from an active position. However, this limit is not imposed by the dignity of the sexual partner but by the internal necessity of having to remain master of oneself. Here, we must again take into account the nature of what the Greeks called *aphrodisia* and the Latins *veneria*, which is translated, clumsily, Foucault notes, as "sexuality."[19] For us, "sexuality" refers immediately to the inner depths of a secret identity. Now, for the Ancients, the *aphrodisia* signify first of all a natural mechanics whose ever-threatening internal violence and excess requires each to have their own regime to impose a measure and rules of use.[20] Mastery of this dangerous dynamic, more than consideration of the other's pleasure, thus requires laying down limits.

While defined as "non-relational," in the sense that sexual pleasure is not thought of as that which, in the confusion of bodies, could be shared out according to a dialectic of gift and abandon, the sexual ethics of the Ancients remains for Foucault entirely determined by respect for social hierarchies. The sexual act is itself only a way of replaying systems of domination on the stage of bodies. This is the sense of the second principle stated, called "principle of socio-sexual isomorphism,"[21] according to which the "good" sexual act, at the level of its form and taking into account the partner's social status, will have to respect general socio-political differences of level. On the basis of this system we can understand why a free man having sexual relations with a male slave, even when he is married, may be positively valued so long as he maintains an active position, but that committing adultery with a married woman is viewed negatively inasmuch she is the property of someone else. Difference of valuation is not deduced from a codification of prohibitions. Differences of valuation are understandable on the basis of the system of the "two principles" isolated by Foucault in his January lectures, which, it needs to be stressed, is constraining because it constitutes a socio-mental armature, even though these constraints do not take the form of juridical prohibitions but are deduced from differentiated valuations. It should be stressed finally that, in this classical ethics, the sexual relationship between married partners is placed at the summit of the hierarchy of values, since it symbolizes and reinforces the prosperity of the household and so, beyond this, of the whole city, but without ever being the object of an exclusive obligation.

2. INVENTION OF THE COUPLE

Artemidorus' text, meticulously examined (essentially Chapters 78, 79, and 80 of Book I), makes it possible to isolate these two principles, which (although the text is from the second century CE[22]) Foucault considers revealing of the system of the *aphrodisia* peculiar to the whole of classical Greek culture. A second series of texts was studied in the lectures of February and March 1981, formed this time by short Stoic treatises on marriage (fragments of Musonius Rufus, Hierocles, and Antipater of Tarsus[23]). According to Foucault they reveal a Hellenistic and Roman model of conjugal behavior that breaks with classical ethics. And, following Paul Veyne in this,[24] Foucault proposes to describe what he calls (11 March) "the invention of the couple," designating by this the moral reconfiguration of the conjugal bond within the matrimonial institution. What will the new prescriptions be? First of all, the strict confiscation of sexuality by the married couple: sexual activity will have to take place solely within marriage (principle of exclusive localization). Then, a finalization of the sexual act by reproduction: sexual activity must be governed by an imperative of procreation, rather than by a pursuit of pleasure (principle of the "de-aphrodization"[25] of the sexual act). Finally, in these Stoic texts we find the deployment of a register of affects (trust, tenderness, mutual affection), both support and consequence of a life shared over time, and going far beyond the range of feelings inspired by simple carnal desire (principle of reciprocal affects).

We are familiar with this model of conjugal perfection. But it has often been denounced as the bitter fruit of a Christian tradition or of bourgeois ideology, inevitably restrictive with regard to the economy of pleasures in the motif of the impurity of an original sinful flesh or of a capitalist productivism hostile to useless expenditure. In the 1981 lectures Foucault wants to demonstrate that the formulation and cultural and practical inscription of this conjugal model, widely supported and relayed by the Stoic moralists, had been asserted since Imperial Rome. This is moreover the sense of the first lecture, which narrates the homage paid by Saint Francis of Sales in his *Introduction to the Devout Life* to the sexual morals of elephants, which are taken as models of conjugal virtue and modesty and which had received analogous praise from pagan naturalists.[26]

The first lecture would thus be a blurring of the borders between the sexual ethics of paganism and Christianity. But this immediately

needs to be qualified. Actually Christian experience does indeed produce a break. However, it does not consist in the promotion of an
austere matrimonial model and a sexuality without pleasure, but in a
restructuring of the relation between the subject, truth, and sexuality.
That said, for Foucault it is a matter of bringing out the three major
consequences of the ethical revolution represented by the Stoic model
of the couple, which could be stated in the following way: desocialization of sexuality; disqualification of homosexuality; "birth" of desire.

The first consequence consists in a desocialization of the matrimonial relationship, and, hence, of sexuality itself—what Foucault designates as the "break" with the "socio-sexual continuum."[27] These
treatises on marriage promote in fact a conjugal bond irreducible to
external social relations characterized by hierarchical and dissymmetric
structures (social logics of domination) or an always somewhat distant
and loose *philia* (general sociability). Now the couple should form a
solid, complete type of unity, more egalitarian than established social
relations and more intense than *philia*. In its Hellenistic and Roman
version, the matrimonial bond tends to become totally heteromorphous to social relations. Sexuality tends to cease being "a metaphor
for the social."[28] What's more, for the Stoics it is nature itself that cuts
out this islet of irreducible conjugality. It is nature that prompts the
formation of attached and faithful couples, not in a transitional way,
as was the case in the classics (see Xenophon's *Oeconomicus*[29]), to give
more foundation to political life, but in order to form a shared existence that has its specific ethical consistency and tends to become its
own end. The sexuality internal to the couple shows the effects of
these transformations: as "consented" pleasure,[30] the wife receives
and accepts the sexual relation rather than submits to it passively.
Moreover, the old assertion of an incompatibility between married life
and philosophical life falls: marriage becomes a necessity for all.[31]

The second consequence of this Stoic promotion of a complete conjugality is a radical disqualification of homosexuality. We have seen that
in the old system that the social position of the partners was more
important than their gender. Homosexuality became problematic only
if the sexual partner of the mature man was a young man from a good
family, and so soon to be called upon to become an active subject in
the city. How, then, can his present passivity be reconciled with his

future status? To resolve this paradox the Greeks thought to develop the homosexual relationship within a pedagogical erotics, offering the boy a relationship that took his abilities and talents into account and adopted his becoming-adult and his education as its aim. Then, in reciprocal trust and mutual respect, initiation into truth and autonomy could justify, conceal, or even more discourage unavowable caresses. This way, the homosexual way of Eros, endeavored to erase the pursuit of purely sexual pleasure, even to the point of wholly renouncing it, as in Platonic erotics.[32] Conversely, the more heterosexual way of Aphrodite took shape, in which the union of bodies and the satisfaction of desires were deduced from a simple natural drive, although the partner's social status still had to be taken into consideration. Now the married couple, in its Roman version, weaves a single fabric, designated by Foucault as "the single chain of love,"[33] a relationship of mutual concord and pleasure of the body. The glorification of the married and proudly faithful couple led to the disappearance of the old duality of Aphrodite and Eros, with the effect that the homosexual relationship is denounced as an abject imposture. For authentic, symmetrical, and respectful concord, and true pleasure without shameful passivity can exist only between husband and lawful spouse. Homosexuality, henceforth, no longer able to lay claim to an Eros that the married couple has confiscated, becomes condemned as a hypocritical and degrading relationship. Foucault finds this whole development in his reading of Plutarch's *Erōtikos*.[34]

The third and doubtless most enigmatic consequence concerns what could be called the birth of desire. During the last lectures of March and the first of April 1981 Foucault develops a strange and strong thesis, of which no echo is found after, neither in *The Care of the Self*, nor in the lectures and interventions of the eighties. When, he explains, absolute fidelity has become a duty of the married man, his masculinity is forced to undergo a "caesura." In the old system, masculinity was equally social and sexual, which authorized all the slippages: the man's statutory superiority authorized him to impose his sexuality on subalterns, and his sexual enjoyment within the couple could always be nourished by a feeling of social domination. Now the new duty of fidelity and the construction of a more equal relationship between man and wife require every married citizen to dissociate "statutory" masculinity and "active" masculinity. In public he will have to govern and dominate others when he is hierarchically superior to them, but desexualizing this social relationship. On the other hand, he will have to reserve his sexual masculinity for his wife, but seeking her consent, taking care that this activity does not take the form of asymmetrical domination. This split within the married man, between his "two sexes"—instead and in place of the old division between the two loves—renders both thinkable and indispensable the split of private and public, which can no longer fall under the same ethical economy.[35] This caesura of masculinity requires the married man to exercise a permanent self-control, very different from what the classics (Plato, Xenophon) thematized as "mastery of self." Insofar as he works to conform to the new matrimonial code, the married man must constantly trace in himself the caesura of his masculinity, between the poles, Foucault says, of the "male" and the "masculine,"[36] and to do this he must bring the techniques of self to bear on the "first pin-prick of desire,"[37] in order to anticipate and better defuse it. For the classics the *aphrodisia* formed an at once natural, paroxysmal, and inseparable bloc of desire-pleasure that from time to time traversed and irradiated the body, calling for specific techniques of self to impose a mastery of self and an ordered distribution of these intensive moments. The married man's technology of self henceforth requires that he gain control of desire at the level of its "budding,"[38] a desire that is thus no longer understood as the mechanical preliminary of pleasure but is crystallized in the form of a "temptation."[39] The bloc of the *aphrodisia* comes apart, and the faithful husband's self-control extracts from it the nugget of desire. Finally, it is on the basis of the married Roman citizen subject to the new conjugal rules that "sexuality" is rendered thinkable as an autonomous stratum of signification. For the Greeks of the classical age the *aphrodisia* were in effect considered either as a natural need like thirst and hunger, requiring like them a dietetics, or as a way of replaying social divisions on the stage of the body. For Hippocrates or Xenophon, to develop a sexual morality was to recodify a natural mechanics or a social logic in the ethical grammar of the body. Henceforth, with the Stoics, through this finicky technique of self-control recommended to married men, sexuality is reconfigured as a permanent dimension of subjectivity, taking the double ethical form of impossible temptation or duty without pleasure.

3. THE PROBLEM OF REDUNDANT DISCOURSE

On the basis of the meticulous study of these Stoic treatises on marriage, often considered insignificant and anecdotal, Foucault describes a decisive tipping point in the history of subjectivity. Now this philosophical testimony is not isolated, as is attested by a whole literature that also praises a strict conjugal sexuality (the poems of Statius or the Younger Pliny's letters).[40] Historians especially observe *in reality* the extension of matrimonial practices and the fixing of strict juridical norms of conjugality, granting the wife new rights and condemning adultery more vigorously. Here Foucault raises the question, almost naive in its immediate formulation, of the exact relationship to be established between the discourse of philosophers advocating the duty of marriage and the reality of social practices in which the extension of the matrimonial rule is clearly visible. Should we say that the philosophical discourse serves as ideological justification of the reality, and that this praise of conjugal union serves only to conceal the loss of old political solidarities with the disappearance of the cities and the establishment of the Empire? Should we say that the discourse endeavors rather to give a rational foundation to a practice, thereby establishing its universal value? Or, more simply, should we recognize that the discourse is after all no more than the testimony, the reflection, or the image of existing social practices?

The discussion of this humble and directly shared proposition draws Foucault into the longest exposition in the history of his Collège de France lectures of his conception of the relations between true discourse and reality (lectures of 11 and 18 March). To start with, Foucault stresses his refusal to find in reality the raison d'être of the discourse that tells the truth of that reality. Truth is an "event"[41] that arises for a given reality. But there is nothing in reality that requires the articulation of a game of veridiction and accounts *in truth* for what is woven in silent reality. The true discourse's striving for *adequacy* to reality cannot and should not represent a sufficient justification, still less the element of self-evidence that prevents us from questioning ourselves. We should really rediscover "epistemic surprise,"[42] the capacity to be surprised by the profusion of true discourse about a reality that does not call for it.

The existence of the mad should not prevent us from being surprised by the emergence of a psychiatric science. The political centralization of powers in the modern period should not prevent our surprise at the birth of a science of the State. And so the spread of matrimonial practice under the Empire should not prevent our surprise at the multiplication of treatises on marriage. The game of veridiction, a term that covers the set of discourses delivered in the name of truth (Foucault is not concerned with giving a verdict on their epistemological rigor), is to be understood as *what happens to reality*, and more precisely *what happens to the subject*, as Foucault noted in his first lecture.[43] That the relation of simple adequacy cannot be accepted as sufficient justification is, moreover, what is demonstrated by the other two major characteristics of games of veridiction that Foucault states: their inutility and their efficiency. Games of veridiction are by and large unproductive in that there is no rational proportion between the profusion of knowledge— once again, the exact sciences are only a tiny part of the set of discourses socially recognized as "true"—and the meagre grasp on reality they allow us. However, for Foucault, this inutility of games of truth is not equivalent to their inefficiency, since they do indeed have effects of reality, or better still of subjectivation. If economic science, to take the example given,[44] proves to be powerless for solving crises, it nonetheless continues to authorize political decisions taken in its name and felt by each. One might even say that psychiatric science maybe does not make possible a better cure for madness, but it feeds into judicial, administrative, and medical decisions and obliges each to construct their identity on the basis of these statements. So like Heidegger, Foucault directly challenges directly the idea that truth-telling is originally constructed as a relation of adequation.[45] This refusal, however, is not intended to give a welcome to the enigmatic murmur of Being, but rather to assert the cacophony of a crowd of games of veridiction that, while giving scarce effective hold on reality, constitute the historical reality on the basis of which the subject's experience of himself is constructed.

4. THE HORIZON OF THE FLESH AND
THE PERMANENC OF THE POLITICAL

The 1981 lectures offer a number of analyses that will be taken up extensively in the second and third volumes of *The History of Sexuality* (*The Use of Pleasure* and *The Care of the Self*). That is to say, for example, the precise study of Artemidorus' *Oneirocritica*,[46] Plutarch's *Erōtikos*,[47] the Hippocratic description of the sexual act,[48] or even the reference to Xenophon's *Oeconomicus* or the treatises on marriage of Musonius Rufus, Hierocles, and Antipater of Tarsus. Moreover, the books published in 1984 offer a more systematic organization (with the tripartition: relationship to the body, the wife, and boys) and general concepts of subjective structuration (precisely the concepts of "use" and "care"). The books are considerably enriched with new textual references, but also divested of some major theoretical or historical perspectives—nothing on the two major principles of classical valuation, the "two sexes" of the married man, games of veridiction, and so on. The ancient studies selected for reading in 1984 are therefore both more complete and more modest, as if Foucault wished to erase some hypotheses that perhaps may be judged too bold, adventurous, or general. The published books are thoroughly precise and informed, but rarely go beyond the meticulous reading of the texts brought together.

Subjectivity and Truth essentially studies ancient techniques of self through the filter of marriage and sexuality. The lectures problematize a Stoic-Roman inflection, whose importance, increased further the following year (*The Hermeneutics of the Subject* in 1982) by the problematization of the "care of self," will oblige Foucault to publish, belatedly, two separate works on Ancient sexual ethics, instead of just one as initially foreseen.[49] The 1981 lectures, moreover, constantly sketch out the broad lines of the Christian experience, and in so doing make perceptible the nervure of the last of the planned volumes of the history of sexuality, *Les Aveux de la chair* (*Confessions of the Flesh*). Foucault regularly announced, on the conclusion of this or that course, that in a subsequent lecture he would study concupiscence and analyze the Christian break, but he never came to leave the soil of Antiquity.[50] It has already been said that the recommendation of an austere conjugality, or even the denunciation of the impure character of sexual pleasure, should not be regarded as Christian inventions. The break takes place at a different point: the elaboration of a technique of confession (*aveu*) that splits the sexual subject's relationship to the truth. Ancient spirituality absolutely accepted that sexual pleasure is an obstacle to the grasp of higher truths (sexuality *or* truth). Christianity splits this relationship by making purifying ascesis, preserved as horizon, depend on a suspicious hermeneutics and exhaustive verbalization of one's hidden desires (truth *of* sexuality). So, in light of Foucault's 1981 reading of the Stoic treatises on marriage, we should not regard Christianity as the first to oblige individuals to adopt an exclusively conjugal sexuality. On the other hand, it did impose an experience of self in which they were obliged to tell another about the truth of their desire.

Like the lectures of 1980, *Subjectivity and Truth* remains impervious to the immediate political scene, occupied in the United States by the liberation in January of the American hostages in Tehran and the start of Reagan's presidency, which would implement the neoliberalism that was the object of Foucault's 1979 lectures, and in France by preparations for a presidential campaign that would be won, in May, by François Mitterand. The study of the ancient arts of living did not, however, mark a withdrawal from political actuality and activism.[51] In November 1980, making no secret of his interest in the study of techniques of self, Foucault indeed announced that they always trace the program of a "politics of ourselves."[52] It is indeed then the study of "governmentality" that is continued, on condition that this is understood as the surface of contact, the historical point of articulation between a government of self and a government of others.[53]

The transcription of the 1981 lectures was based on a recording made by Jacques Lagrange and deposited in the Collège de France library. There are some breaks due to defects in the tape or turning cassettes over. A few passages are difficult to hear. The preparatory manuscript presents few differences from the lectures delivered. Some passages, however, were abandoned, especially at the end of lectures. They have been restored as additions.

F.G.

1. Michel Foucault, *Le Courage de la vérité. Cours au Collège de France, 1983-1984*, ed., F. Gros (Paris: Gallimard-Seuil, "Hautes Études," 2009); English translation Graham Burchell, *The Courage of Truth. The Government of Self and Others II. Lectures at the Collège de France 1983-1984*, English series editor Arnold I. Davidson (Basingstoke: Palgrave Macmillan, 2011).
2. Michel Foucault, *Leçons sur la volonté de savoir. Cours au Collège de France, 1970-1971*, ed., D. Defert (Paris: Gallimard-Seuil, "Hautes Études," 2011; English translation Graham Burchell, *Lectures on the Will to Know. Lectures at the Collège de France 1970-1971*, English series editor Arnold I. Davidson (Basingstoke: Palgrave Macmillan, 2013).
3. Michel Foucault, *Sécurité, Territoire, Population. Cours au Collège de France, 1977-1978*, ed., M. Senellart (Paris: Gallimard-Seuil, 2004); English translation Graham Burchell, *Security, Territory, Population*, English series editor Arnold I. Davidson (Basingstoke: Palgrave Macmillan, 2007). Lecture of 15 February 1978.
4. Michel Foucault, *Du gouvernement des vivants. Cours au Collège de France, 1979-1980*, ed., M. Senellart (Paris: Gallimard-Seuil, "Hautes Études," 2012); English translation Graham Burchell, *On the Government of the Living. Lectures at the Collège de France, 1979-1980*, English series editor Arnold I. Davidson (Basingstoke: Palgrave Macmillan, 2014). Lecture of 12 March 1980.
5. *Du gouvernement des vivants*, p. 6; *On the Government of the Living*, p. 5.
6. Even more precisely, "reflexive truth acts"; ibid., Fr., pp. 79-80; Eng., pp. 81-82.
7. Ibid., Fr., p. 79; Eng., p. 81.
8. The question posed in 1980 is indeed that of "the government of men through the manifestation of truth in the form of subjectivity"; ibid., Fr., p. 79; Eng., p. 80.
9. Ibid., Fr., p. 223, p. 226, p. 253, p. 256, p. 258; Eng., p. 227, p. 230, p. 258, p. 262, p. 264.
10. Lectures edited by H.-P. Fruchaud and D. Lorenzini in M. Foucault, *L'Origine de l'herméneutique de soi. Conférences prononcées à Dartmouth College, 1980* (Paris: Vrin, "Philosophie du présent," 2013); English translations Graham Burchell, *About the Beginning of the Hermeneutics of the Self. Lectures at Dartmouth College, 1980* (Chicago: University of Chicago Press, 2015).
11. See for example, *L'Origine de l'herméneutique de soi*, pp. 37-38; *About the Beginning of the Hermeneutics of the Self*, pp. 24-25; *Mal faire, dire vrai. Fonction de l'aveu en justice*, ed., F. Brion and B. E. Harcourt (Louvain-la-Neuve: Presses universitaires de Louvain, 2012) pp. 12-13; English translation Stephen W. Sawyer, *Wrong-Doing, Truth-Telling* (Chicago: University of Chicago Press, 2014) pp. 23-24; "Sexuality and Solitude," in *EW*, 1, p. 177.
12. See above, lecture of 14 January for the distinction *bios/zōē* and lecture of 25 March for the contrast Greek *bios*/Christian subjectivity.
13. See *L'Origine de l'herméneutique de soi*, p. 38; *About the Beginning of the Hermeneutics of the Self*, p. 25.
14. M. Foucault, "Histoire de la sexualité," in: *La Volonté de savoir* (Paris: Gallimard, 1976) vol. I, ch. II; English translation Robert Hurley, *The History of Sexuality. Volume 1: An Introduction* (New York: Pantheon, 1978) Part Two.
15. See above, lectures of 28 January and 4 February, as well as *Mal faire, dire vrai*; *Wrong-Doing, Truth-Telling*, lectures of 13 and 20 May 1981.
16. See above, lecture of 4 February, p. 99.
17. See above, lectures of 28 January, p. 75 sq.
18. See above, lectures of 21, 28 January, 4 February, and 4 and 25 March.
19. See above, lecture of 1 April, p. 283: "This notion of *aphrodisia* was basically characterized precisely by the fact that in Greek thought there did not exist a sort of uniform and continuous category permitting the definition of something like sexuality, or what Christians called the flesh. The Greeks knew neither sexuality nor the flesh."
20. On the nature of the *aphrodisia*, see above, lectures of 25 February and 1 April, and more generally *L'Usage des plaisirs* (Paris: Gallimard, 1984); English translation Robert Hurley, *The Use of Pleasure* (New York: Pantheon Books, 1985).
21. See above, *passim*, in particular lectures of 28 January and 4 and 25 March.
22. Foucault resolves this paradox by asserting that Artemidorus' text above all proposes to take stock of a centuries old tradition that precedes it.
23. This was studied in the lecture of 11 February. A more complete version is found in *Le Souci de soi* (Paris: Gallimard, 1984) in the chapter: "La femme," pp. 173-216; English translation

Robert Hurley, *The Care of the Self* (New York: Pantheon Books, 1986), Part Five: "The Wife," pp. 145-185.

24. P. Veyne, "La famille et l'amour sous le Haut-Empire romain," in P. Veyne, *La Société romaine* (Paris: Seuil, 1991) p. 108: "Under the Empire, there is no longer any question of it being understood that dissension may rule between husband and wife, since henceforth the very functioning of marriage is supposed to rest on harmony and the rule of the heart. A new idea is born: the 'couple' of master and mistress of the household."

25. The expression appears in the lecture of 25 March, p. 258. In *Le Souci de soi*, p. 213; *The Care of the Self*, p. 182, Foucault speaks rather of "dehedonization."

26. The example of the sexually virtuous elephant is found in *L'Usage des plaisirs*, pp. 23-24; *The Use of Pleasure*, pp. 17-18 (see also, "Sexuality and Solitude," p. 179).

27. See above, lecture of 4 February, pp. 102-104.

28. See above, lecture of 21 January, p. 60.

29. This text is presented in the lecture of 11 February, p. 126 (see also *L'Usage des plaisirs*, "La maisonnée d'Ischomaque," pp. 169-183; *The Use of Pleasure*, "Ischomachus' Household," pp. 152-165).

30. On the idea of the spouse's "consent" (*kharis*), see above, lecture of 4 March, p. 193-194.

31. On marriage as an "absolute" duty, see above, lecture of 4 February.

32. These theses are set out in the lecture of 28 January, above, and at greater length in *L'Usage des plaisirs*, chapters "Érotique" and "La veritable amour" (pp. 205-269); *The Use of Pleasure*, Parts Four and Five: "Erotics" and "True Love," pp. 187-246.

33. See above, lecture of 4 March, p. 192.

34. See ibid., p. 176 sq.

35. See above, lecture of 25 March, p. 262.

36. See above, lecture of 1 April, p. 275.

37. Ibid., p. 288.

38. Ibid.

39. Ibid.

40. See above, lecture of 11 March, pp. 212-216.

41. Ibid.

42. See above, lecture of 18 March, pp. 236-237, footnote *.

43. See above, lecture of 7 January, p. 11: "When in a culture there is a true discourse about the subject, what is the subject's experience of himself and what is the subject's relationship to himself in view of the fact of this existence of a true discourse about him?"

44. See above, lecture of 25 March, p. 236-237, footnote *.

45. The 1971 *Leçons sur la volonté de savoir; Lectures on the Will to Know* on ancient Greece could already broadly be understood as a close discussion of Heidegger's theses on truth as *adaequatio* (on this point see D. Defert in the "Situation du cours," pp. 273-275; "Course Context," pp. 276-278).

46. See *Le Souci de soi*, "Rêver de ses plaisirs," pp. 13-50; *The Care of the Self*, "Dreaming of One's Pleasures," pp. 3-36.

47. Ibid., Fr., pp. 224-242; Eng., pp. 193-210.

48. See *L'Usage des plaisirs*, "L'acte, la dépense, la mort," pp. 141-156; *The Use of Pleasure*, "Act, Expenditure, Death," pp. 125-140.

49. Still in March 1983, Foucault speaks of a single volume entitled *L'Usage des plaisirs* (on the plan of this first version of the history of ancient sexuality, the manuscript of which is kept in the Bibliothèque Nationale de France, see D. Defert, "Chronologie," *DE*, I, p. 61/"Quarto," vol. I, p. 86).

50. See, for example, the end of the lecture of 11 February: "This marks the break between marriage and other social relations, the beginning of the heteromorphism that will characterize marriage within the system of social relations and that Christianity will exploit in a way we will try to see later" (above p. 141) or of 25 February: "Subjectivity, truth, and desire, this is what is formed, in the way I have just told you, when one moves from the Greek experience of the *aphrodisia* to the Christian experience of the flesh. But this is the overall movement, the point of arrival that I will try to show you at the end of this series of lectures" (above, p. 158).

51. In his "Chronologie," Daniel Defert notes that in March 1981 Foucault "refuses to sign belated petitions supporting the election of François Mitterand as President of the Republic, in accordance with an oft-reasserted principle that an intellectual is not an electoral spiritual director." But otherwise, in June "he takes part in Geneva, with Bernard Kouchner and Yves Montand, in the creation of the International Committe against piracy, for the defense of the *boat people*. He supports the right of intervention in international politics in the name of the 'right of the governed'" (*DE*, I, p. 59/ "Quarto," vol. I, p. 82).
52. *About the Beginning of the Hermeneutics of the Self*, p. 76.
53. Ibid., pp. 25-26: "The contact point, where the way individuals are driven by others is tied to the way they conduct themselves, is what we can call, I think, government."

This page intentionally left blank

INDEX OF CONCEPTS AND NOTIONS

Compiled by Nicola Lennon

Page numbers followed by n refer to end of chapter notes

abesse (separations) 225n
ablution 16
abortion 150
absolute duty 111, 314n
absolute power 291n
abstinence
 from meat, wine, entertainments 246n
 from sex 150-1, 168n, 266
Abyss, the 246n
active agency 302
activity, principle of 75, 84-90, 92,
 98, 102-3, 135, 181-3, 257-63, 267-74,
 282-3, 303
adaequatio, truth as 314n
Adam and Eve 21n
Admetus and Alcestis, narrative
 of 136-7
adultery 4, 6, 15, 62-3, 73n, 81-2, 161-2, 206,
 263-4, 304, 309
 definition of 75
 law against (*lex de adulteriis*) 206-7
advice for existence 27
advice for life 27
advice on conduct 295
afterlife, the 296
agapēsis (love, friendship, affection)
 164, 197, 202n
agennes (dishonour) 181, 199n
agreeable consent see *kharis*
akarpon (without fruit) 191, 201n
alcohol 169n, 268n
 abstention from 246n
 prohibition of 24n
 drunkenness 191, 201n, 290n
Alexandria 233, 273
Alexandrians 17, 21n
algebra 237n
allegorical dreams 55, 71-2n

allegory 4-5, 15, 55
ambition 285
anagkaion 72n
anagkaiotatos 223
anagkē dē prōton sunduazesthai
 (necessary union of male and
 female) 144-5n
analogy, domain of 55
anangkaioterōn (the most necessary) 112
androsunē 275
aneu gunaikos (the life without wife, celibate
 life) 112
aneu karitōn (without grace) 194, 196-7
anger 27, 29, 44n, 285, 296
animal fables 4, 8
 see also elephant, fable of the
animals 2, 4, 130-1, 139, 200n
 see also bees; elephants; fish; lions; ta de
 sunagelastika
anthropology 12, 27
anthropomorphism 21n
anthrōpos gar tē phusei sunduastikon
 mallon ē politikon (man's inclination to
 live as a couple, even more than in a
 city) 145n
anti-Semitism 39, 41, 45n
 see also Judaism
anxiety (*anxia*) 224n, 296
ap'allēllōn hēdonēs (mutual pleasure) 199n
aphilois (without friendship) 200n
aphrodisia (the pleasures of sex) 21n, 68,
 75-97, 100-1, 110, 123, 148-69, 175-7,
 180-2, 196, 198, 203-4, 207, 222, 250-1,
 254-5, 257-60, 263, 266-7, 269-73, 276,
 282-91, 295, 297, 298, 299n, 304, 305,
 308, 313n, 314n
aphrodisiac 21n
aphrodisiac dissymmetry 93-4

apo nekrou kai apo aphrodisiōn (cleanse oneself of any contact) 150
apolausis (enjoyment) 186
apologism 229, 230
apparatus (*dispositif*) 48-9, 289
appropriateness, principle of 83
aretē (virtue) 181, 186, 187, 196
aristocracy 215, 261, 267n, 272n, 273, 277, 295-6
    service 290n
    see also competitive aristocracies; monarchy
armies 139, 146n
art of dying 28, 30
    see also death
art of government 294
    see also governmentality
art of memory 28
art of rhetoric 28, 30
art of tombs 218n
Artemision (Temple of Artemis) 149, 168n
arts of living 25-35, 42, 44n, 49-50, 213-14, 274-6, 280, 312
    see also askēsis; mathēsis; meletē
asceticism 177, 246n, 274
    ascetic Platonic erotics 96n
Asia 70n
askeinai 69n
askēsis (ascesis) 25, 33, 246n, 311-12
assassination 290n
atelēs (without purpose) 191
Augustan period 206-7
autocracy 240
autonomy 107-8
aversion 236n
avid desire (an *epithumia*, a *pothos*) 129

bainesthai kai paidoporeisthai (to indulge in being covered and begetting) 96n
baptism 5-6, 246n
barbarian invasions 119n
beatitude 31
beauty 45n, 96n, 199n, 200n, 290n
bees 128-9, 144n, 184
beggars 71n
behavioral reactions 285
benevolence 267n
    see also *eunoia*
bereavement 28
    see also death; widowhood
bestiaries 21n
bigamy 211
biopoetics 34n
biopolitics 34n
bios (life) 34, 44n, 251-4, 268n, 302, 313n
bios chrēmatistikos (life of wealth) 253

bios theōrētikos (theoretical life) 151-2
bioūn (way of living) 34
birth 21n, 150
blabē (injury or setback) 57, 62
boat people 315n
bodily fluid 153-4, 169-70n
body (*sōma*) 146n, 151, 168n
botany 20n
bourgeois ideology 305
bourgeois morality 250, 296
brigandage 201n
brothels 72n
    see also *ergasterion*; prostitution
building, house/boat 139-40
bureaucracy 273-4

cadavers 150
    see also body (*soma*); death
caesura 249, 260, 267n, 275, 283, 286, 307-8
canon law 119n
capitalism 25, 40-2, 159, 250, 296, 305
Cappadocian Fathers 17
capriciousness 290n
care of the self 146n, 307, 311
care, concept of 311
caritas (mutual affective bond) 213
Carolingian Empire 256
Cartesianism 268n
caste 207, 213
castissimus ardor (chaste ardor) 214, 219, 224n
castration 280, 291n
casuistry 259
catastasis 116-17
Charlatans 70n
chastity 2, 6, 25, 168n, 172n, 214, 229, 246n
children, 'begetting' 96n, 121n, 164, 172n, 247n
choice of loves 298
chrematistics 252
Christian Late Roman Empire 44n
Christianity
    arts of living 24n
    ascetic and rigorist approach 24n
    animal fables 15
    conjugal behavior 6
    literature 15
    marriage 2
    morality 15-18, 36-7, 101, 177, 255-6
    organization as a Church 17
    origins of 14, 17
    plurality of accounts 17
    texts and manuscripts 4-5
    sexual ethics 15
    sexual morality 16-18, 25, 42
    see also confession; Jesus Christ; Judeo-Christianity; New Testament
Church, the 236n
cities 69n, 142, 144n, 240-1, 273, 304

citizenship 80, 121n
civilization 11, 28, 39, 98
classical science 48
code, valorization and illusion of the 98-9
codification 230, 232, 245
    and the flesh 99-102
coitus 155, 169n, 170n
comitial disease 170-1n
community (*koinōnia*) 126
    desire for 129
competitive aristocracies 272-4, 290n
    see also aristocracy
complaisance (kindness, obligingness) 193, 201n
complex machinery (*dispositif*) 23n
concord (*concordia*) 213, 218
concubines 61, 211
concupiscence 14, 23n, 288, 296
confession (*aveu*) 14, 53, 302, 311
confession (*confession*) 23n
confiscation of sexuality 305
conjugal conduct 2, 6
conjugal fidelity 25, 244, 249, 259, 279
conjugal virtue 4, 9, 305
conjugality, indissoluable 15
conjugaux (conjugal) 144n
consent (spouse's) 314n
constraint 57, 62, 72n, 79, 237n
contemplative life 29, 36, 109
continence 246n
control over desire (*ekgrateian pros epithumian*) 268n
Corinthians, doctrine of Saint Paul to the 2, 20n
cosmology 9, 246n
'couple or the herd' 130-1
couple, the 212, 215, 223n, 234, 305, 314n
coups d'état 291n
courage 45n
crime 11-14, 239, 290n
criminology 12
cruelty 290n
cultural prohibition 303
Cynicism 43, 69n, 112-17, 120-2n, 151, 160, 234
    see also *to kunizein*
Cyrene, Sacred Laws of 149, 168n

dead letter 234
de-aphrodization of sexual acts 258, 305
death 13-14, 28, 30, 31n, 55, 61, 63, 72n, 73n, 151, 218n, 219, 298
    sexual proximity and 154-6, 158, 168n
    see also bereavement; illness; tomb; widowhood
debauchery 103, 201n, 276-82, 290n, 291n
debitum conjugale (conjugal debt) 231, 247n
debt 57, 72n, 79

dehedonization 314n
delinquency 294
Delphi, ancient site in Greece 150, 168n
Demiurge 246n
deployment (*dispositif*) 23n
depravity 206, 291n
desiderium (desire, miss) 215-17, 225n
desire 165, 167, 236n, 241, 265-8, 285-9
    birth of 306
    'first pin-prick of' 308
    sexual 21n, 23-4n, 152, 157-8, 193, 285
    see also *desiderium; epithumian iskhurian; supplicium*
desocialization of sexuality 306
destiny 47-8, 55
Devil, the 6
devoutness 7
di'holōn genesthai tēn krasin (fusion of all their elements) 146n
di'holōn krasis (the complete mixture, total fusion) 139-40
diatribes 105
dietetics 302, 308
diexagōgē (to live to the end) 132
difference of valuation 304
dikaios (the just and unjust) 141
dimorphism 258-9
direction 302
discourse of truth 26, 167, 222, 244-5
discourse, functions of 244
discretion 25
divine principles 246n
divorce 17, 218n, 279
document (*documentum*) 15
documentary 237
Dodona, Hellenic oracle 150
dogma 236n, 246n
domestic life 56-7
    see also family; household
doubling of self 276
dowry 210, 211
dreams
    dreams-rêves 47, 53-4
    dreams-songes 47, 54, 55, 59, 70n, 71n
    interpretation of, see oneirocriticism
dual relationships see *sunduasmos; sunduastikos*
dualism 246n

eating 268n
economic postulate 238
economic science 310
economy of pleasures 36
education 236n
effectiveness (and ineffectiveness) 236n, 244, 245
effeminacy (effeminatus) 96n, 187, 201n, 291n

effeminate womanizer, paradox of 75, 90
egalitarianism 243
egkrateia (mastery, continence,
    self-control) 24n, 152, 266, 287
ego (res cogitans) 268n
Egypt, ancient 205-6, 210-11, 233
ei gamēteon (should one marry?) 105-8
eis aretēn (virtue) 199n
ejaculation 153, 257
ek diestōtōn (distinct elements) 139, 146n
ek sunaptomenōn (joined parts) 139, 146n
ekdosis (system of exchange in marriage) 209
elephant, fable of the 1-9, 14-16, 20-4n, 25-8,
    35-7, 104, 228-30, 305, 314n
elitism 267n, 271-2, 276-7
emperors of the Roman Empire 279-80
empodion (handicap) 127
en koinōnia pasa philia estin (shared
    interest) 145n
end (terme) 254
endurance 268n
England 281
    kings of 282
engraved art 120n
engrosser (to get a boy pregnant) 94
enhypnion 71n
enjoyment (apolausis) 200n
enjoyment (jouissance) 192
Enkratites 17, 24n
ensemble (whole, continuum) 256, 260
enupnia 53-5
    see also dreams-rêves
Epicurean mechanism 190
Epicureans 43, 69, 112, 121n
Epidaurus, sanctuary of (Argolis,
    Peloponnese) 150, 168n
epilepsy 154-5, 159, 170n, 171n
epimeleia (care, concern for the other) 186
epimeleia heautou (care of oneself) 294-6
epimeleisthai (care for oneself, care for the
    other) 186, 200n
Epirus, ancient Greek state 168n
epistemic surprise 236n, 237, 309
epistemology 254, 310
epithumia (desire of one sex for the other) 130,
    186, 192-3, 196, 266, 286-8
epithumiai (desires) 200n
epithumian iskhurian (violent desire) 142n
equality 259, 264-5, 268n, 275
    see also homonoia
equivalence principle 60
ergasterion (workshop/brothel) 57, 61, 72n
erōs (taking the other into account as
    subject) 93-4, 204, 250, 267n, 289n
    see also love
erōs-aphrodisia (synthesis of love and sexual
    pleasures) 204

erotic, principle of the 93
eroticization 166
erotics 96n, 314n
Erōtikos (the dialogue on love) 176-98, 199n
ethics
    ancient 72n
    classical 177, 302-5
    ethical perception 93, 103-4, 160
    sexual 15, 51, 85, 97, 103-4, 175-6, 195,
        230-1, 303-6, 311
Eucharist 24n
eunoia (benevolence) 137n
eustathēs (firm, well-balanced life) 131
evil 121n, 246n
evil spirits 22n
evolution 25
excess 65
exemplification 92
exercise of power 302
exercises 44n
exile 28
extra-marital affairs 15, 81
    see also adultery 81

fables, see animal fables; elephant, fable of the
faithfulness 264, 308
Fall of man 5, 21n
family 3, 56-7, 68-9, 84, 106-8, 119n, 207,
    240, 242, 262-3
fate 225n
fatherhood 56-7, 63, 66-8, 84, 121-2, 246n
favourability 121n
favourable values 76-7
feminine principle 246n
femininity 278
fidelity 211-12, 243-4, 249, 259, 279, 307
fish 131, 144n, 229n, 255, 256
flesh, the 119n, 199n, 229n, 255-6, 282-3, 296
forbidden fruit 5
force/ power of love (amoris vis) 7, 22n
fortune 56, 72n, 296
France 37n, 41, 281, 282, 312, 315n
Frankfurt period (1797-1800) 44n
French Revolution 18-19
friendship 133-4, 144n, 145n, 298
    see also homologia; kath'homologian;
        koinōnikais; philia; philia hetairikē;
        suggēnikē; tēn te suggenikēn kai tēn hetairikēn

gamein (marrying) 114, 120n
games of truth 221-2, 236-7n
Genealogy of Morality 40
general juridification of human behavior 303
general sociability 306
geneseōs eneken (procreation) 172n
genetics 237n
genitals, warming of the 152, 169-70n

genos (species, race) 128, 142n
Germany 41
gēroboksoi (feeders and nurses of the old) 126
gifts 72n
  see also *ousia*
gloria (glory) 215, 225n, 252, 268n
Gnosticism 229, 246n
God 142n
  delegate of 281
  mediating position with man 281
  messenger of 121-2n
  New Testament representations 246n
  Old Testament representations 246n
  reverence to 121-2n
  word of 6
godparents 119n
goneōn epimeleisthai (honoring parents) 114
government of men 302, 313n
government of self 296
governmentality 281, 288, 294-5, 301-2, 312, 315n
grace 290n
  see also *kharis*
gratitude 267n
Great Games 268n
Greek antiquity 6-9, 30, 33, 51, 53, 57, 65, 70n, 76, 168n, 193, 205

hagneia (debauchery, impurity) 6
  see also debauchery
happiness 27, 31, 34, 217
harmfulness 236n
hate 236n
health 55
Hebrew 230
hēdonē (pleasure) 163, 164, 186-7, 196-7 200n
  see also aphrodisia
hēdonen periesti karpousthai kai apolausin hōras
  kai sōmatos (sensual pleasures and fleeting
  pleasures of the body) 200n
Hegelian movement 41
Hellenistic period
  aphrodisia 76, 100, 123
  arts of living 23n, 30, 35
  attitude of mistrust 148
  bees 144n
  Christianity and 26-7, 227, 255-6
  conjugal behaviour 305
  culture 9
  diversity of sexual acts 65
  government of self 296
  love 196
  marriage documentation 209-10
  matrimonial practice, spread of 227, 232, 233, 249-50, 306
  monarchies, development of 208, 273
  philosophy 69, 158

political power 240
population, daily life of 205
(sexual) morality 177, 296
sexuality prohibition 98
societies 14
technique of life 295
technology of self 295
thought 9
treatises on anger 44n
urban development 208
veridiction 245
women 180
henōmena (beings, absolutely unitary) 139
herds see *sunagelastikoi*
heresy 38, 246n
hermeneutics 312
  of the self 23n
heterogeneity 184-6, 258, 262-3, 276
heteromorphism 80-2, 136, 306, 314n
heterosexuality 80, 82, 307
Hexameron 21n
hierarchization, principle of 83, 273
High Middle Ages 260
  see also Middle Ages
higher truths 311
Hippocratic Oath 86, 96n
historical consciousness 39
historical studies 13, 240
historico-philosophical approach 11
historiography 42
history of sexuality 288, 303
history of subjectivity 303, 309
homelessness 121n
homilia (meeting or assembly) 78, 129
homilies 119n, 124-5
homologia (friendship, concord) 134, 145n
homomorphism 80-1
homonoia (equality) 137n
  see also equality
homosexuality 27, 80, 82, 90, 94, 96n, 172n, 199n, 256
  disqualification of 306-7
hormē (natural impulse, female) 184, 186, 199n
hōs gunē katathumios tō gegamēkoti (desired wife
  for the man who has married her) 145n
hostility 121n
household (*oikos*) 69n, 133, 135, 142, 187, 210, 261, 298, 304, 314n
hubris (violence) 6
hugron douton kai oikouron (enervating and housebound) 200n
hugros (moist) 187
human history 236n
human nature 69n
human practice 236-7n
human sciences 27

humanity 69n
  history of 5-6
humors 153-4
hunting 168n
hupeikhō (to give way, make concessions, consent) 194
hymen 225n

idea morum (a sort of model of conduct) 15
ideality 241
ideology 9, 241-2, 245, 305, 309
illness 11-14, 29, 55, 63, 170n, 239, 282, 294
illusion of the code 303
image 225n
imago (image) 216
immorality 161
Imperial period 158, 205, 240
imperial power 291n, 30
incest 27, 62-7, 74n, 100, 119n
  strict Germanic code of 260
individual rights 60-1
infanticide 180
infibulation 291n
injury 62
insanity *see* madness
inscribed (*se marque*) 239
insomnia 216
institutions 236n, 237n
insularization of marriage 124
insult 290n
integral amalgamation 146n
integral union 146n
International Committee against piracy 315n
international politics 315n
intimacy 3
introspection 302
islanders 70n
isolation of sexual activity 16
isomorphism 75, 77-82, 84, 90-3, 97, 102, 181-3
  see also socio-sexual isomorphism
Italy 70n
Ithaca 71n

Judaism 39-40, 230
  see also anti-Semitism
Judeo-Christianity 25, 39-42, 156
jurgio (quarrel) 224n
jurisdiction 245

kairos (the occasion) 52
kata peristasin (is of the realm of circumstance, conjuncture) 112, 114
kata phusin (according to nature) 141, 142n, 185
katathumios (present in the heart of) 136
kath'homologian (friendship contract) 145n
kēdosunē (one caring for the other) 137n

kharis (grace, obligingness, agreeable consent) 193-7, 204, 213
khrēsis aphrodisiōn (practice of sexual acts) 284, 295
kindness 164, 267n
  see also complaisance; kharis
kings, power and characteristics of 281-2
kinship 14, 119n, 134
  see also family; *suggeneîke*
koina einai panta (everything considered common) 145n
koinōnia (the community of existence) 113, 118, 129, 145n, 189, 191, 193, 196
koinōnikais (community friendships) 145n
koinōnos (common) 146n
krase (affective relations of marriage) 203
krasis (complete fusion or blend) 138-40, 148, 231, 261
  see also di'holōn krasis
kunizein 115-16

language 30
  see also art of rhetoric
Late Antiquity 4-5
latent relation of connection 237n, 239
Latin antiquity 6
learning 32
lesbianism 65-6, 74n
letters of Pliny 213-20, 225n, 309
liberation 246n
licentiousness 268n
limitation of involuntary impulses 36
lions 200n
literature 27
localization, principle of 82-3, 160, 163, 305
logic 237-8, 243
logicist postulate 238
logicist principles 243
logicist dodge 235-6
logos 223, 242-3, 245
loss of substances (*eis auton apousias poïēsetai*) 72n
love (*erōs*) 22n, 69n, 105, 121n, 146n, 169n, 172-3n, 184, 190-2, 201n, 225n, 236n, 246n, 262, 290n
  of boys (*paidikos erōs*) 166n, 186, 193, 195, 199n, 200n, 298
  matrimonial 140
  of women 166n, 187, 200n, 298
  pederastic 166, 181-4
  see also *Erōtikos* (the dialogue on love)
Lucian's *Amores* 298
lust 290n, 291n

madness 11-14, 26, 239, 282, 294
mandrake (*dudha'im*) 21n
manifestation of the truth 302
Marcionism 229, 246n
Marcionite doctrine 246n

marriage
  as a handicap 127
  classical ends of 127
  comparative advantages and
    disadvantages 105-7, 110-11, 119n
  hyper-valorization of 75, 81, 83, 91,
    102-5, 124
  insularization of 124
  married life 297-8
masculinity 275, 280, 307-8
  masculine sexual acts 257-8
master/disciple relationship 32
mastery of oneself 36
mastery of self 36, 265-6, 275-6, 279, 293,
  308
mastery of the passions 36
masturbation 65-7, 74n
mathematical physics 18-19
mathēsis (teaching) 25, 33, 48, 69n
matter 246n
meat, eating 24n
medical regimes 297
medicine 20n, 29, 55, 156
  as art of living 44n
  physicians on sexual acts 159
meditation 33, 44n
  see also *meletē*
meletan 69n
meletē (meditation, reflection on) 25, 33
memory 31n
  see also art of memory
menstruation 150, 171n, 230
meta gamou (with marriage) 112, 114
metaphysics of subjectivity 268n
Middle Ages
  arts of living 31
  Christianity 260
  codification of sexual activity 99, 165-6
  effeminacy 90
  incest 99, 260
  fables 1
  family groups 119n
  general juridification of human
    behaviour 303
  homosexuality 256
  penal practices 119n
  Physiologus 4, 21n, 228
  religious power 281
military practice 236n
mineralogy 20n
mira concordia (remarkable, admirable
  concord) 213
miracles 282
mixis (mixture of elements that remain
  distinct) 138
mixture see *di'holōn krasis; ek diestōtōn; ek
  sunaptomenōn; krasis; mixis*
moderation 6, 8, 65, 89

modesty 6-7, 16, 22n, 25, 79, 165, 173n, 265n,
  290n, 298, 305
Monadism 246n
monarchy 209, 240, 273, 281
  see also aristocracy; kings
monasticism 177
monogamy 3, 15, 25, 228, 244, 249, 259
monotheism see Stoicism
moral philosophy 82, 158-9
moral value 7-8, 52, 60, 63, 65, 76
moralists 29
moralization 206-7
morals, virtues and 2
mother earth 63
motherhood 63-5, 74n
mutual affection 305
mutual assistance 135
mutual respect (*timē*) 164, 202n
mysticism 246n

natural acts 284
natural analogy 58
natural history 15, 20n
natural impulses see *hormē; orexis*
natural sciences 2
natural vessel, theory of the 85-6
naturalism 4-6, 15-16, 25, 130-1, 228, 305
naturalness 67-8, 84-8
neoliberalism 312
Neo-Platonism 246n
  Neoplatonist mysticism 38
New Testament 246n
nocturnal visions 53, 54-5
  see also dreams
nomos 59, 62-3, 83-4
non-relational sexuality 303-4
nursery schools 236n

obedience 302
objectivation 284-8, 294
obligation 12-13, 237n, 241, 304
obligingness 202n
  see also *compaisance; kharis*
Oeconomicus (Xenophon) 124-7
Oedipus Complex 63-4
offensa (sulk) 225n
oikoiros (housebound) 187
oikos (household) 69n, 133, 135, 142, 187, 210,
  261, 298, 304, 314n
old age 126, 142n, 225n
Old Testament 246n
oneirocriticism (dream interpretation) 49-77,
  97, 160, 182n, 257, 297, 303
oneiroi 54-5, 72n
  see also *dreams-songes*
oneiros 71n
ontological astonishment 236n, 237
ontological status 31-3, 36

opinion 236n
oral sex 64-5, 100
orexis (natural impulse, male) 184, 199n
organic unity 138-40
Orphism 151, 160n, 168n
ouden kōlusei 115
ousia (substance) 57, 72n

paganism
  Christianity and 6, 15-17, 26, 37-42, 239,
    249, 255, 305
  conceptual origins 37-42
  ethics 18, 36, 68
  heresy and 38
  naturalism 4-6, 15, 305
  pagan Antiquity 43, 49, 229
  reflection and practice 228-9
  (sexual) morality 16-17, 37-8, 42, 101,
    176-7, 229
  veridiction 245
  watershed 44n
paidagōgein (mastery over) 264, 268n
paideia (teaching of the truth) 110
paiderastein meta philosophias 119n
paidikon 211
paidopoieisthai (having children) 114
paidopoiïa (purpose of marriage) 163-4
panacea 21n
panegryric metaphor (*panegyris*) 252-3, 268n
para nomon (foreign to the law more than
  contrary to the law) 62
para phusin (outside nature) 64, 84, 185, 199n
Paradise 5, 21n
paradox 98
parents, care of 121n
paroxysmality 152, 288, 298, 308
Parsifal theme 282
particle physics 237n
partners 86
passion 7, 169n, 282, 290n
  see also *pothon iskhuron*
passions, the 246n
passive pleasure, discrediting of 75, 88-9
passivity 62, 88-9, 98, 103-4, 199n, 278-9,
  282, 306-7
pathos 260
Pauline theology 246n
pedagogical erotics 75, 307
pedagogise (*pédagogiser*) 264
pedagogy 27, 92-5, 185, 295-6, 298
pederasty 175, 183-4, 187-8
  see also love: pederastic
penal practices 119n
penance 246n
penetration (*perainein*) 66-8, 72-4n, 75,
  84-90, 257
penitential techniques 302

per mutuam caritatem (mutual affective
  bond) 213
perception 99
perfection, principle of 83
Peri gamou (treatise on marriage) 130, 138
peri ton bion (whose object is life) 251
  see also techniques of life
peristasin moi legeis (particular
  circumstance) 121n
peristasis (save circumstances) 114, 117
philia (friendship) 185, 189-90, 192, 195-7,
  199n, 200n, 306
philia hetairikē (friendship of
  comradeship) 134-7, 164
philian dia kharitos (friendship and
  obligingness/kindness) 201n
philian kai aretēn (friendship and virtue) 199n
philosophical life, the 302
philosophikōs paiderastein (love boys
  philosophically) 111
phusis (nature) 59, 185-6
Physiologus 15, 21n, 229
piracy 315n
pistis (trust) 197
Platonic erotics 307
pleasure 88-9, 94, 96n, 103-4, 120n, 144n,
  135, 241, 267, 288-9
  of the body 171n
  devalorization of 103-4
  voluptuous 159
  see also *aphrodisia*
Plutarch's *dialogue on love* 298
poems of Statius 213-14, 217-18, 220, 309
polis (political body) 132
political activism 312
political history of truth 237n
politics 30, 252
politics of ourselves 312
politikos (political) 132
polymorphism 237n, 238
polytheism 39
positive knowledge 302
positivist approach 10-11
possessions 121n, 139-40, 146n, 161
pothon iskhuron (powerful passion) 142n
power
  conceptions of 294
  distribution of 273
  political 240, 272, 276-8, 290n, 295,
    309-10
practical tolerances 303
praetorians 291n
pregnancy 171n, 230
  see also abortion; birth; menstruation
prescriptive discourse 221, 233-4, 295
principle of exclusive localization 305
principle of the naturalness of penetration 47

private life 261-2, 274
procreation 16
procreationism 16, 246-7n, 305
prodigality 144n
productivism 305
proēgoumena: politeuesthai (being a citizen,
    leading the citizen's life) 114
proēgoumenon (principle act) 111-12, 114, 116,
    120n, 124
proēgoumenōs (primordially) 69n, 112, 113-17
proegoumenōs prattetai (actions with value in
    themselves) 120n
professional apprenticeship 30-1
prohibition 58n, 98, 99
property 121n
prophecy (truth-telling) 71n, 150
pros aretēn (exhort to virtue) 200n
prostitution 57, 61, 72n, 73n, 114, 161, 263,
    277-8, 280, 291n
    see also brothels; *ergasterion*; working girls
prudence 172n
psychiatric science 309-10
psychiatry 12
psychoanalysis 90
puberty 114, 171n
public life 30, 36n, 261-2, 274
public service 122n
publicity 30
pure expenditure 236n
purification 3, 16, 25, 96n, 103, 150-1, 155,
    157, 168n, 171n
    confessional 14
    washing and 1-2
Pythagoreanism 8, 38, 43, 45n, 151,
    160n, 161

racism 45n
    see also social racism
rape 22n, 72n, 206, 291n
rarity 16
rationality (*raisonnable*) 9, 69n, 244-5
rationalization 276
    of a code 254-5
    of behavior 242-5
reading grid 303
reality 55, 237-8, 240-1, 250, 310
reason 26
reciprocal affects, principle of 298, 305
reduction of the act 287
redundant discourse (*discours en trop*) 227-35,
    243-4, 309-10
reflection 245
reflexive truth acts 313n
Reformation/Counter-Reformation 3-4
regime of existence 29
regime of the body 29
regime of the soul 29

register of affects 305
relationship to ourselves 12
religion and marriage 206
religious acts 149-50
religious consciousness 41-2
religious prohibition 155-6
remarriage 17, 219
Renaissance period 20n, 29-30
representative reduplication 233
repressive hypothesis 303
reproduction, human biological 5-6, 25, 64-5,
    246n, 298, 305
republican liberty 291n
reputation 31n
requies in labore (seek rest in work) 216
resistance 266
respect (*timē*) 197, 298
resurrection of the body 246n
rhetorical exercises 105
    see also art of rhetoric
right of the governed 315n
rite of the taurobolium 291n
ritualistic acts 149-50
Roman period
    aphrodisia 76, 100-1, 123, 148-9, 176-7, 196,
        204-5, 276, 283
    aristocracy 261, 271-2, 295-6
    arts of living 23-4n, 27, 30-1, 35
    attitude of mistrust 148-9
    bees 144n
    codification of sexual relations 210, 227,
        245
    conjugal behaviour 305-8
    'debauched prince' 291n
    ensemble, continuum 256
    history 26
    governmentality 301
    imperial society 111, 273
    incest 260
    juridical-moral rules 38n
    marriage 211, 223n, 232, 249-50,
        305-8
    medicine 29
    moral principles 206
    philosophy 45n, 69
    political power 240, 272-3, 277-81
    Roman matron 214
    Romanization of Egypt 206, 210
    sexual conduct/ethics 51, 53, 65, 91, 98,
        119n, 176-7, 206, 296
    Stoicism and 311
    technique of life 295
    technology of self 295
    treatises on anger 44
    roof (*stegos*) 126
    ruin 28
    rules of conduct 27

sacrifice 149, 168n
sage 33, 36, 42n, 69n, 266
sagittal relation of representation 237n, 239
salvation 296
scansion (fundamental division) 251, 255, 256
schemas of existence 27-8
scholastic penance 119n
science 236n, 237n, 238-9
scouting 121n
scripture 4-5
secrecy 16, 22n, 25
self-control 36, 266, 268n, 274, 291n, 308
self-examination 33, 285
self-knowledge 293-4
self-mastery see mastery of self
self-restraint 45n
self-sacrifice 302
Semitism 230
    see also anti-Semitism; Judaism
senatorial class 277, 291n
separation of the sexes, principle of 83
serpents 21n
servitude 291n
sex-status 263n, 275, 284
shame 201n, 264, 278
shared interest see en koinōnia pasa
    philia estin
si qui forte casus inciderit (if the opportunity
    arises) 112
silence 246n
sin 290n
single chain of love 307
slavery 61-2, 66, 68, 72-4n, 86-7, 96n, 145n,
    162-3, 172n, 187, 200n, 263, 268n, 279,
    291n
sleep 168n, 268n
social class 243, 272, 276
social domination 307-8
social life 30
social logics of domination 306
social practices 233-4, 309
social racism 45n
social status 56, 60, 62, 67, 181, 194, 214,
    263, 272, 291n, 304, 307
social stereotypes 27
social virtues 2
socialism 41
sociality, dual and plural 131-2
socio-sexuality
    continuum 47, 306
    correlation 274
    heteromorphism 80
    isomorphism 75, 80, 102, 104-5, 124,
        257-60, 267-70, 272, 279, 304
socius (companion) 218
sōma (body) 57

sophōn polis (city of wise men) 115-16,
    120n, 121n
sōphrosunē (a wisdom) 6, 265n
soteriology 246n
sovereignty 60, 82, 89-90, 242,
    280-1
spectacle 252
Speculum majus 21n
spiritual direction 295, 301, 315n
spiritualism 246n
splitting of self 276
'spouse or the friend' (Aristotle)
    133-8
sterility 21n
Stoicism
    adversaries of 43
    arts of living 42-3, 176
    Christian sexual morality 25
    on marriage 115, 137-8, 141
    mixture 138
    monotheism 38
    philosophical form 43, 50
    sexual ethics 304
    texts inspired by 141, 176
stoikheion (marriage rooted in the
    household) 69n
strategic knowledge 236n
strict economy in sexual relations
    15-16
structuration of sexuality 303
struggle 252
subjectivation of sex 275, 288
subjective structuration 311
subjectivity / subjectification 9-12, 19, 158,
    239, 253-4, 267, 282, 284
suggenikē (friendship, kinship)
    134, 145n
sumbiosis (shared life) 198
sumplokē (entanglement) 78, 79
sunagelastikoi (beings living in a herd or
    group) 130-1
sunduasmos (the connection between two
    elements) 131, 133
sunduastikos (dual bond) 131, 132
sunduazein (to live as a couple) 131
sunduazesthai (first union) 132, 172n
sunduazetai (animal couples) 144n
suneinai (to practice the sexual act)
    129, 142n
sunousia (sexual intercourse) 78, 189, 191,
    193, 200n
sunousiai aphrodisiōn (coming together) 76
    see also aphrodisia
supplementary discourse 242-3
supplementary game 238-9
supplicium (desire) 216-17, 225n

surrealism 56, 72n
suzein (live together) 129, 142n
syndyasticism 132-4
Syria 24n

ta d'hēnōmena kai sumphuē (whole of a single nature) 146n
ta de kata peristasin (actions performed according to their relations) 120n
ta de sunagelastika (animal justice) 144n
ta pros ton bion allēloin summēkhanasthai (to arrange the affairs of their life together and reciprocally) 129-30, 142n
taboo 100
teaching 32, 33
technique of existence 295
technique of life 295
techniques for living 251
techniques of communication 302
techniques of domination 302
techniques of life (*tekhnai peri ton bion*) 254, 256, 259, 267
techniques of production 302
techniques of self 293, 302, 312
technologies of life 256
technologies of subjectivity / subjectivation 256-7
technology of the self 94-5, 256, 266, 270-1, 277-83, 295-6
tekhnai (techniques) 34-5, 251, 254-5, 287
tekhnai peri bion (techniques of life) 34-5, 269-70
tekhnē (technique) 251
teleios 69n
telos (this end, this purpose) 191
temperance 25, 65
temptation 308
tēn te suggenikēn kai tēn hetairikēn (friendship between family and friends) 145n
tenderness 305
tēs th'homilias kai tēs koinōnias (intercourse and common life) 143n
test of the soul 302
theology 27
theon sebein (honoring God) 114
theoreomatic dreams 71n
theōria 109
thought 246n
thrift 45n
to genos hēmōn aidion ē (eternal race) 143n
to kunizein (living cynically) 115
to pros ton bion (all the things concerning life) 130
tois axiois epimeleias (worthy of care) 200n
tolerance 98
tomb (*sēma*) 168n
topos 178

torture (*tormentum*) 216-17, 224n
training *see* professional apprenticeship
tranquility (*vita tranquilla*) 27, 31, 120n
tree of temptation 21n
tribadism 74n
true/false game 236-9, 245, 309-10
trust (*pistis*) 164, 202n, 305
truth acts 302
truth of sexuality 312
typology 25, 135
tyrannicide 290n

unitary games 238-9
United States
    liberation of American hostages in Tehran (1980) 312
universal truth value 11, 12
universe, origins of the 246n
unprofitable game 238-9
urbanization 267n, 295
use of pleasures 291n
use, concept of 311
usefulness (*sumpheron*) 69n, 236n
useless expenditure 305
utilitarianism 238
utility 144n, 236n
utopia 234

Valentinians 229
validation 267n
valorization 13, 99, 258, 260, 267-75, 297-8, 303
value system 267, 270-1
vanity 290n
veneria (activity, sexuality) 283, 304
vengeance 201n
veridical discourse 221
veridiction 13, 221, 237-9, 245, 309-11
verification 236
veritable document (*documentum*) 2
villages 145n
violence 201n, 284, 290n, 298
violence *see epithumian iskhurian*
virginity 17, 117, 119n, 149, 150, 291n
virility 262
virtue 2, 4, 9, 199n, 298
    see also eis aretēn; philian kai aretēn
virtuousness 9

Wars of Religion 236n
washing 1-2, 6
water, baptismal 5-6
watershed (search) 37, 44n, 189
wealth 72n, 74n, 80, 268n
    see also ousia
Weberian discourse analysis 244

what happens to reality 310
what happens to the subject 310
whole knowledge (*savoir*) 28n
widowhood 219
    see also bereavement
willingness 267n
wisdom 33, 44n
    fall of 246n
womb 153
women, juridical status of 243

work (*labores*) 224n
'working girls' 73n
    see also prostitution

youth 278
youths 263

zēn (property of living) 34
zōē 268n, 302, 313n
zoology 2, 9, 20n

INDEX OF NAME

Compiled by Nicola Lennon

Page numbers followed by n refer to end of chapter notes

Aelian (Claudius Aelianus) 6-8, 15, 22n, 26, 228
Aeschines 277, 291n
Albertus Magnus (Saint Albert the Great) 4, 20n
Alcibiades 265-6, 287
Aldrovandi, Ulisse 2-3, 15-16, 20n, 24n, 228
Ambrose, Saint 4, 21n
Antipater of Tarsus 42, 69n, 137-8, 141, 142n, 178, 230, 297, 305, 311
Antisthenes 152
Antoninus Pius 225n
Apollo, god of music, truth and prophecy 168n
Apollo Mystes 70n
Apollonius of Tyana 160, 171n
Aphrodite, goddess of love 22n, 201n, 202n, 307
Apuleius 155, 171n
Archidamus 200n
Aretaeus of Cappadocia 155, 170n
Ariès, Philippe 172n
Aristotle 8-9, 22n, 96n, 105, 131-7, 140-1, 144n, 145n, 161, 163, 172n, 261
Artaud, Antonin 291n
Artemidorus of Ephesus 49-69, 70n, 71n, 72n, 74n, 75-86, 96n, 97-8, 101, 119n, 160, 257, 259, 269, 297, 299n, 303, 305, 311, 313n
Artemis, goddess of hunting 168n
Arthur, legendary king of England 282
Asclepius, god of medicine 168n
Athenaeus 297
Augustine, Saint 23n, 39-40, 159, 167, 171n, 230-1, 246n
Augustus, 1st Emperor of the Roman Empire 291n
Aulus Gellius 170n

Aurelius Augustinus 44n
Autobulus, son of Plutarch 200n

Bailey, D. S. 255, 268n
Beauvais, Vincent de 4
Béjin, André 172n
Bellovacensis, Vincentius 21n
Bersez, J. 21n
Bloch, Marc 281, 291n
Blumenkranz, B. 44n
Bollack, J. 120n
Bonhöffer, Adolf Friedrich 111, 120n
Boswell, J. 223n, 255-6, 268n
Bourgeois, B. 44n
Bréhier, Émile 111, 144n
Broudehoux, J.-P. 223n
Brown, Peter 37, 44n
Buffon, Georges-Louis Leclerc, Comte de 2-3, 5, 20n
Burkert, W. 168n

Caelius Aurelianus 154, 170n
Caillemer, E. 223n
Caligula, 3rd Emperor of the Roman Empire 290n
Celsus 155, 170n
Charles VI, king of England 282
Chevallier, P. 23n
Chiron, P. 170n
Chrysippus 43
Chrysostom, John (Archbishop of Constantinople) 105, 119n, 124-5, 143n
Cicero 112, 120n, 144n, 159, 252, 268n
Claudius, 4th Emperor of the Roman Empire 279, 290n
Clement of Alexandria 17, 21n, 112, 120n, 229-30, 246n

Cohen, J. 44n
Commodus, son of Marcus Aurelius 225n
Crates 113, 121n, 151
Crook, J.A. 223n
Crouzel, H. 122n

Daphnaeus, son of Archidamus 189, 193, 200n
Daremburg, C. 171n
Decidiana, Domitia 224n
Defert, Daniel 23n, 315n
Democritus 112, 154, 170n
Derrida, Jacques 70n
Descartes, Renée 48, 70n, 268n
Diogenes Laertius 112-13, 120n, 121n, 168n, 169n, 252, 268n
Dover, K.J. 96n

Elagabalus, 25th Emperor of the Roman Empire 279-81, 291n
Epictetus 33, 44n, 69n, 105, 111-16, 119n, 120n, 121n, 160, 199n, 230, 266, 268n, 287, 291n
Epicurus 112, 140, 169n
Eros, god of love 307
Euripides 146n
Eustathius of Antioch 21n

Farge, Arlette 294, 299n
Faustina, daughter of Antoninus Pius 225n
Ferri, Silvio 168n
Festugière, André-Jean 49, 53-4, 70n, 72n, 74n, 268n
Feuerbach, L. 45n
Feyereisen, P. 119n
Florisoone, Charles 20n
Foucault, Michel
    Lectures at the College de France 23n, 45n, 119n, 313n
    Lectures at Dartmouth College 119n, 313n
    On the Government of the Living 23n, 246n, 313n
    Security, Territory, Population 313n
    The Archaeology of Knowledge 247n
    The Care of the Self 313n
    The Courage of Truth 301, 313n
    The Hermeneutics of the Subject 44n, 200n, 291n
    The History of Sexuality 20n, 23n, 310, 313n
    The Order of Things: An Archaeology of the Human Sciences 20n
    The Use of Pleasure 72n, 96n, 199n, 201n, 268n, 291n, 313n
    Wrong-Doing, Truth-Telling, The Function of Avowal in Justice 119n

Francis of Sales, Saint 2-3, 7, 15-16, 20n, 37, 228, 305
Frend, W.H.C. 44n
Freud, Sigmund 49, 70n,167, 287n

Galen of Pergamon 55, 70n, 154, 155, 169n, 170n, 297
Galilei, Galileo 18
Gessner, Conrad 4, 20n
Gregory of Nyssa 105, 119n
Gros, Frédéric 301n

Hadot, P. 199n
Hegel, Georg Wilhelm Friedrich 40-1, 44n
Heidegger, Martin 268n, 310
Hephaestus, son of Hera and god of blacksmiths 201n
Hera, goddess of women and marriage 194, 201n
Heraklides Ponticus 252-3
Héritier-Augé, Françoise 74n
Herodian 291n
Herodotus 8, 23n, 165, 172n
Hesiod 201n, 214, 224n
Hesychius of Alexandria 150, 168n
Hierocles 42, 69n, 111-12, 114, 120n, 130-3, 141, 178, 222, 225n, 297, 305, 311
Hipparchia 113
Hippocrates 86, 96n, 153, 155, 169n, 171n, 308, 311
Homer 54
Humbert, G. 223n

Iamblichus 268n
Irenaeus of Lyon 24n
Ixion, king of the Lapiths 194

Jerome, Saint 172n
Jesus Christ 246n, 281-2
Juba II of Mauretania 22n
Jupiter, god of sky and lightning 220, 225n
Justin 21n, 24n, 229, 246n

Kant, Immanuel 48, 70n
Kepler, Johannes 18
Kouchner, Bernard 315n

Labarrière, J.L. 144n
Lactantius 119n
Lahlou, J. 22n
Lannoy, J.-D. de 119n
Libanius 105, 119n
Lucian of Somosata 185, 200n, 277, 290n, 298, 299n
Lucretius 169n

Macrobius 154, 170n
Marcion 246n
Marcus Aurelius 144n, 219, 225n, 278, 285, 287, 291n
Marx, Karl 41
Maximus of Tyre 277, 290n
Mitterand, François, 21st President of France 312, 315n
Montand, Yves 315n
Musonius Rufus 42, 69n, 105, 111, 127-33, 136-8, 141, 143n, 144n, 145n, 146n, 160-2, 172n, 178, 230, 263-4, 268n, 271, 297-8, 299n, 305, 311

Nephele, goddess (Nuee, cloud) 194
Nero, 5th Emperor of the Roman Empire 279, 290n
Newton, Isaac 18
Nietzsche, Friedrich 45n, 48, 70n, 268n

Origen Adamantius 21n, 117, 122n
Ovid 150, 168n

Paul, Saint 2, 20n
Paulus Aegineta 155, 171n
Periander, tyrant of Ambracia 290n
Philo of Alexandria 21n, 144n
Philodemus of Gadara 44n
Philostratus 160, 171n
Pindar 194, 201n
Plato 96n, 119n, 163, 168n, 172n, 178, 185, 190-1, 196, 199n, 265-6, 268n, 286, 291n, 294, 299n, 307-8
Pliny the Elder 7-8, 15, 22n, 26, 37, 155, 170n, 228, 262
Pliny the Younger 212-17, 223n, 224n, 309
Plutarch 43, 44n, 96n, 105, 139-41, 146n, 149, 160, 163-5, 168n, 172n, 173n, 176-84, 187-97, 199n, 200n, 204n, 230, 271, 277, 290n, 297-8, 299n, 307, 311
Pohlenz, M. 111, 120n
Pomeroy, S.B. 223n
Porphyry 144n
Protogenes 184-6
Proust, Marcel 188, 216
Pythagoras of Samos 151, 160, 252

Quintus Sextius the Elder 43, 45n

Rabbow, Paul 44n
Reagan, Ronald, 40th President of the United States 312
Renan, Ernest 225n
Rousseau, Jean-Jacques 17

Rufinus of Aquileia 21n
Rufus of Ephesus 154, 159, 170n, 171n, 297

Sabinus 154, 170n
Sappho 194
Schopenhauer, Arthur 48, 70n
Scribonius Largus 155, 170n
Seneca 42, 44n, 45n, 69n, 161, 163, 172n, 230, 271, 285
Sextius (Quintus Sextius the Elder) 45n
Socrates 265-6, 286-7
Solinus, Gaius Julius 6, 22n
Solon 164
Sophocles 64
Soranus 297
Statius 212-15, 217-19, 223n, 224n, 225n, 309
Stekel, W. 96n
Stobaeus 127, 144n, 146n
Suetonius 276, 290n, 291n
Sulla (Lucius Cornelius Sulla Felix) 290n

Tacitus 213, 223n, 276, 290n
Tatian the Assyrian 24n
Temkin, O. 170n, 171n
Tertullian 167n, 246n, 287n
Theodosius I 24n
Tiberius, 2nd Emperor of the Roman Empire 290n
Tibullus, Albius 150-1, 168n
Timarchus 278
Titus Livy 22n
Tralles, Alexander von 155, 171n

Valentinus 246n
Vatin, Claude 119n, 223n, 233-4, 247n
Venus, goddess of love 168n, 169n, 217-18, 225-6n
Veyne, Paul 43, 45n, 172n, 205, 212, 223n, 271-2, 290n, 305, 314n
Voelke, A.J. 199n
Voltaire (François-Marie Arouet) 17

Weber, Max 41-2, 45n, 244

Xenophon 105, 124-8, 130-2, 140-1, 143n, 144n, 261, 265-6, 268n, 306, 308, 311

Yates, Francis 28, 44n

Zeno of Citium 43
Zeus, god of sky and thunder 121n, 168n, 172n, 194, 201
Zucker, Arnaud 21n